Advances in Intelligent Systems and Computing

Volume 231

Series Editor

Janusz Kacprzyk, Warsaw, Poland

Vladimir Trajkovik · Anastas Mishev
Editors

ICT Innovations 2013

ICT Innovations and Education

Editors
Vladimir Trajkovik
Faculty of Computer Sciences and
    Engineering
Ss Cyril and Methodius University
Skopje
Macedonia

Anastas Mishev
Faculty of Computer Sciences and
    Engineering
Ss Cyril and Methodius University
Skopje
Macedonia

ISSN 2194-5357   ISSN 2194-5365   (electronic)
ISBN 978-3-319-01465-4   ISBN 978-3-319-01466-1   (eBook)
DOI 10.1007/978-3-319-01466-1
Springer Cham Heidelberg New York Dordrecht London

Library of Congress Control Number: 2013943577

Printed on acid-free paper

Springer is part of Springer Science+Business Media (www.springer.com)

Preface

The ICT Innovations conference is the primary scientific action of the Macedonian Society in Information and Communication Technologies (ICT-ACT). The conference provides a platform for academics, professionals, and practitioners to interact and share their research findings related to basic and applied research in ICT.

The ICT Innovations 2013 conference gathered 160 authors from 17 countries reporting their scientific work and novel solutions in ICT. Only 26 papers were selected for this edition by the International Program Committee, consisting of 222 members from 49 countries, chosen for their scientific excellence in their specific fields.

ICT Innovations 2013 was held in Ohrid, at the Faculty of tourism and hospitality, in the period September 12–15, 2013. The special conference topic was ICT Innovations and Education. The conference also focused on variety of ICT fields: Quality of Experience in Education, Language Processing, Semantic Web, Open Data, Artificial Intelligence, Robotics, E-Government, Bioinformatics, High Performance Computing, Software Engineering and Theory, Security and Computer Networks. The conference was supported by HE D-r Gjeorge Ivanov, the President of the Republic of Macedonia.

We would like to express sincere gratitude to the authors for submitting their works to this conference and to the reviewers for sharing their experience in the selection process. Special thanks to Bojana Koteska and Igor Kulev for their technical support in the preparation of the conference proceedings.

Ohrid,
September 2013

Vladimir Trajkovik
Anastas Mishev
Editors

Organization

ICT Innovations 2013 was organized by the Macedonian Society of Information and Communication Technologies (ICT-ACT).

Conference and Program Chairs

Vladimir Trajkovik   Faculty of Computer Science and Engineering,
University of Ss Cyril and Methodius, Skopje
Anastas Mishev   Faculty of Computer Science and Engineering,
University of Ss Cyril and Methodius, Skopje

Program Committee

Senthil Kumar A.V.   Bharathiar University, India
Rocío Abascal-Mena   Universidad Autónoma Metropolitana -
Cuajimalpa, Mexico
Ariffin Abdul Mutalib   Universiti Utara, Malaysia
Mouhamed Abdulla   Concordia University, Canada
Nevena Ackovska   Ss. Cyril and Methodius University, Macedonia
Seyed-Abdollah Aftabjahani   Georgia Institute of Technology, Georgia
Syed Ahsan   Technische Universität Graz, Austria
Zahid Akhtar   University of Cagliari, Italy
Abbas Al-Bakry   Babylon University, Iraq
Ahmad Al-Zubi   King Saud University, Jordan
Zharko Aleksovski   Philips Research, Eindhoven, Netherlands
Azir Aliu   South East European University, Macedonia
Giner Alor Hernandez   Instituto Tecnologico de Orizaba, Mexico
Adel Alti   University of Setif, Algeria
Luis Alvarez Sabucedo   Universidade de Vigo. Depto. of Telematics, Spain
Hani Alzaid   Queensland University of Technology, Australia
Suzana Andova   Technical University of Eindhoven, Netherlands

Ljupcho Antovski   Ss. Cyril and Methodius University, Macedonia
Ezendu Ariwa   London Metropolitan University, United Kingdom
Goce Armenski   Ss. Cyril and Methodius University, Macedonia
Hrachya Astsatryan   Institute for Informatics and Automation Problems,
                                       National Academy of Sciences of Armenia,
                                       Armenia
Tsonka Baicheva   Institute of Mathematics and Informatics, Bulgaria
Verica Bakeva   Ss. Cyril and Methodius University, Macedonia
Valentina Emilia Balas   Aurel Vlaicu University of Arad, Romania
Antun Balaz   Institute of Physics Belgrade, Serbia
Lasko Basnarkov   Ss. Cyril and Methodius University, Macedonia
Ildar Batyrshin   Mexican Petroleum Institute, Mexico
Genge Bela   Petru Maior University, Romania
Orlando Belo   University of Minho Campus de Gualtar, Portugal
Marta Beltran   Rey Juan Carlos University, Spain
Ljerka Beus-Dukic   University of Westminster, United Kingdom
Radu Bilba   Universitatea George Bacovia, Romania
Gennaro Boggia   DEE - Politecnico di Bari, Italy
Slobodan Bojanic   Universidad Politécnica de Madrid, Spain
Mirjana Borštnar Kljajić   Faculty of Organizational Sciences, Slovenia
Dragan Bosnacki   Eindhoven University of Technology, Netherlands
Klaus Bothe   Institute of Informatics, Humboldt University,
                                       Berlin, Germany
Zaki Brahm   Research Unit (URPAH) Tunisia
Francesc Burrull   Universidad Politecnica de Cartagena, Spain
Jose-Raul Canay Pazos   Universidade de Santiago de Compestela, Spain
Kalinka Regina Castelo Branco   USP, Brasil
Nick Cavalcanti   UFPE, United Kingdom
Ruay-Shiung Chang   National Dong Hwa University, Taiwan
Somchai Chatvichienchai   Department of Information and Media Studies,
                                       University of Nagasaki, Japan
Jenhui Chen   Chang Gung University, Taiwan
Hsing-Chung Chen   Asia University, Taiwan
Qiu Chen   Tohoku University, Japan
L. T. Chitkushev   Boston University, USA
Ivan Chorbev   Ss. Cyril and Methodius University, Macedonia
Ping-Tsai Chung   Long Island University, New York, USA
Betim Cico   Polytechnic University, Tirana, Albania
Deepak Dahiya   Jaypee University of Information Technology, India
Ashok Kumar Das   International Institute of Information Technology,
                                       India
Danco Davcev   Ss. Cyril and Methodius University, Macedonia
Antonio De Nicola   ENEA, Italy
Zamir Dika   Southeast University, Macedonia
Vesna Dimitrova   Ss. Cyril and Methodius University, Macedonia

Ivica Dimitrovski   Ss. Cyril and Methodius University, Macedonia
Ciprian Dobre   University Politehnica of Bucharest, Romania
Martin Drlik   Constantine the Philosopher University in Nitra,
                               Slovakia
Hua Duan   Shanghai Jiaotong University, China
Saso Dzeroski   Jozef Stefan Institute, Slovenia
Victor Felea   "Al.I.Cuza" University of IASI, Romania
Sonja Filiposka   Ss. Cyril and Methodius University, Macedonia
Simon Fong   University of Macau, Hongkong
Samuel Fosso Wamba   School of Information Systems & Technology,
                               Wollongong University, Australia
Neki Frasheri   UPT Tirana, Albania
Kaori Fujinami   Tokyo University of Agriculture and Technology,
                               Japan
Elena G. Serova   GSOM & St. Petersburg State University, Russia
Suliman Mohamed Gaber   USM, Malaysia
Slavko Gajin   RCUB, Serbia
Andrey Gavrilov   Novosibirsk State Technical University, now
                               Visiting Professor of Kyung Hee University,
                               Korea
Amjad Gawanmeh   Khalifa University, United Arab Emirates
Sonja Gievska-Krilu   Ss. Cyril and Methodius University, Macedonia
Dejan Gjorgjevik   Ss. Cyril and Methodius University, Macedonia
Danilo Gligoroski   NTNU, Norway
Katie Goeman   KU Leuven, Belgium
Abel Gomes   Univeristy of Beira Interior, Department of
                               Computer Science and Engineering, Portugal
Jorge.Marx Gómez   Oldenburg University, Germany
David Guralnick   Kaleidoscope Learning, New York, USA
Marjan Gusev   Ss. Cyril and Methodius University, Macedonia
Zoran Hadzi-Velkov   Ss. Cyril and Methodius University, Macedonia
Tianyong Hao   City University of Hong Kong, Hong Kong
Natasa Hoic-Bozic   University of Rijeka, Depart. of Informatics,
                               Croatia
Fu-Shiung Hsieh   Chaoyang University of Technology, Taiwan
Yin-Fu Huang   Department of Computer Science & Information
                               Engineering, National Yunlin University of
                               Science and Technology, Taiwan
Yo-Ping Huang   National Taipei University of Technology, Taiwan
Chi-Chun Huang   Department of Information Management, National
                               Kaohsiung Marine University, Taiwan
Ladislav Huraj   University of SS Cyril and Methodius in Trnava,
                               Slovakia

Augostino Marengo	University of Bari, Italy
Ninoslav Marina	University for Information Science and Technology "St. Paul the Apostle", Macedonia
Jasen Markovski	Eindhoven University of Technology, Netherlands
Smile Markovski	Ss. Cyril and Methodius University, Macedonia
Cveta Martonovska	Goce Delchev University, Macedonia
Darko Matovski	MAN, UK
Yu Song Meng	National Metrology Centre, A*STAR, Singapore
Marcin Michalak	Silesian University of Technology, Poland
Dragan Mihajlov	Ss. Cyril and Methodius University, Macedonia
Marija Mihova	Ss. Cyril and Methodius University, Macedonia
Aleksandra Mileva	Goce Delchev University, Macedonia
Anastas Mishev	Ss. Cyril and Methodius University ,Macedonia
Igor Mishkovski	Ss. Cyril and Methodius University, Macedonia
Kosta Mitreski	Ss. Cyril and Methodius University, Macedonia
Pece Mitrevski	St. Kliment Ohridski University, Macedonia
Irina Mocanu	Politehnica University, Romania
Ammar Mohammed	Koblenz University, Germany
Radouane Mrabet	ENSIA, Morocco
Irena Nancovska Serbec	University of Ljubljana, Faculty of Education, Slovenia
Phuc Nguyen Van	Asian Institute of Technology and Management, Vietnam
Viorel Nicolau	"Dunarea de Jos" University of Galati, Romania
Alexandru Nicolin	Horia Hulubei National Institute of Physics and Nuclear Engineering, Romania
Novica Nosovic	University of Sarajevo, Banja Luka
Florian Nuta	Danubius University of Galati, Romania
Abel Nyamapfene	University of Exeter, United Kingdom
Eleonora Pantano	University of Calabria, Italy
João Paulo Papa	Universidade Estadual Paulista, Brasil
Jehan-Francois Paris	Department of Computer Science, University of Houston, USA
Peter Parycek	Danube University Krems, Austria
Shushma Patel	London South Bank University, United Kingdom
Paul Peachey	University of Glamorgan, United Kingdom
Predrag Petkovic	Faculty of Electronic Engineering, University of Niš, Serbia
Antonio Pinheiro	Universidade da Beira Interior, Portugal
Giuseppe Pirlo	Bari University, Italy
Florin Pop	University Politehnica of Bucharest, Romania
Zaneta Popeska	Ss. Cyril and Methodius University, Macedonia
Andreja Pucihar	Ecenter, Slovenia
Selvakumar R.K.,	Anna University Tirunelveli, India

Vladimir Radevski	South East European University, Macedonia
Jiwat Ram	University of Adelaide, Australia
Dejan Rancic	Elektronski Fakultet, Nish, Serbia
Danda Rawat	Eastern Kentucky University, United States of America
Manjeet Rege	Rochester Institute of Technology, United States of America
Andreas Riener	Johannes Kepler University Linz, Institute for Pervasive Computing, Austria
Jatinderkumar Saini	Gujarat Technological University, India
Suresh Sankaranarayanan	University of WestIndies, Jamaica
Venkat Sastry	Defence College of Management and Technology, Cranfield University, UK
Vladimír Siládi	Matej Bel University, Slovakia
Manuel Silva	ISEP, Portugal
Dr. Dharm Singh	MP University of Agri & Tech Udaipur, India
Brajesh Kumar Singh	FET, R.B.S. College, Bichpuri, India
Ana Sokolova	University of Salzburg, Austria
Yeong-Tae Song	Towson University, USA
Michael Sonntag	Johannes Kepler, University Linz, Austria
Dejan Spasov	Ss. Cyril and Methodius University, Macedonia
Bratislav Stankovic	University for Information Science and Technology "St. Paul the Apostle", Macedonia
Georgi Stojanov	American University of Paris, France
Radovan Stojanovic	University of Montenegro, Montenegro
Igor Stojanovic	Faculty of Computer Science
Stanimir Stoyanov	University of Plovdiv, Bulgaria
Chandrasekaran Subramaniam	Kumaraguru College of Technology, Coimbatore, India
Chang-Ai Sun	University of Science and Technology, China
Kenji Suzuki	The University of Chicago, USA
Irfan Syamsuddin	State Polytechnic of Ujung Pandang, Indonesia
Anas Tawileh	Cardiff University, UK
Sudeep Thepade	MPSTME, SVKM's NMIMS University, Mumbai, India
Ousmane Thiare	University Gaston Berger of Saint-Louis, Senegal
Dimitar Trajanov	Ss. Cyril and Methodius University, Macedonia
Ljiljana Trajkovic	Simon Fraser University, Canada
Vladimir Trajkovik	Ss. Cyril and Methodius University, Macedonia
Igor Trajkovski	Ss. Cyril and Methodius University, Macedonia
Chidentree Treesatayapun	Cinvestav, Mexico
Yuh-Min Tseng	Department of Mathematics, National Changhua University of Education, Taiwan
Goran Velinov	Ss. Cyril and Methodius University, Macedonia

Nguyen Quoc Bao Vo	Posts and Telecommunications Institute of Technology, Vietnam
Sanja Vranes	Mihajlo PupIn Institute, Serbia
Krzysztof Walkowiak	Wroclaw University of Technology, Polland
Wan Adilah Wan Adnan	Universiti Teknologi MARA, Malaysia
Shuai Wang	New Jersey Institute of Technology, USA
Santoso Wibowo	CQUniversity, Australia
Michal Wozniak	Wroclaw University of Technology, Polland
Shuxiang Xu	University of Tasmania, Australia
Tolga Yalcin	Ruhr-University Bochum, HGI, EMSEC, Germany
Chao-Tung Yang	Department of Computer Science, Tunghai University, Taiwan
Wuyi Yue	Konan University, Japan
George Z. Chen	Heriot-Watt University, UK
Mazdak Zamani	Universiti Teknologi, Malaysia
Zoran Zdravev	Goce Delchev University, Macedonia
Katerina Zdravkova	Ss. Cyril and Methodius University, Macedonia
Xiangyan Zeng	Fort Valley State University, USA
Defu Zhang	Xiamen University, China
Dawid Zydek	Idaho State University, USA

Organizing Committee

Ana Madevska Bogdanova	Ss. Cyril and Methodius University, Macedonia
Goran Velinov	Ss. Cyril and Methodius University, Macedonia
Zoran Zdravev	Goce Delchev University, Macedonia
Lidija Goracinova	FON University, Macedonia
Azir Aliu	South East European University, Macedonia
Cvetko Andreevski	St. Kliment Ohridski University, Macedonia

Technical Committee

Sonja Filiposka	Ss. Cyril and Methodius University, Macedonia
Vesna Dimitrova	Ss. Cyril and Methodius University, Macedonia
Bojana Koteska	Ss. Cyril and Methodius University, Macedonia
Igor Kulev	Ss. Cyril and Methodius University, Macedonia

Contents

Technology Enhanced Learning – The Wild, the Innocent and the E Street Shuffle

Vladan Devedzic

Faculty of Organizational Sciences, University of Belgrade, Serbia
devedzic@fon.rs

Abstract. Although Technology Enhanced Learning (TEL) is still developing and attracts a lot of R&D attention, initiatives and funding worldwide, it has been around for quite some time and allows for a critical assessment. Once just "somewhere at the intersection of pedagogy and learning technology", today TEL spans many other fields and phenomena, like social and organizational processes, computer games, knowledge management, standardization, policy making in various sectors, sustaining the impact of learning, and efforts to overcome digital divide. Still, not everything goes smoothly. As reports and surveys indicate, there are many challenges ahead, still waiting to be tackled.

Keywords: Technology enhanced learning, TEL, pedagogy, ICT in education.

1   Introduction

The term *Technology Enhanced Learning* (*TEL*) is self-explanatory and easily understood intuitively. There are also some definitions around, and all of them are good. So, for instance, in this text it is convenient to adopt the one from [41] where TEL is defined as "any online facility or system that directly supports learning and teaching. This may include a formal virtual learning environment (VLE), an institutional intranet that has a learning and teaching component, a system that has been developed in house or a particular suite of specific individual tools." Note also that Wikipedia stipulates that "*E-learning* refers to the use of electronic media and information and communication technologies (ICT) in education. E-learning is broadly inclusive of all forms of educational technology in learning and teaching. E-learning is inclusive of, and is broadly synonymous with *multimedia learning, technology-enhanced learning* (TEL), *computer-based instruction* (CBI), *computer-based training* (CBT), *computer-assisted instruction or computer-aided instruction* (CAI), *Internet-based training* (IBT), *Web-based training* (WBT), *online education, virtual education*, virtual learning environments (VLE) (which are also called *learning platforms*), *m-learning*, and *digital educational collaboration*. These alternative names emphasize a particular aspect, component or delivery method." [43]

TEL encompasses and spans many areas, fields, processes and phenomena. A quick look at TEL researchers' and practitioners' current topics of interest (e.g., the topics covered by the ECTEL 2013, one of the TEL community's most widely

V. Trajkovik and A. Mishev (eds.), *ICT Innovations 2013*,
Advances in Intelligent Systems and Computing 231,
DOI: 10.1007/978-3-319-01466-1_1, © Springer International Publishing Switzerland 2014

recognized annual conferences, `http://www.ec-tel.eu/`) immediately reveals this diversity:

- Technological underpinning: large-scale sharing and interoperability, personalization, user modeling and adaptation, context-aware systems, social computing, Semantic Web, mobile technologies, serious games and 3D virtual worlds, network infrastructures and architectures for TEL, sensors and sensor networks, augmented reality, roomware, ambient displays and wearable devices, data mining and information retrieval, recommender systems for TEL, learning analytics,…
- Pedagogical underpinning: problem- and project-based learning / inquiry based learning, computer-supported collaborative learning, collaborative knowledge building, game-based and simulation-based learning, story-telling and reflection-based learning, learning design and design approaches, communities of learners and communities of practice, teaching techniques and strategies for online learning, learner motivation and engagement, evaluation methods for TEL,…
- Individual, social & organizational learning processes: cognitive mechanisms in knowledge acquisition and construction, self-regulated and self-directed learning, reflective learning, social processes in teams and communities, social awareness, trust and reputation in TEL, knowledge management, organizational learning,…
- Sustainability and scaling of TEL solutions: Massive Open Online Courses (MOOCs), open educational resources (OER), learning networks, teacher networks, cloud computing in TEL, bring your own device (BYOD), orchestration of learning activities, learning ecologies, learning ecosystems, fitness and evolvability of learning environments, business models for TEL,…
- Learning contexts and domains: schools of the future, promoting learning and employability within disadvantaged groups and communities, applications of TEL in various domains, formal education: initial (K-12, higher education), post-initial (continuing education), workplace learning in small, medium and large companies, networked enterprise settings as well as public and third sector organizations, distance and online learning, lifelong learning,, vocational training, informal learning, non-formal learning, ubiquitous learning,…
- TEL in developing countries: ICT inclusion for learning, digital divide and learning, generation divide and learning, education policies, rural learning,…
- TEL, functional diversity and users with special needs: accessible learning for all, visual, hearing and physical impairments, psycho-pedagogic support for users, educational guidance for tutors, adapted learning flow, content and monitoring process, standards about accessibility and learning,…

There are also new, emerging topics, because the field of TEL is spreading out fast. However, without going into details, TEL is still about *learning* in the first place (or learning and teaching, precisely); that's the L in TEL. It is still about how learning is enhanced, assisted and supported by *technology*, the T in TEL. But it is definitely not just about T, nor it is just about L – there is an E as well, since TEL stands for technology *enhanced* learning. The often neglected and misinterpreted E is there to abridge the T and the L, and is largely multifaceted, complex, and brings the necessary rich, yet subtle, complementing, elegant low-key tone to the field.

2   Technology – The Wild

To many people, technology is seductive. Just think about the fascination each new version of iPhone creates in minds of so many people. In one of his inspiring, exciting talks, given at the AIED'99 conference in Le Mans, France, Elliot Soloway exclaimed about young learners: "What they want is *technology*! Technology!!!" And given the momentum of technological development, it is no wonder that technology greatly influences (and often drives) progress in many disciplines and fields. In TEL, the idea of the T part is to provide technology-rich learning environments to support and enhance teaching and learning, in terms of enabling more motivating, more engaging and more effective learning experiences. These environments are diverse and range from virtual learning environments, to learning management systems, to personal learning environments, to universally-available tools such as social software, virtual worlds, on-line games and many more, to real-world spaces that complement the use of modern technologies in education [23].

2.1   Learning Technology – A Very Brief Overview

A *Learning Management System* (*LMS*) is a software application for the administration, documentation, tracking, reporting and delivery of e-learning education courses or training programs [14]. LMSs range from systems for managing training and educational records to software for distributing online or blended/hybrid college courses over the Internet with features for online collaboration. Platforms like Moodle, BlackboardLearn, Sakai, and many more are popular LMSs used to manage university/college, corporate, and other categories of learning delivery and administration.

Intelligent Tutoring Systems (ITSs) are computer systems that use methods and techniques of artificial intelligence to improve the processes of computer-based teaching and learning [11]. They aim to provide immediate and customized instruction or feedback to learners, usually without intervention from a human teacher. There are many examples of ITSs being used in both formal education and professional settings, in which they have demonstrated their capabilities and limitations.

A *Virtual Learning Environment* (*VLE*) is an e-learning education system based on the Web that and includes a content management system and enables virtual access to classes, class content, tests, homework, grades, assessments, and other external resources (such as appropriate Website links) [21]. In a VLE, learners and teachers typically interact through threaded discussions, chat, Web conferencing systems such as Adobe Connect, and Web 2.0 tools.

There is much hype around about *using Web 2.0 and social media in learning* [22]. Using Facebook and Twitter in the classroom, engaging students through blogging, encouraging them to listen to topic-related podcasts etc. are but a few ideas on how these widely used technologies should be applied in TEL.

Personal Learning Environments (*PLEs*) are systems that help learners take control of and manage their own learning in terms of setting their own learning goals, managing their learning activities, the content they use, the learning process, and communication with others [2]. PLEs are based on the idea that learning is a continuous process, seeks to provide supporting tools takes place in different contexts and situations, and will not be provided by a single learning provider. Technically, PLEs represent an integration of a number of Web 2.0 technologies like blogs, wikis, RSS feeds, Twitter, Facebook, etc. around the independent learner.

E-book technologies offer new ways of interacting with massively shared, adaptive and dynamic books, such as alternative versions of text, sharing annotations, crowd authoring (where textbooks are produced by students, for students), embedding graphs and simulations showing live data, using tools such as timers and calculators to support structured learning and formative assessment, co-reading (where readers are automatically put in contact with others currently reading the same page) and many more [36].

Digital badges (*DBs*) are means for identifying, measuring, validating and recognizing skills, competencies, knowledge, and achievements of learners regardless of their origin (formal, non-formal, informal), as well as a way of accrediting non-formal learning [24]. Badges may be awarded by authorities, by peers, or may be automatically assigned on completion of certain tasks. Mozilla Foundation has developed an infrastructure to award, manage and validate DBs.

M-learning (or *mobile learning*) *technologies* focus on learning across contexts and learning with mobile devices, i.e. learning that happens when the learner is not at a fixed, predetermined location, but takes advantage of the learning opportunities offered by mobile technologies [32].

Learning analytics (*LA*) involves the collection, analysis and reporting of large datasets about learners and their contexts in order to improve learning and the environments in which learning takes place [38]. LA systems research also focuses on allowing real-time analysis of disparate data in order to show when a learner is making good progress or is struggling. LA is often combined with *information visualization techniques* to provide effective, interactive exploration of TEL data, discover learning patterns and help learners with recommendations [13].

Massive Open Online Courses (*MOOCs*) use cloud computing, instructional design technology, various IT platforms and other modern technology to enable large-scale interactive participation and open access to education via the Web [36]. Any person who wants to take such a course can, with no limit on attendance. In addition to traditional course materials such as videos, readings, and problem sets, MOOCs provide interactive user forums that help build a learning community. Coursera, Udacity and edX are popular MOOC platforms and providers. *Open Educational Resources* (*OERs*) [5] are a similar concept, but unlike MOOCs provide only learning resources, not any certification of competence.

A *serious game* a game designed for a primary purpose other than pure entertainment [39]. There are specific serious gaming environments for education that include specific virtual worlds and simulations. Until recently, they were used widely outside of formal education systems; their use within schools is less common, but is getting more momentum.

2.2   Examples

Certain technology may be designed primarily for learning/teaching, but much of the existing technology designed for other purposes is used in education as well. The Top 100 Tools for Learning Website (http://c4lpt.co.uk/top100tools/) maintains a list of tools widely used in education, and at the top positions one usually sees various Google tools, Twitter, YouTube, Skype and the like.

Paperista. Paperista (http://www.uzrok.com/paperista/) is a recently developed Web-based visualization and exploration tool that enables researchers and students interested in learning analytics and educational data mining to explore a dataset about research publications in these two related fields, Fig. 1. The tool provides multiple views, thus allowing users to observe and interact with topics from these fields and understand their evolution and relationships over time.

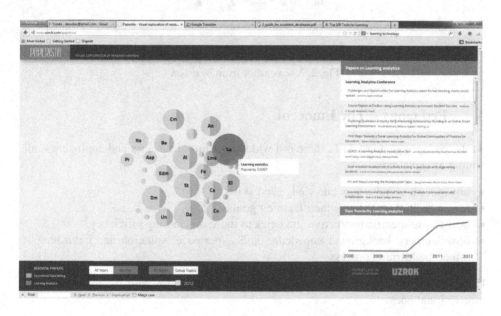

Fig. 1. A screenshot from *Paperista*

Neuroph. Neuroph (http://neuroph.sourceforge.net/) is lightweight Java neural network (NN) framework to for developing common neural network architectures. It includes a well designed, open source Java library with a small number of basic classes that correspond to basic NN concepts. It is used as a NN educational tool at a number of universities worldwide.

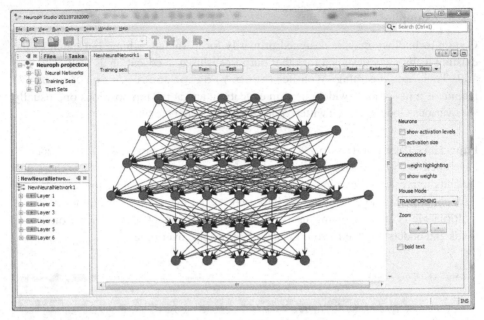

Fig. 2. A screenshot from *Neuroph*

3   Pedagogy – The Innocent

At the pedagogy end of TEL, there is a wide spectrum of theories and approaches, all aiming at explaining:

- how people acquire new knowledge and skills
- how to help them achieve their learning goals
- how to use specific instructive strategies to increase learning efficiency
- how learners' background knowledge and experience, situation and environment affect the learning process

Pedagogical theories are often used in TEL in conjunction with psychological and motivational ones.

3.1   Pedagogical Theories/Models/Approaches – A Very Brief Overview

Pedagogical theories/approaches usually imply specific teaching and learning models. For example, the traditional *transmission model* assumes that there is a body of knowledge that learners need to acquire, typically from lectures or direct instruction, or from interacting with it [9]. This roughly corresponds to making learning materials available on the corresponding course Website or through the school's LMS. In the *constructivist model*, attention is on the individual and learners' background is taken into account in their process of gradually developing understandings of and beliefs about elements of the domain [35]. The teacher facilitates group dialogue that explores an element of the domain with the purpose of leading to the creation and shared understanding of a topic, and encourages further reading, exploration of a corresponding Web site, and development of learners' meta-awareness of their own understandings and learning processes. She/He provides opportunities for learners to determine, challenge, change or add to existing beliefs and understandings through engagement in tasks that are structured for this purpose, possibly working collaboratively in a shared wiki space. Thus knowledge is developed/built/constructed gradually, as the result of interplay between external stimuli and internal interpretation. The goal of the learner is to acquire the "correct" interpretation, solving the given problem on her/his own and then self-assessing and self-reflecting on the produced outcome [30]. Discovery, hands-on, experiential, collaborative, project-based, and task-based learning are all instructional approaches that base teaching and learning on constructivism.

Sociocultural learning theories take greater account of the important roles that social relations, community, and culture play in cognition and learning [42]. They draw heavily on the work of Vygotsky [40] that emphasizes that learning is embedded in social events, and social interaction plays a fundamental role in the improvement of learning. Sociocultural learning theories provide a strong theoretical support for *collaborative learning*, where learners at various performance levels work together in small groups toward a common goal.

Not all TEL is instruction-based, but when it *is*, then instructional design matters. *Instructional design* is the systematic process by which instructional materials are designed, developed, and delivered [20]. It is the practice of creating instructional experiences that make the acquisition of knowledge and skill more efficient, effective, and appealing [27]. It assumes determining the current state and needs of the learner, defining the end goal of instruction, and creating some intervention to assist achieving the goal and satisfying the learner's needs. There are many instructional design models but many are based on the generic ADDIE model/framework [29] that represents instruction in five phases: analysis, design, development, implementation, and evaluation. Note that the final step should lead to revising instruction according to learners' reactions to the content, activities, and assessment. This, in turn should facilitate adopting an iterative cycle of educational design and practice.

The term *learning design* is often used interchangeably with instructional design and denotes the process of planning and structuring a sequence of *learning activities* leading to completion of one or more *learning goals*. In TEL, learning activities can include working with LMSs/VLEs, searching the Web, finding and evaluating resources, listening to podcasts, solving problems, collaborating with peers, constructing a wiki, manipulating data, using relevant discussion forums and social media, etc. These activities facilitate communication, shared understanding of the learning domain, community building and learners' reflection on what they have learned. Learning goals should be SMART – specific, measurable, achievable, relevant and time-oriented [28].

3.2   Recent Trends

A great impact of the use of technologies for learning is in *informal* and *workplace learning* [3]. In such contexts, social software is widely used not only for information seeking and communication purposes, but also for informal learning that: takes place in small episodes, in response to problems or issues, or is driven by the interests of the learner; is sequenced and controlled by the learner; is heavily contextual in terms of time, place and use; is cross-disciplinary and often takes place in communities of practice.

Many of the above mentioned theories, approaches and models did not take the full advantage and the affordances of the new on-line, interconnected, digital realm. In *Learning 2.0*, "learners' familiarity with Web2.0 technologies opens up a completely new space for and style of learning, focusing on collaborative knowledge building; shared assets; breakdown of distinction between knowledge and communication." [6] Learning 2.0 enables learners to control and track individual learning using a set of Web 2.0 tools and systems that enable collaboration (wikis, blogs, social bookmarking tools, virtual worlds such as Second Life, etc.). Learning 2.0 doesn't replace the conventional means of learning; it augments the conventional methods of learning with a set of Web 2.0 tools and systems. Knowledge (as meaning and understanding) is socially constructed. Learning takes place through conversations about content and grounded interaction about problems and actions.

Connectivism views learning as a primarily network-forming process that focuses on individuals exploring the knowledge situated externally from people, i.e. on the Web [37]. Connectivism emphasizes the role of the social and cultural context in learning. In a way, it is similar to Vygotsky's "zone of proximal development" (ZPD) [40] and to Bandura's social learning theory [4] that proposes that people learn through contact. Siemens calls connectivism a learning theory for the digital age, since it reflects how technology affects how people live, how they communicate and how they learn.

Another popular pedagogical theory in the TEL community is *activity theory* and the related *expansive learning* [15]. It contextualizes the interaction between humans and computers indicating that "(a) Contents and outcomes of learning are not merely knowledge in texts and the heads of students but new forms of practical activity and artifacts constructed by students and teachers in the process of tackling real-life projects or problems – it is learning what is not yet known. (b) Learning is driven by genuine developmental needs in human practices and institutions, manifested in disturbances, breakdowns, problems, and episodes of questioning the existing practice. (c) Learning proceeds through complex cycles of learning actions in which new objects and motives are created and implemented, opening up wider possibilities for participants involved in that activity. This perspective on teaching and learning highlights the potential impact of new tools as vehicles for transforming activity procedures." [15]

There are many other pedagogic approaches, models and theories to choose from. Scaffolding [16], curriculum development and rhizomatic knowledge [10], metacognition [8], *bricolage* [44], *framework of knowledge ecologies* [33] and *learning styles* [8] are just some of them. See [3] and [36] for comprehensive overviews.

3.3   Examples

iCare model. A popular learning/pedagogical model in TEL is the iCare model [19]. iCare stands for:

- Introduce the topic: not just the learning outcomes
- Connect: not just content
- Apply: through activities, exercises
- Reflect: through discussions
- Extend: provide supplemental or advanced material

Simultaneously, iCare is a toolkit for TEL, helping staff in educational institutions in designing courses for online learning. Entire courses can be constructed according to iCare principles. They can be applied to each module/lesson in a course; typically, a different Web page is associated with each iCare element. For example, a course with six modules might have thirty Web pages associated with it, five pages per module. In a typical module, the Introduction element (Web page) may include the learning outcomes, prerequisites, readings, and necessary equipment/software for module activities. The Connect element should provide the necessary concepts, principles, facts, processes etc. related to the module. Scaffolds can be provided as well, allowing learners to discover the parts of the module for themselves (which is often recommended and makes learning more efficient).

The Apply element should offer challenges and activities that allow learners to apply and practice the knowledge they gained in Connect to tasks and problems from the real-world. Web quests, simulations, games, quizzes and the like can be used to scaffold specific activities. The Reflect is self explanatory: activities, questions, self-evaluations etc. are provided to help students apply metacognitive processes as they articulate what they've learned/experienced from Connect to Apply sections of the module. The Extend section offers opportunities to individualize learning experiences with additional, optional learning materials and activities.

Peer-to-Peer Learning and Peeragogy. Peer-to-peer (P2P) learning [31] and its recent derivative called peeragogy [1] promote learners' interactions with other learners to attain learning goals. They investigate conditions and opportunities for learning through existing social networks, facilitated peer learning, and two-way (or more) communication. Central to P2P learning is the idea that every learner can also be a "teacher", and that learning can be individual-, group- and community-driven. A range of techniques that self-motivated learners are investigated for applicability and suitability in different learning contexts, as well as the challenge of peer-producing a useful and supportive context for self-directed learning.

4   Enhancing L with T – The E Street Shuffle

> *Our primary focus is on the **enhancement** of learning and*
> *teaching: this drives our approach. Technology supports this*
> *enhancement goal, and is therefore a factor in the development*
> *of effective learning, teaching and assessment strategies.* [18]

There are many ways for E in TEL to manifest itself, but there is also a lot of room for improvement to this end. It is exactly in E where TEL spans to and intertwines with many other disciplines, aspects, processes and phenomena, and where interdisciplinary knowledge, skills, experiences and practices are the most essential. It is here that diversity grows exponentially, in terms of the aspects of TEL relevant in educational institutions, at workplaces, at home, and elsewhere. And it is also here that challenges for future development of TEL are the most abundant.

4.1   L ↔ T – A 2-Way Street?

"If you build it, they will come."
No. Perhaps it works in movies and in fairytales, but not in real-world TEL. The T end of the TEL community is notorious for failing to consider TEL in light of, and subservient to, the human cognitive system [12]. They tend to see TEL practice as technologically rather than pedagogically driven. However, it is not what technology learners use that counts, but what they take from it. In Dror's words: "How the learners interpret, understand and internalize the learning material via technology is what the focus should be on... Too many TEL are over focused on the technology rather than on the learners' cognition." Once learners are given learning resources through an LMS or another technology, the learning process just begins – there is no guarantee that they will use it. They typically need motivation, as well as some meta-cognition ("to know what they know" and "to know what they need to know"). Likewise, learning must be challenging to learners, and TEL technology must take these simple facts into account. TEL technology should also incite commitment to learning.

On the other hand, Attwell and Hughes note that "it is not clear at all how pedagogical theories have impacted the actual use of technology in learning, education and training, other than they offer ways to make sense of how technologies fit in the picture" [3]. In their survey, they have found almost no survey data on the impact of technology on pedagogy. If any impact, that would be in the informal and workplace learning domain, not in curriculum driven education. They also note that most of the existing studies of that impact come from researchers, or from their institutions or funding agencies, not from the new practices that are more related to informal learning and the workplace.

Price and Kirkwood have come to a similar conclusion: "while the use of technology *may* enhance learning, the *evidence* supporting these claims is tangential, as is the evidence illustrating changes in the practices of higher-education teachers" [34]. They explain why: technologies and tools are transient, can often be used in many different ways, and if they work in one context (group, faculty or institution) it does not necessarily imply that they will work in another. In order to be able to claim a certain level of impact of technology on TEL, they suggest to first collect and share evidence (reports, case studies, etc.) with the community, providing details such as [25]: What was the teaching and learning concern or issue being addressed by the intervention? Why did you need to engage with it? How was the pre-existing situation to be improved? What was the topic/discipline and at what level? What technology was used and why? What evidence was used to drive or support the design of the intervention? What was the design of the intervention? What was the context within which it was used? How did the intervention relate to assessed activities (formative or summative)? How many students were involved? What was the nature of the evaluation undertaken and/or the evidence gathered? What was the impact of the intervention (on students' learning/on teaching practice/on the activities of others)? How successful was the intervention at addressing the issue identified at the outset? In other words, higher education teachers need to devise and design activities to promote learning and to use technologies in ways that enable students to achieve desired educational ends [12].

Many other researchers and studies have addressed the role of E in bridging the gap between T and L. What follows is a summary of just a minor part of such contributions, coming from various authors and indicating a rich array of prospects:

- The TPACK framework (Technological Pedagogical Content Knowledge) describes the kinds of knowledge – technology, pedagogy, and content – needed by a teacher for effective pedagogical practice in a TEL environment [26]. TPACK includes 7 different knowledge areas: Content Knowledge (CK), Pedagogical Knowledge (PK), Technology Knowledge (TK), Pedagogical Content Knowledge (PCK), Technological Content Knowledge (TCK), Technological Pedagogical Knowledge (TPK), and Technological Pedagogical Content Knowledge (TPCK).
- Gulz et al. argue that TEL focuses on the tasks and content of the material to be learned, but lacks off-task interaction opportunities so essential for learners and learning processes [17].
- JISC suggests the following approaches [23]: reviewing how course design and validation can be informed by technology; investigating more flexible and creative models of delivery through technology; supporting release of OERs; exploring how learners experience and participate in technology-rich learning; researching more carefully into the competencies required for learning in a digital age and the support available to learners; exploring how designing learning and the development of planning tools can be informed by effective pedagogic practice; enabling connectivity to information and to others; knowledge-sharing and co-authoring across multiple locations; opportunities for reflection and planning in personal learning spaces; rapid feedback on formative assessments; participation in communities of knowledge, inquiry and learning; learning by discovery in virtual worlds; development of skills for living and working in a digital age.
- Barra and Usman advocate that TEL should be largely profiled to help in assessment and to support adaptive release / conditional activities (allowing academics to release content (lecture notes, online quizzes, assignment dropbox) on VLEs to learners only after certain conditions are met) [5].

- Technologists may want to take a more agile way of working with pedagogists, in terms of exploiting metaphors used in the *agile paradigm* of software development [7]. Agile software development enables having working software prototypes early in the project, and upgrading them iteratively and incrementally in very short cycles. Agile development also stresses rigorous testing of all software under development and thus implicitly increases the software quality. Transposed to the world of TEL, it means that technologists should work more intensively with pedagogists, on very short cycles, using pedagogists as domain experts and "customers" to collaborate with in a kind of "requirements engineering game". It can help both parties – technology developers can insist on continuity of this process to eliminate their reluctance ("Will it matter?") and blindness ("Let's build it, they will surely come!"), whereas pedagogists can make their beliefs about technology more realistic.

4.2   A Survey from UK (2012)

A recent survey of TEL trends in UK [41] shows some shifts from T and L being at the opposing ends of TEL towards more integration between the two. The survey shows that the major reasons why a higher education institution (HEI) in UK uses/considers to use TEL are to enhance the quality of learning and teaching and improve access to learning for students off campus. More and more academic staff are knowledgeable of TEL and capable of delivering it. More importantly, the proportion of *Web supplemented modules* (those that use the Web just to *support* module delivery) has steadily decreased over the years, and the proportion of *Web dependent modules* (those where interaction with content and communication through the Web is essential) is increasing. Optimization of services for mobile devices by institutions (access to *library services, email* and *course announcements, timetabling information, access to course materials* and *personal calendars* for iPhone, iPad and Android devices) are more and more mobile-enabled. Enthusiasm of teachers (especially tutors) for TEL is increasing, and so are the number of distance learning courses and the number of distance students. Teaching *about* using TEL at HEIs and workplaces alike is also getting popular. Subject areas that make more extensive use of TEL include medicine, nursing, health; management, accountancy, finance, business; education; and social sciences, psychology, law, teaching etc. Among the most supported TEL software tools at HEIs are plagiarism detection tools, e-submission tools, e-assessment tools, e-portfolio tools, wikis and blogs.

5   Conclusions

In order for TEL to develop more efficiently, future VLEs should support mobile learning, personalization of learning content, capturing general attitudes of learners towards learning, focusing on enrichment of student experiences and understanding learner behavior and interactions through learning analytics and big data.

However, we the gap in technology related skills required by teaching and learning professionals cannot be bridged by qualifications alone or by initial training [3]. A programme of opportunities for continuing professional development related to TEL is also needed to enable people to remain up to date.

Pedagogists and technologists should talk more.

Acknowledgments. The work presented in this paper is partially supported by The Ministry of education, science and technology development of the Republic of Serbia through the project No. III47003, Infrastructure for Technology Enhanced Learning in Serbia.

References

1. Alexander, B., et al.: The peeragogy handbook (2013)
2. Attwell, G.: Personal Learning Environments – the future of eLearning? eLearning Papers 2(1), 1–7 (2007)
3. Attwell, G., Hughes, J.: Pedagogic Approaches to Using Technology for Learning: Literature Review (2010)
4. Bandura, A.: Social Learning Theory (1977)
5. Barra, P., Usman, S.: Technology enhanced learning and assessment (2013), http://www.heacademy.ac.uk/assets/documents/disciplines/bios ciences/2013/Barra&Usman.pdf
6. Beetham, H., McGill, L., Littlejohn, A.: Thriving in the 21st century: Learning Literacies for the Digital Age (LLiDA project). The Caledonian Academy, Glasgow Caledonian University, Glasgow (2009)
7. Cockburn, A.: Agile software development: The cooperative game, 2nd edn. Addison-Wesley (2006)
8. Coffield, F.: Just suppose teaching and learning became the first priority. Learning and Skills Network (2008)
9. Cohen, D.: revolution in one classroom: The case of Mrs. Oublier. Educational Evaluation and Policy Analysis 12(3), 311–344 (1997)
10. Cormier, D.: Rhizomatic education: Community as curriculum. Innovate 4(5) (2008)
11. Devedzic, V.: Semantic Web and Education. Springer, Heidelberg (2006)
12. Dror, I.E.: Technology enhanced learning: The good, the bad, and the ugly. Pragmatics & Cognition 16(2), 215–223 (2008)
13. Duval, E.: Attention Please! Learning Analytics for Visualization and Recommendation. In: Proceedings of LAK11: 1st International Conference on Learning Analytics and Knowledge, pp. 9–17. ACM (2011)
14. Ellis, R.K.: Field Guide to Learning Management Systems. ASTD Learning Circuits (2009)
15. Engeström, Y.: Activity Theory and Individual and Social Transformation. In: Perspectives on Activity Theory: Learning in Doing: Social, Cognitive & Computational Perspectives, pp. 19–38 (1999)
16. Feden, P., Vogel, R.: Education. McGraw-Hill, New York (2006)
17. Gulz, A., Silvervarg Flycht-Eriksson, A., Sjödén, B.: Design for off-task interaction – rethinking pedagogy in technology enhanced learning. In: Proceedings of the 10th IEEE International Conference on Advanced Learning Technologies, pp. 204–206. IEEE (2010)

18. HEFCE (Higher Education Funding Council for England): Enhancing Learning and Teaching Through the Use of Technology: A revised approach to HEFCE's strategy for e-learning (2009)
19. Hoffman, B., Ritchie, D.C.: Teaching and learning online: Tools, templates, and training. Technology and Teacher Educational Annual (1998)
20. Instructional Design Central: What is Instructional Design (2012),
    http://www.instructionaldesigncentral.com/htm/
    IDC_instructionaldesigndefinitions.htm
21. JISC: Briefing Paper 1: MLEs and VLEs Explained (2007),
    http://www.jisc.ac.uk/whatwedo/programmes/
    programme_buildmle_hefe/mle_lifelonglearning_info/
    mle_briefingpack/mle_briefings_1.aspx
22. JISC: Celebrating 10,000 followers... and our resources to help engage students through social media (2012),
    http://www.jiscs.ac.uk/news/stories/2012/08/socialmedia.aspx
23. JISC: Effective Practice in a Digital Age: A guide to technology-enhanced learning and teaching. Higher Education Funding Council for England (HEFCE) (2009),
    http://www.jisc.ac.uk/media/documents/publications/
    effectivepracticedigitalage.pdf
24. Joseph, B., et al.: Six ways to look at badging systems designed for learning (2012),
    http://www.olpglobalkids.org/content/six-ways-look-badging-
    systems-designed-learning
25. Kirkwood, A., Price, L.: Enhancing learning and teaching through technology: a guide to evidence-based practice for academic developers (2011)
26. Koehler, M.J., Mishra, P.: Introducing TPCK. In: Colbert, J.A., Boyd, K.E., Clark, K.A., Guan, S., Harris, J.B., Kelly, M.A., Thompson, A.D. (eds.) Handbook of Technological Pedagogical Content Knowledge for Educators, pp. 1–29 (2008)
27. Merrill, M.D., Drake, L., Lacy, M.J., Pratt, J.: ID2_Research_Group: Reclaiming instructional design. Educational Technology 36(5), 5–7 (1996)
28. Meyer, J.: What would you do if you knew you couldn't fail? Creating S.M.A.R.T. Goals. Attitude Is Everything: If You Want to Succeed Above and Beyond. Meyer Resource Group, Incorporated (2003)
29. Morrison, G.R.: Designing Effective Instruction, 6th edn. John Wiley & Sons (2010)
30. Nicol, D.: The foundation for graduate attributes: Developing self-regulation through self and peer assessment. The Quality Assurance Agency for Higher education. London, UK (2010)
31. O'Donnell, A.M., King, A.: Cognitive perspectives on peer learning. Lawrence Erlbaum, UK (1999)
32. O'Malley, C., et al.: Guidelines for learning/teaching/tutoring in a mobile environment. MOBIlearn (2003), http://www.mobilearn.org/download/
    results/guidelines.pdf
33. Pata, K.: Revising the framework of knowledge ecologies: How activity patterns define learning spaces? In: Lambropoulos, N., Romero, M. (eds.) Educational Social Software for Context-Aware Learning: Collaborative Methods & Human Interaction, pp. 1–23. IGI Global, Hershley (2009)
34. Price, L., Kirkwood, A.: Technology enhanced learning – where's the evidence? In: Proceedings of ascilite 2010 conference, Sydney, Australia (2010)
35. Richardson, V.: Constructivist Pedagogy. Teachers College Record 105(9), 1623–1640 (2003)

36. Sharples, M., et al.: Innovating Pedagogy 2012: Exploring new forms of teaching, learning and assessment, to guide educators and policy makers. Open University, Milton Keynes (2012)
37. Siemens, G.: Connectivism: A learning theory for the digital age. International Journal of Instructional Technology and Distance Learning 2(1), 3–10 (2005)
38. Siemens, G.: What Are Learning Analytics? (2010),
    http://www.elearnspace.org/blog/2010/08/25/
    what-are-learning-analytics/
39. Ulicsak, M., Wright, M.: Games in Education: Serious Game (2010),
    http://media.futurelab.org.uk/resources/documents/
    lit_reviews/Serious-Games_Review.pdf
40. Vygotsky, L.S.: Mind and society: The development of higher psychological processes. Harvard University Press (1978)
41. Walker, R., Voce, J., Ahmed, J.: 2012 Survey of Technology Enhanced Learning for higher education in the UK. Universities and Colleges Information Systems Association, Oxford (2012)
42. Wang, L.: Sociocultural Learning Theories and Information Literacy Teaching Activities in Higher Education. Reference & User Services Quarterly 47(2), 149–158 (2007)
43. Wikipedia: E-Learning (2013), http://en.wikipedia.org/wiki/E-learning
44. Wiseman, B.: Introducing Levi-Strauss and Structural Anthropology. Totem Books (2000)

30. Siemens, G.: Connectivism: A learning theory for the digital age. International Journal of Instructional Technology and Distance Learning (2005)

Studies in Bioengineering and Medical Informatics: Current EU Practices and Western Balkan Initiative

Goran Devedžić

University of Kragujevac, Faculty of Engineering, Serbia
devedzic@kg.ac.rs

Abstract. Scientific and engineering fields of Bioengineering (BE) and Medical (Health) Informatics (MI) are recognized as two key challenges within essential research and innovation strategies in EU and other leading regions and countries worldwide. Many universities are redesigning existing or developing new study programs to provide education and training for tomorrow's biomedical engineers, physicians, and researchers. European Commission is responding to the evident needs by promoting actions through the variety of scientific (e.g. Framework Program and Horizon 2020) and educational (e.g. Tempus, Erasmus, etc.) strategies. The Education, Audiovisual and Culture Executive Agency (EACEA) granted regional Tempus project *"Studies in Bioengineering and Medical Informatics - BioEMIS"* aiming at introducing dedicated study programs at Western Balkan countries, namely Serbia, Montenegro, and Bosnia and Herzegovina. The following, state of the art analysis of European study programs in BE and MI provides basic guidelines for creating new curricula.

Keywords: Bioengineering, Medical (Health) Informatics, Study Programs, Western Balkan.

1 Introduction

New Information and Communication Technologies (ICT) have tremendous impact in all areas of human life and work. They are followed by fascinating breakthroughs in medicine and health systems. To support those trends and create symbiosis of engineering and biomedical knowledge [1-3], higher education is remodeling and redesigning existing and opening new study programs that will meet educational and professional challenges of future engineers, physicians, and researchers. The fields of Bioengineering (BE) and Medical (Health) Informatics (MI) are recognized as two out of a few key challenges within essential research and innovation strategies in EU and in other leading regions and countries worldwide [4-8]. In that course and following the actual educational, training, scientific, industrial and health sectors' needs, University of Kragujevac (Serbia) initiated in 2011/2012 dedicated curriculum development and engaged the major universities in the Western Balkan (WB) region (namely, in Serbia, Montenegro, and Bosnia and Herzegovina) to join in a systematic action of BE&MI study programs creation. The initiative resulted in Tempus project

V. Trajkovik and A. Mishev (eds.), *ICT Innovations 2013*,
Advances in Intelligent Systems and Computing 231,
DOI: 10.1007/978-3-319-01466-1_2, © Springer International Publishing Switzerland 2014

"Studies in Bioengineering and Medical Informatics - BioEMIS", funded by European Commission's the Education, Audiovisual and Culture Executive Agency (EACEA).

The fields of BE&MI are deeply rooted in the culture of inter- and multidisciplinary research and collaboration. During the past decade BE&MI have matured as a profession, and its diversity has been increasingly recognized. Nearly every university worldwide wanting to be at the edge of a technological progress offers a curricula in BE&MI at master and doctoral levels, and numerous offer bachelor level degrees, as well. It is clear that modern medicine would not exist without modern diagnostic and therapeutic equipment and instruments that are the result of synergy of medical/engineering knowledge. The influence of ICT in health and medicine is even more critical.

However, only a few universities in WB region offer modules or courses that are related to BE. Those are usually incorporated in the traditional study programs, e.g. electrical engineering or mechanical engineering, without accreditation and qualification identification. In the field of MI there are almost no such attempts, while local Medical Faculties are traditionally more or less closed to the multidisciplinary knowledge. In contrast to the educational missing links, research activities, projects and scientific papers are quite evident.

On the other hand, health industry together with telecommunication is the most growing in the region. Many clinical centers, public or private, companies for production support and services increased their need for dedicated professionals in BE&MI. For highly skilled graduates, dedicated education is crucial for deep understanding of the specific clinical needs, assisting in medical equipment and devices utilization and maintenance, and designing medical information systems. Therefore, the aim of the BioEMIS project is to transfer good EU practices in this field to WB and to introduce dedicated profiles of education at local universities. Having in mind relatively small geographic distance, common language, common culture, common history and compatible education systems, one of the key objectives is creation of compatible study programs at the most feasible levels.

This study involves the short (comparative) overview of the current EU strategies for tomorrow's bioengineers education, presented in the following section. Third section brings the thorough analysis of the existing BE&MI study programs at European universities. Basic guidelines for curriculum development in the field of MI are provided as well. Conclusions summarize current EU practices and reflections to the Western Balkan initiative.

2   BE&MI Education and Research Strategies for 2020 – A Short Overview

World Wide Web and Internet technologies brought remarkable changes in the area of education. Not only in the way we teach, but also in the way we communicate the knowledge and train on skills [9]. Generally, two decades ago, many approaches were not available, many research issues were not current, many technologies did not exist, many nowadays "well known" achievements were not discovered yet. Accelerated changes in the education process come from the evident influence of ICT and Internet technologies. Educational paradigm shifted and fundamental scientific breakthroughs happened. Research and innovation priorities placed in EU developmental strategies [10] made a starting point and created basic environment that specified the key axes of BE&MI educational development, at least within the short-term framework – for 2020.

EU strategy for smart, sustainable and inclusive growth [10] comprehensively targets these reinforced priorities by strengthening the knowledge and innovation policy is one of the cornerstones. This policy emphasizes education, R&D and innovation as the main features of the set of crucial initiatives that underpin the strategy as a whole. The aim of the EU and the Commission is "to speed up the development and deployment of the technologies needed to meet the challenges identified (health and ageing, being one of these). The first will include: 'building the bio-economy by 2020', 'the key enabling technologies to shape Europe's industrial future' and 'technologies to allow older people to live independently and be active in society' " [10]. In particular, EU insists on collaborative, emerging, enabling, and converging approaches immanent to inter- and multidisciplinary research areas. For instance, such approaches interface ICT, nanotechnologies, mathematics, physics, biology, chemistry, earth system sciences, medical and health sciences, material sciences, neuro- and cognitive sciences, social sciences or economics [12]. This is, obviously, in accordance to the stimulant policy of development and innovation in ICT sector to achieve fully operable digital society [10-12]. Having these (interconnected) strategic priorities and pillars set, it is clear that the area of BE&MI has the strategic importance for the 21st century [13].

Such setting assumes permanent development, redesign and modification of study programs that provide essential and specific knowledge for future engineers and physicians working in inter- and multidisciplinary environments. The aim is to enable better understanding of the health and disease determinants oriented towards, among other goals, disease prevention and treatment, diagnosis improvement, effective screening methods, enhancement of health data usage and intensive deployment of ICT in healthcare practice – using Key Enabling Technologies [12], [14]. These highly multidisciplinary technologies involve micro- and nanotechnology, biotechnology, computation, advanced materials and advanced manufacturing systems as outcome of the synergy of life, physical, and engineering sciences. Many of the achievements of human endeavor, particularly in the last two decades, are the result of merging distinct technologies and disciplines into a unified whole. This new research model and a "blueprint for innovation", which provokes and implies the new educational model as well, is called convergence [15].

Just like in Europe, in the other leading regions and countries worldwide the trend is to treat biomedical engineering, as a separate discipline [16-17], [19], [30-33]. It is estimated that "biomedical engineering is the fastest growing engineering discipline with a projected employment growth of 72% by 2018" [16]. However, the situation at higher education is not unified and still varies to the great extent. North American higher education area offers a broad spectrum of BE&MI study programs at all three levels at the most of universities, while a number of others are hurriedly preparing for the degrees in this field [1-2], [6], [20-21], [28-30]. Pointing out the important role and significance of so-called "nano-bio-info convergence", i.e. nanotechnology, biotechnology and information technology convergence for the 21st century advances, the policy makers point to the strategic importance of BE&MI research and innovation pathways.

Developed Asian-Pacific countries promote BE&MI studies through dedicated study programs, usually under alliance with national biomedical engineering societies, respecting governmental priorities [17-19], [22-25], [28-29], [31], [45-46]. Although this region is the homeland of some of the most developed economies, uneven overall development and isolated rural locations are additional motivating factors for deployment of telemedicine services aimed to enhance diagnosis, therapy and continual medical education, in addition to the mainstream focuses. In that sense last decade was particularly intensive.

The situation in the Latin America and the Caribbean is somewhat different in the sense that BE&MI curricula are not well defined and need more detailed plan of education and investments [32-35]. Studies show that key strategic targets are overall improvement of healthcare system and accreditation of health services and institutions. Thus the demand for BE&MI professionals is significantly increased since underlying actions and objectives, such as vital signals monitoring systems development, surgical micro instrumentation and minimally invasive procedures development, prostheses and implants development, embedded software development, to mention a few, are inconceivable without dedicated workforce. Latin American Regional Council on Biomedical Engineering (CORAL) set essential regional priorities, among which are fostering, promotion and encouraging of research work, study programs, publications and professional activities, coordination of the activities between (regional) universities, health institutions, industries, societies, and consultative services for any national or international agency in all matters related to Biomedical Engineering in the region [34-35].

Obviously, worldwide developments, promotion, and reinforcements in the field of BE&MI show increased pace and urge for additional engagement and efforts in all areas, including higher education. To underpin underlying strategic actions, in the global economy context, the essential decisions in the European Higher Education Area are already made: "European Commission has increasingly emphasized the role of universities in contributing to the knowledge society and economy: Europe must strengthen the three poles of its knowledge triangle: education, research and innovation. Universities are essential in all three" [11].

Tempus project "Studies in Bioengineering and Medical Informatics - BioEMIS" has the consortium of 24 partner institutions. Among the partners are 13 universities (4 from EU: University of Birmingham (UK), University Pierre and Maria Currie – Sorbonne (France), Tampere University of Technology (Finland), and University of Maribor (Slovenia); 4 from Serbia: Universities of Kragujevac, Belgrade, Nis, and University of Defense; 4 from Bosnia and Herzegovina: Universities of Banja Luka, East Sarajevo, Mostar, and Bihac; and one from Montenegro: University of

Montenegro). Another 11 partners are Ministry of education, medical chambers, clinical and rehabilitation centers, and medical industry companies from the Western Balkan region.

3 Analysis of the Existing BE&MI Study Programs at European Universities

In order to develop study programs in BE&MI at all three university levels in WB countries, BioEMIS Tempus project team from University of Kragujevac (Serbia) conducted thorough analysis of the similar study programs offer in Europe. The objective of the analysis is to create the foundation for harmonized future inter-institutional cooperation, students' and faculty staff mobility, integrated programs of study, training, and research [36]. During the last decade dozens of papers have been published on this topic. However, the most of them considered the field of BE&MI at national level. For instance, Proceedings of the Institution of Mechanical Engineers, Part H: J. Engineering in Medicine dedicated their Volume 223(4) to Biomedical Engineering Education [37-44], as well as IEEE Engineering in Medicine and Biology Magazine - Volume 26, Issue 3 [23-24], [36], [53-55], Annals of Biomedical Engineering - Vol. 34, No. 2 [56-65], IEEE Transactions on Biomedical Engineering [1-2], IEEE Reviews in Biomedical Engineering [3], IEEE Engineering in Medicine and Biology Society Conferences [18], [32-34], Methods of Information in Medicine [67-70], and many other journals, conferences, and reports [63], [66]. In addition, educational projects BIOMEDEA (2004) and Curricula Reformation and Harmonization in the field of Biomedical Engineering (Tempus - JP 144537-2008) brought significant advances in setting guidelines, standards and procedures for harmonization of education and generation of qualifications in BE&MI [26], [71], [51-52]. BIOMEDEA with more than 60 partners consortium aims at contributing to the realisation of the European Higher Education Area (EHEA) in medical and biological engineering and science (MBES) in the sense of biomedical and clinical engineering education, accreditation, training and certification, while Curricula Reformation and Harmonization in the field of Biomedical Engineering with 24 partners consortium aims to create generic programs for graduate and postgraduate studies in biomedical engineering.

Education systems in Western Balkan countries underwent global changes and improvements through Bologna Process. However, wider actions for BE&MI study programs development at WB universities did not take place. Therefore, BioEMIS conducts joint and systematic action aimed at fulfilling WB universities', health, and industrial needs in this strategic field of research, knowledge and innovation.

3.1 Criteria of the Analysis

The main objective of the analysis is to identify core courses that define the broad field of BE&MI, which will be the basis for study programs development. In that order we analyzed different European study programs containing relevant courses. It should be noted that although the mainstream in this area is to develop integrated BE&MI study programs, many universities still tend to offer these study programs within other ones, e.g. engineering study programs. This analysis is based on:

1. Core BE & MI study programs
2. Variety of study programs that contain BE & MI related disciplines/courses
   (a) Computational Biology
   (b) Computational Physics
   (c) Computational Chemistry
   (d) Applied Mathematics
   (e) Electrical/Electronic Engineering
   (f) Mechanical Engineering
   (g) Medicine
   (h) Dentistry
   (i) Sports Engineering / Physical Education and Sports
   (j) Computer Science / Computer Engineering
   (k) Life Sciences
   (l) <various alternative programs>

The courses appear under different titles, depending on the focus within the broad field of interest. The frequency analysis is based on generic titles that reflect the core (essence) of the disciplines and needs. For instance, Medical Information Systems, Information Systems and Databases, Information Systems in Healthcare, Biomedical Information Systems, Information Systems in Biomedicine, and similar titles appear more or less for the same content. Medical Information Systems is chosen as generic course title. The same approach is applied for all courses that are identified as similar in content.

At the bachelor level, there is a set of compulsory courses of a fundamental importance, namely mathematics, biology, physics, chemistry, ICT, foreign (English) language, as well as a set of general biomedical courses, such as anatomy, physiology, biochemistry. These are not explicitly shown in the following diagrams and charts. The same holds for placement (internship), project, and final work (thesis).

Syllabi for the same generic title differ, as well. Although naturally overlapped in the contents to the great extent, general classification can be made as follows:

1. general/unified – for all entry qualifications
2. partly general/ unified, i.e. set of compulsory courses and the set of electives (mainly depending on entry qualifications)
   (a) pro medical
   (b) pro dental
   (c) pro engineering
   (d) pro biological
   (e) ...

Accordingly, BE&MI study programs may have the form of: (a) "pure" BE, (b) "pure" MI, and (c) BE&MI, where "pure" stands for mainly engineering or ICT oriented study programs, with negligible representation of counter field major disciplines. However, the analysis shows that even within the "pure" study programs, specializations provide specific orientations, i.e. needs of students: engineer, medical doctor, natural scientist, mathematician, ICT engineer, etc. To fulfill essential requirements of BE&MI study program and enable research and innovative dimension, curriculum should contain variety of courses to cover necessary advances in mathematics, engineering and ICT, such as geometrical modeling, data bases and artificial intelligence in medicine, statistics, image processing, as well as advanced disciplines of natural sciences like biochemistry, biophysics, medical optics, measurements, cell biology, etc.

3.2   Universities

The analysis included 221 European universities, spread over 30 countries. The selection of the universities has been done according to the indices provided by world known ranking studies and portals, namely [47-50]:

- the Academic Ranking of World Universities
  (http://www.shanghairanking.com/), and
- Leiden Ranking (http://www.leidenranking.com/)

Not all European universities were analyzed for study programs containing BE&MI related courses. The main lead in universities selection is based on Leiden Ranking list of European universities [48], due to their indicators, but the Shanghai list [47] and search results of portals [49-50] were used for additional selection. Note that Leiden Ranking list contains 214 universities (ranking 2013), where not all of them offer BE&MI related study programs. The results include 94 Leiden ranked universities. The distribution of analyzed universities is shown in Figure 1.

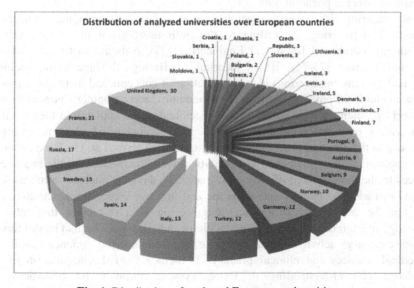

Fig. 1. Distribution of analyzed European universities

3.3   Biomedical Technology

Overall analysis tends to comprehend broad field of BE&MI from different aspects, respecting basic definitions of the constituent subfields. In its widest meaning the field BE&MI stems from the global field of Biomedical Technology (Fig.2). It should be noted that in some publications Biomedical Technology and Bioengineering are used interchangeably, lacking, in general, establishment of clear boundaries.

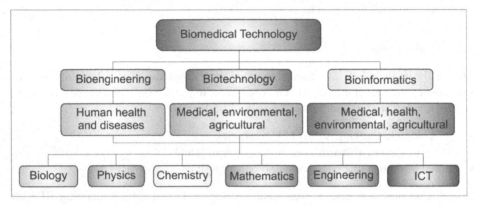

Fig. 2. Global field of Biomedical Technology

Three major subfields are bioengineering, biotechnology, and bioinformatics. All three span almost over complete biomedical and health issues of living organisms, including humans, animals, flora, and environment. Fundamental principles, approaches, methodologies, and technology are interlaced and converging in numerous distinct application domains.

Bioengineering is broader discipline than often interchanging concept of Biomedical engineering. According to the definitions given in Encyclopedia of Agricultural, Food, and Biological Engineering [27], National Institutes of Health [75], and Internet [73-74], Bioengineering or Biological Engineering (including biological systems engineering) is the application of concepts and methods of biology (and secondarily of physics, chemistry, mathematics, and computer science) to solve real-world problems related to the life sciences and/or the application thereof, using engineering's own analytical and synthetic methodologies and also its traditional sensitivity to the cost and practicality of the solution(s) arrived at. Or, Bioengineering is the application of experimental and analytical techniques based on the engineering sciences to the development of biologics, materials, devices, implants, processes and systems that advance biology and medicine and improve medical practice and health care. On the other side, Biomedical engineering is a discipline that advances knowledge in engineering, biology and medicine, and improves human health through cross-disciplinary activities that integrate the engineering sciences with the biomedical sciences and clinical practice. It includes: (a) the acquisition of new knowledge and understanding of living systems through the innovative and substantive application of experimental and analytical techniques based on the engineering sciences, and (b) the development of new devices, algorithms, processes and systems that advance biology and medicine and improve medical practice and health care delivery [72], [74]. Casual distinction could be made in the terms that bioengineering is oriented towards health and diseases issues of all living organisms, while biomedical engineering focuses more towards corresponding issues of the humans.

Biotechnology is a vast field of science, and hence difficult to concisely define. Nevertheless, for the purpose of this review definition presented in Merriam-Webster's Medical Dictionary [76] reflects the complexity of the field concerned with medicine, health, environment, and agriculture: Biotechnology is the body of knowledge related to the use of organisms, cells or cell-derived constituents for the purpose of developing products which are technically, scientifically and clinically useful; alteration of biologic function at the molecular level (i.e., genetic engineering) is a central focus. It is further divided into three main parts: (a) green biotechnology – agricultural processes, (b) red biotechnology – related to the health care processes, and (c) white biotechnology – related to the industrial and environmental processes [77].

Bioinformatics in its narrow, original and straight forward sense is focused on creating and analyzing datasets of macromolecular structures, genome sequences, and the functional genomics, which combines elements of biology and computer science. Accordingly, (molecular) bioinformatics is defined as "conceptualizing biology in terms of molecules (in the sense of Physical Chemistry) and applying "informatics techniques" (derived from disciplines such as applied maths, computer science and statistics) to understand and organize the information associated with these molecules, on a large scale. In short, bioinformatics is a management information system for molecular biology and has many practical applications" [78]. In addition, bioinformatics, although distinctive and independent discipline, is quite often accepted in much broader sense of Biomedical Informatics or Medical (Health) Informatics, which in turn is also considered as different and independent discipline. Following the National Institutes of Health definition, bioinformatics focuses on "research, development, or application of computational tools and approaches for expanding the use of biological, medical, behavioral or health data, including those to acquire, store, organize, archive, analyze, or visualize such data" [79]. That is, "the field that concerns itself with the cognitive, information processing, and communication tasks of medical practice, education, and research, including the information science and the technology to support these tasks..., and deals with biomedical information, data and knowledge – their storage, retrieval and optimal use for problem solving and decision making" [80].

All these definitions clearly emphasize fundamentals rooted in science, i.e. biology, chemistry, physics, mathematics, as well as engineering and ubiquitous ICT. Understanding BE&MI as the converging synthesis of medicine and life sciences, engineering, and technology (Fig.3) directs curriculum development towards balanced mixture of courses that will support integration of converging and enabling disciplines to meet the aforementioned challenges of education, research, and innovation.

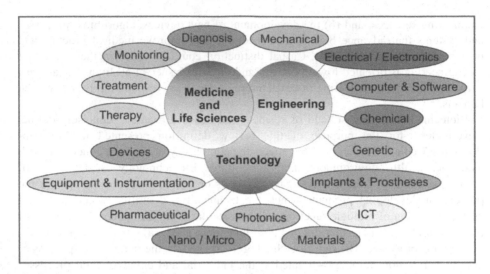

Fig. 3. Synthesis of Medicine and Life Sciences, Engineering, and Technology

EU and other European universities have the rich experience, good practice and yet evolving approaches in educating professionals of BE&MI. The following generic sets of courses at bachelor and master levels present state of the art in BE&MI study programs offer, with an emphasis on curriculum of Medical Informatics.

3.4   Curriculum in Medical (Health) Informatics

As previously mentioned, the synthesis of engineering, medical, and biological systems generates different forms of curricula, depending on the major field of interest and application. Rough classification into general BE&MI, "pure" BE, and "pure" MI classes is merely conditional. However, the analysis shows that "pure" MI study programs are significantly present at European universities, although less than general BE&MI and "pure" BE study programs. The following analysis is based on the generic titles of the courses, due to huge variety of alternatives. Compulsory fundamental science, humanities, and literacy courses are not explicitly shown.

"Pure" MI study programs offer relatively small selection of courses. The main reason is the nature of ICT disciplines and overall involvement in the field of biomedical science, although significance and intensity are more than obvious. At the bachelor level there are only 6 generic courses directly oriented towards biomedical applications, namely (Fig.4) [81]:

1. Health Services Management and Policy
2. Introduction to Bioinformatics
3. Health Information Systems
4. Medical Data Bases / EHR
5. Data Structures and Algorithms
6. Introduction to Scientific/Biomedical Programming.

These core generic courses are almost evenly distributed among European MI curricula, making them the essential for the MI professionals. Common three years bachelor level curricula provide just the basics for future professionals aiming at working in an interdisciplinary environment. Such education prepares the ITC professionals in two ways: to directly engage with public health and hospital/clinical information services, and to proceed to the next levels of education.

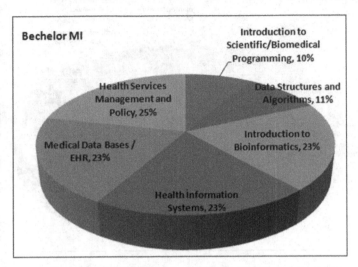

Fig. 4. Distribution of MI bachelor courses at European universities

The set of generic MI courses at master level covers advanced knowledge innovative disciplines. There are 14 courses in total (Fig.5):

1. Bioinformatics
2. Geometrical Modeling (in Medicine)
3. Navigation and Robotic Systems in Medicine / Computer Aided Medical Procedures / Computer Assisted Surgery;
4. Biomedical Statistics
5. Biomedical Data Analysis and Processing
6. Medical Information Systems
7. Computer Graphics (in Medicine)
8. Artificial Intelligence in Medicine
9. Computer Aided Diagnosis
10. E-learning in Medicine
11. Programming in the Biomedical Engineering
12. Augmented Reality in Medicine
13. Telemedicine
14. E-health.

One should note that 0% frequency for E-health and Telemedicine courses means actually "less than 1%". These courses are taught as separates disciplines at some universities. Same courses are quite often included as topics within more common courses like Medical Information Systems and Computer Aided Diagnosis for the former two, and Geometrical Modeling (in Medicine), Navigation and Robotic Systems in Medicine / Computer Aided Medical Procedures / Computer Assisted Surgery, Computer Aided Diagnosis for the latter.

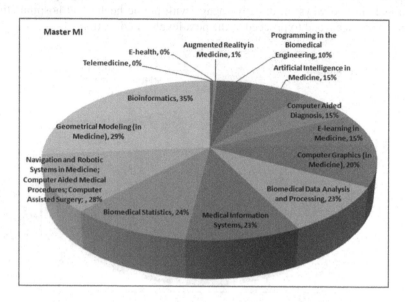

Fig. 5. Distribution of MI master courses at European universities

Larger diversity of courses offered at master level implies not only the need for more advanced knowledge and specialization, but also preparation for further research work at doctoral level. This offer is the reflection of the main axes of development within key European strategies of research and innovation.

International Standard Classification of Occupations [82] classifies biomedical engineers in the professionals' Unit Group 2149: Engineering Professionals not elsewhere classified. On the other side, explicit classification for medical (health) ICT specialists does not exist. This implies that they are generally classified in professional's Sub-major Group 25: Information and Communications Technology Professionals, and consequently in Unit Group 25x9: <Information and Communications Technology Professionals> not elsewhere classified. However, market analyses reports that job opportunities for MI professionals show an unprecedented demand in the last few years [83]. Availability and utilization of ICT products forced, especially during last decade, public health and clinical workforce to change the way they acquire, store, retrieve, and use medical and health data and information. Proper support of specialized ICT professionals and clinical engineers become the inevitable for coping with expectations and overall progress. Unfortunately, in the WB countries there are no specific regulation and legislative in the area of BE&MI. Tempus project BioEMIS tends to create the links at the educational and professional level.

4   Conclusions

Higher education systems in Western Balkan countries underwent thorough changes and improvements through Bologna Process. However, the field of BE&MI that is recognized as the one of the most prominent in the decade to come demonstrates insufficient engagement. Tempus project "Studies in Bioengineering and Medical Informatics - BioEMIS" aims at introducing BE&MI study programs at all three university levels in three countries. So far, the only university at the Western Balkan that offer accredited study program in biomedical engineering is University of Belgrade. Some teaching and mostly research activities and the results in the field of BE&MI are quite evident at most of the universities. Therefore, the leading universities at WB region joined at systematic action of development a set of dedicated bachelor, master, doctoral, and specialization study programs.

Analysis of study programs offered at 221 European universities yielded basic guidelines for study programs development at the different levels. Those guidelines, as well as standards and recommendations, are used at WB universities to transfer of "know-how" from EU and other European universities. Following the regional needs in this field and respecting the principle of complementarity, we created targeted plan that is ambitious, but realistic and achievable. We will introduce 4 PhD programs and modernize one in BE&MI (3 years, 180ECTS), introduce 5 M.Sc. programs and modernize one (2 years, 60ECTS), and introduce 3 specializations (1 year, 30ECTS).

Proposed systematic action at WB universities in the full alignment with the key strategic priorities and underpinning developmental pillars. Bioengineering problems (including medical) are among the most important that engineers will be solving in the future. The overall benefits are not only for the universities and the students, but for the WB community in general.

Acknowledgments. This study is supported by TEMPUS project 530417-TEMPUS-1-2012-1-UK-TEMPUS-JPCR, "Studies in Bioengineering and Medical Informatics", funded by EU Commission and project "Application of Biomedical Engineering in Preclinical and Clinical Practice" (grant No.III-41007) funded by the Serbian Ministry of Education, Science and Technological Development. The author would like to express his appreciation to R&T assistants of Faculty of Engineering (University of Kragujevac, Serbia), Suzana Petrovic, Andjelka Stankovic, Sasa Cukovic and Bogdan Dimitrijevic for their extraordinary contribution in collecting data and information about European study programs in BE&MI. Special thanks goes to Radovan Stojanovic, Zlatko Bundalo and Duncan Shepherd for their encouragement and support, and to Nenad Miric and Dragutin Rajic for their invaluable advices.

References

1. Lee, R.C.: Convolving Engineering and Medical Pedagogies for Training of Tomorrow's Health Care Professionals. IEEE Trans. on Biomedical Engineering 60, 599–601 (2013)
2. He, B., Baird, R., Butera, R., Datta, A., George, S., Hecht, B., Hero, A., Lazi, G., Lee, R.C., Liang, J., Neuman, M., Peng, G.C.Y., Perreault, E.J., Ramasubramanian, M., Wang, M.D., Wikswo, J., Yang, G.-Z., Zhang, Y.-T.: Grand Challenges in Interfacing Engineering With Life Sciences and Medicine. IEEE Trans. on Biomedical Engineering 60, 589–598 (2013)
3. Saranummi, N.: Connected Health and Information Technology for Health. IEEE Reviews in Biomedical Engineering 6, 21–23 (2013)
4. European Commission: Work Programme 2013: Cooperation – Health,
   `ftp://ftp.cordis.europa.eu/pub/fp7/health/docs/`
   `fp7-health-wp-2013_en.pdf`
5. European Commission: Specific Programme Implementing Horizon 2020 - The Framework Programme for Research and Innovation (2014-2020),
   `http://ec.europa.eu/research/horizon2020/pdf/proposals/`
   `com%282011%29_811_final.pdf`
6. National Academy of Engineering: The Engineer of 2020: Visions of Engineering in the New Century, `http://www.nap.edu/catalog/10999.html`
7. National Academy of Engineering: Grand Challenges for Engineering,
   `http://www.engineeringchallenges.org/Object.File/Master/`
   `11/574/Grand%20Challenges%20final%20book.pdf`
8. National Academy of Sciences, National Academy of Engineering, Institute of Medicine: Reshaping the Graduate Education of Scientists and Engineers,
   `http://www.nap.edu/catalog/4935.html`
9. Vest, C.M.: Educating Engineers for 2020 and Beyond. National Academy of Engineering Annual Meeting and The Bridge 36(2), 38–44 (2005),
   `https://engineering.purdue.edu/~engr116/`
   `ENGR19500H_fal/General_Course_Information/Common/`
   `Educating_Engineers_Vest_2005.pdf`
10. European Commission: EUROPE 2020 - A strategy for smart, sustainable and inclusive growth, `http://eur-lex.europa.eu/LexUriServ/LexUriServ.do?uri=`
    `COM:2010:2020:FIN:EN:PDF`
11. European Centre for Strategic Management of Universities: Higher Education Governance Reforms Across Europe, `http://www.utwente.nl/mb/cheps/publications/`
    `Publications%202009/C9HdB101%20MODERN%20PROJECT%20REPORT.pdf`
12. European Commission: Regulation of the European Parliament and of the Council - establishing Horizon 2020 - The Framework Programme for Research and Innovation (2014-2020), `http://eur-lex.europa.eu/LexUriServ/`
    `LexUriServ.do?uri=COM:2011:0809:FIN:en:PDF`
13. House of Commons' Science and Technology Committee: Bioengineering - Seventh Report of Session 2009–10. The House of Commons, London (2010),
    `http://www.publications.parliament.uk/pa/cm200910/`
    `cmselect/cmsctech/220/220.pdf`
14. United Nations: The UN Platform on Social Determinants of Health - Health in the post-2015 development agenda: need for a social determinants of health approach,
    `http://www.worldwewant2015.org/file/300184/download/325641`

15. Sharp, P.A., et al.: The Third Revolution: The Convergence of the Life Sciences, Physical Sciences, and Engineering. Massachusetts Institute Of Technology (2011), http://dc.mit.edu/sites/dc.mit.edu/files/MIT%20White%20Paper%20on%20Convergence.pdf

16. Barabino, G.: A Bright Future for Biomedical Engineering (Editorial). Annals of Biomedical Engineering 41(2) (2013)

17. Fleming, J., Iyer, R.M., Shortis, M., Vuthaluru, H., Xing, K., Moulton, B.: Biomedical Engineering Curricula: Trends in Australia and Abroad. World Transactions on Engineering and Technology Education 10, 23–28 (2012)

18. Lithgow, B.J.: Biomedical Engineering Curriculum: A Comparison Between the USA, Europe and Australia. In: Proceedings of 23rd Annual Conference – IEEE/EMBS, Istanbul, Turkey (2001)

19. Matsuki, N., Takeda, M., Yamano, M., Imai, Y., Ishikawa, T., Yamaguchi, T.: Effects of unique biomedical education programs for engineers: REDEEM and ESTEEM projects. Advances in Physiology Education 33, 91–97 (2009)

20. National Academy of Engineering: Educating the Engineer of 2020: Adapting Engineering Education to the New Century, http://www.nap.edu/catalog.php?record_id=11338

21. Katona, P.G.: Biomedical Engineering and The Whitaker Foundation: A Thirty-Year Partnership. Annals of Biomedical Engineering 34, 904–916 (2006)

22. Wang, F.: Valuation of Online Continuing Medical Education and Telemedicine in Taiwan. Educational Technology &Society 11, 190–198 (2008)

23. Poon, C.C.Y., Zhang, Y.-T.: Perspectives on High Technologies for Low-Cost Healthcare - The Chinese Scenario. IEEE Engineering in Medicine and Biology Magazine 5, 42–47 (2008)

24. He, B.: Biomedical Engineering in China - Some Interesting Research Ventures. IEEE Engineering in Medicine and Biology Magazine 5, 12–13 (2008)

25. Roth, F., Thum, A-E.: The Key Role of Education in the Europe 2020 Strategy. CEPS Working Document No.338, http://www.ceps.eu/book/key-role-education-europe-2020-strategy

26. Nagel, J.H.: Biomedical Engineering Education in Europe – Status Reports. BIOMEDEA, http://www.biomedea.org/Status%20Reports%20on%20BME%20in%20Europe.pdf

27. Dooley, J.H., Riley, M.R., Verma, B.: Biological Engineering: Definition. In: Encyclopedia of Agricultural, Food, and Biological Engineering, pp. 120–123. Taylor and Francis, New York (2010)

28. Canada Biomedical Sciences and Engineering University Programs, http://www.canadian-universities.net/Universities/Programs/Biomedical_Sciences_and_Engineering.html

29. Biomedical Engineering Education Programs, http://www.cmbes.ca/index.php?option=com_content&view=article&id=53&Itemid=189

30. Humphrey, J.D., Coté, G.L., Walton, J.R., Meininger, G.A., Laine, G.A.: A new paradigm for graduate research and training in the biomedical sciences and engineering. Advances in Physiology Education 29, 98–102 (2005)

31. Singh, K.: Biomedical Engineering Education Prospects in India. In: 13th International Conference on Biomedical Engineering, IFMBE Proceedings, vol. 23, pp. 2164–2166. Springer (2009)

32. Schneider, B., Schneider, F.K.: The Role of Biomedical Engineering in Health System Improvement and Nation's Development. In: 32nd Annual International Conference of the IEEE EMBS, pp. 6248–6251. IEEE (2010)

33. Monzon, J.E.: The Challenges of Biomedical Engineering Education in Latin America. In: Proceedings of the 27th Annual IEEE Conference on Engineering in Medicine and Biology, pp. 2403–2405 (2005)

34. Monzon, J.E.: CORAL and its role in Latin America over the last decade. In: Proceedings of the 25th Annual International Conference of the IEEE EMBS, vol. 4, pp. 3454–3456. IEEE (2003)

35. Consejo Regional de Ingeniería Biomedica Para América Latina, http://www.coralbiomedica.org/

36. Nagel, J.H., Slaaf, D.W., Barbenel, J.: Medical and Biological Engineering and Science in the European Higher Education Area - Working Toward Harmonization of Biomedical Programs for Mobility in Education and Employment. IEEE Engineering in Medicine and Biology Magazine 26, 18–25 (2007)

37. Slaaf, D.W., van Genderen, M.H.P.: The fully integrated biomedical engineering programme at Eindhoven University of Technology. Proceedings of the Institution of Mechanical Engineers, Part H: Journal of Engineering in Medicine 223, 389–397 (2009)

38. Baselli, G.: Biomedical engineering education at Politecnico di Milano: Development and recent changes. Proceedings of the Institution of Mechanical Engineers, Part H: Journal of Engineering in Medicine 223, 399–406 (2009)

39. Joyce, T.: Currently available medical engineering degrees in the UK. Part 1: Undergraduate degrees. Proceedings of the Institution of Mechanical Engineers, Part H: Journal of Engineering in Medicine 223, 407–413 (2009)

40. Joyce, T.: Currently available medical engineering degrees in the UK. Part 2: Postgraduate degrees. Proceedings of the Institution of Mechanical Engineers, Part H: Journal of Engineering in Medicine 223, 415–423 (2009)

41. Theobald, P., Jones, M.D., Holt, C.A., Evans, S.L., O'Doherty, D.M.: Medical engineering at Cardiff University. Part 1: Undergraduate programmes of study. Proceedings of the Institution of Mechanical Engineers, Part H: Journal of Engineering in Medicine 223, 425–430 (2009)

42. Theobald, P., O'Doherty, D.M., Holt, C.A., Evans, S.L., Jones, M.D.: Medical engineering at Cardiff University. Part 2: Postgraduate programmes of study. Proceedings of the Institution of Mechanical Engineers, Part H: Journal of Engineering in Medicine 223, 431–435 (2009)

43. Cunningham, J.L.: Development of degrees in medical engineering at the University of Bath. Proceedings of the Institution of Mechanical Engineers, Part H: Journal of Engineering in Medicine 223, 437–441 (2009)

44. Wallen, M., Pandit, A.: Developing research competencies through a project-based tissue-engineering module in the biomedical engineering undergraduate curriculum. Proceedings of the Institution of Mechanical Engineers, Part H: Journal of Engineering in Medicine 223, 443–448 (2009)

45. Kikuchi, M.: Status and future prospects of biomedical engineering: a Japanese perspective. Biomedical Imaging and Intervention Journal 3(3), e37 (2007)

46. Sato, S., Kajiya, F.: The Present State and Future Perspective of Biomedical Engineering in Japan, http://www.dtic.mil/dtic/tr/fulltext/u2/a409923.pdf

47. Academic ranking of world Universities, http://www.shanghairanking.com/

48. CWTS Leiden Ranking, http://www.leidenranking.com/

49. Study Portals: The European Study Choice Platform,
    http://www.studyportals.eu/
50. QS World University Rankings, http://www.topuniversities.com/
    university-rankings/world-university-rankings/2012
51. Tempus project JP 144537-2008: Curricula Reformation and Harmonization in the field of
    Biomedical Engineering, http://projects.tempus.ac.rs/
52. Magjarevic, R., Lackovic, I., Bliznakov, Z., Pallikarakis, N.: Challenges of the Biomedical
    Engineering Education in Europe. In: 32nd Annual International Conference of the IEEE
    EMBS, pp. 2959–2962 (2010)
53. Iakovidis, I., Le Dour, O., Karp, P.: Biomedical Engineering and eHealth in Europe -
    Outcomes and Challenges of Past and Current EU Research Programs. IEEE Engineering
    in Medicine and Biology Magazine 26, 26–28 (2007)
54. Maojo, V., Tsiknakis, M.: Biomedical Informatics and HealthGRIDs: A European
    Perspective - Past and Current Efforts and Projects in the Synergy of Bionformatics and
    Medical Informatics. IEEE Engineering in Medicine and Biology Magazine 26, 34–41
    (2007)
55. Siebes, M., Viceconti, M., Maglaveras, N., Kirkpatrick, C.J.: Engineering for Health - A
    Partner in Building the Knowledge Economy of Europe. IEEE Engineering in Medicine
    and Biology Magazine 26, 53–59 (2007)
56. Lerner, A.L., Kenknight, B.H., Rosenthal, A., Yock, P.G.: Design in BME: Challenges,
    Issues, and Opportunities. Annals of Biomedical Engineering 34, 200–208 (2006)
57. Neuman, M.R., Kim, Y.: The Undergraduate Biomedical Engineering Curriculum:
    Devices and Instruments. Annals of Biomedical Engineering 34, 226–231 (2006)
58. Paschal, C.B., Nightingale, K.R., Ropella, K.M.: Undergraduate Biomedical Imaging
    Education. Annals of Biomedical Engineering 34, 232–238 (2006)
59. Lutchen, K.R., Berbari, E.J.: White Paper: Rationale, Goals, and Approach for Education
    of Biosystems and Biosignals in Undergraduate Biomedical Engineering Degree
    Programs. Annals of Biomedical Engineering 34, 248–252 (2006)
60. Hammer, D.A., Waugh, R.E.: Teaching Cellular Engineering. Annals of Biomedical
    Engineering 34, 253–256 (2006)
61. Brinton, T.J., Kurihara, C.Q., Camarillo, D.B., Pietzsch, J.B., Gorodsky, J., Zenios, S.A.,
    Doshi, R., Shen, C., Kumar, U.N., Mairal, A., Watkins, J., Popp, R.L., Wang, P.J.,
    Makower, J., Krummel, T.M., Yock, P.G.: Outcomes from a Postgraduate Biomedical
    Technology Innovation Training Program: The First 12 Years of Stanford Biodesign.
    Annals of Biomedical Engineering, pp. 1–8 (2013)
62. Allen, R.H., Acharya, S., Jancuk, C., Shoukas, A.A.: Sharing Best Practices in Teaching
    Biomedical Engineering Design. Annals of Biomedical Engineering, pp. 1–11 (2013)
63. Bruzzi, M.S., Linehan, J.H.: BioInnovate Ireland—Fostering Entrepreneurial Activity
    Through Medical Device Innovation Training. Annals of Biomedical Engineering, pp. 1–7
    (2013)
64. Louie, A., Izatt, J., Ferrara, K.: Biomedical Imaging Graduate Curricula and Courses:
    Report from the 2005 Whitaker Biomedical Engineering Educational Summit. Annals of
    Biomedical Engineering 34, 239–247 (2006)
65. Oden, M., Mirabal, Y., Epstein, M., Richards-Kortum, R.: Engaging Undergraduates to
    Solve Global Health Challenges: A New Approach Based on Bioengineering Design.
    Annals of Biomedical Engineering 38, 3031–3041 (2010)
66. Viik, J.: Biomedical Engineering Education in Finland. Finnish Society for Medical
    Physics and Medical Engineering,
    http://www.lfty.fi/PDF/bme_in_finland.pdf

67. Mantas, J., Ammenwerth, E., Demiris, G., Hasman, A., Haux, R., Hersh, W., Hovenga, E., Lun, K.C., Marin, H., Martin-Sanchez, F., Wright, G.: Recommendations of the International Medical Informatics Association (IMIA) on Education in Biomedical and Health Informatics. Methods of Information in Medicine 2, 105–120 (2010)

68. López-Campos, G., López-Alonso, V., Martin-Sanchez, F.: Training Health Professionals in Bioinformatics - Experiences and Lessons Learned. Methods of Information in Medicine 3, 299–304 (2010)

69. Otero, P., Hersh, W., Luna, D., González Bernaldo de Quirós, F.: A Medical Informatics Distance Learning Course for Latin America - Translation, Implementation and Evaluation. Methods of Information in Medicine 3, 310–315 (2010)

70. Berner, E.S., McGowan, J.J.: Use of Diagnostic Decision Support Systems in Medical Education. Methods of Information in Medicine 4, 412–417 (2010)

71. Nagel, J.H.: Criteria for the Accreditation of Biomedical Engineering Programs in Europe. BIOMEDEA, http://www.biomedea.org/Documents/Criteria%20for%20Accreditation%20Biomedea.pdf

72. Imperial College London: Definition of Biomedical Engineering, http://www3.imperial.ac.uk/pls/portallive/docs/1/51182.PDF

73. Wikipedia: Biological Engineering, http://en.wikipedia.org/wiki/Bioengineering

74. Sloan Career Cornerstone Center: Bioengineering Overview, http://www.careercornerstone.org/pdf/bioeng/bioeng.pdf

75. National Institutes of Health: Collection Development Manual: Bioengineering, http://www.nlm.nih.gov/tsd/acquisitions/cdm/subjects10.html

76. Merriam-Webster's Medical Dictionary: Definition of Biotechnology, http://ghr.nlm.nih.gov/glossary=biotechnology

77. Bartoszek, A., Bekierska, A., Bell-lloch, J., de Groot, T., Singer, E., Woźniak M.: Managing innovations in biotechnology (2006), http://dugi-doc.udg.edu/bitstream/handle/10256/4289/1Memoria.pdf?sequence=1

78. Luscombe, N.M., Greenbaum, D., Gerstein, M.: What is Bioinformatics? A Proposed Definition and Overview of the Field. Methods of Information in Medicine 4, 346–358 (2001)

79. National Institutes of Health: NIH Working Definition of Bioinformatics and Computational Biology, http://www.bisti.nih.gov/docs/CompuBioDef.pdf

80. Bernstam, E.V., Smith, J.W., Johnson, T.R.: What is biomedical informatics? Journal of Biomedical Informatics 43, 104–110 (2010)

81. Devedžić, G., Stojanović, R., Bundalo, Z., Shepherd, D., Petrović, S., Stanković, A., Ćuković, S.: Developing Curriculum in Bioengineering and Medical Informatics at Western Balkan Universities. In: 2nd Mediterranean Conference on Embedded Computing, MECO 2013, Budva, Montenegro (2013)

82. International Standard Classification of Occupations - ISCO-08: Volume 1 - Structure, Group Definitions And Correspondence Tables. International Labour Organization, Geneva, Switzerland, http://www.ilo.org/wcmsp5/groups/public/—dgreports/—dcomm/—publ/documents/publication/wcms_172572.pdf

83. Jobs for the Future and Burning Glass Technologies. A Growing Jobs Sector: Health Informatics. Report (2012), http://www.jff.org/sites/default/files/CTW_burning_glass_publication_052912.pdf

Integrating Computer Games in Primary Education for Increased Students' QoE

Toni Malinovski[1], Marina Vasileva[2], and Vladimir Trajkovik[1]

[1] Faculty of Computer Science and Engineering,
Ss. Cyril and Methodius University, Skopje, Republic of Macedonia
tmalin@nbrm.mk, trvlado@finki.ukim.mk
[2] Primary School Ss. Cyril and Methodius, Skopje, Republic of Macedonia
vasileva_marina@yahoo.com

Abstract. Continuous improvement of the learning environment and pedagogical methodology are tasks that every educational program has to endure. This study explores integration of computer games in the primary educational program, their proper alignment with the learning curriculum, while creating powerful learning environment which increases student's effectiveness and motivation to learn. The research methodology follows user-oriented approach while evaluating different variables in a computer game enhanced class, which impact the overall students' Quality of Experience (QoE). A case study is conducted with a traditional and non-traditional class in several primary schools in Macedonia, students' feedback is collected through surveys and the data set is analyzed according to the proposed methodology. Path analyses model is presented illustrating the relationships among several relevant variables in the class with a computer game and their connection to the students' QoE.

Keywords: Computer Game, Primary Education, QoE, Learning, Evaluation, Path Analyses Model.

1 Introduction

The rapid development of the emerging technologies in the recent years has increased their implementation in different areas of modern society. Information and Communication Technology (ICT) provides resources and different tools, which extend opportunities for creation, management and distribution of information among involved parties. Therefore ICT may highly contribute in the field of education, where communication of knowledge and information is extremely relevant, while creating a powerful learning environment which can improve students' learning experience and enhance teaching methods. These new technologies can provide benefits for teachers to increase effectiveness and flexibility, support easier planning and lessons preparation, while adapting the learning content to new ideas. On the other hand, students' motivation can be increased while learning in a technology-enhanced setting and student-centered environment, compared to the traditional classroom.

V. Trajkovik and A. Mishev (eds.), *ICT Innovations 2013*,
Advances in Intelligent Systems and Computing 231,
DOI: 10.1007/978-3-319-01466-1_3, © Springer International Publishing Switzerland 2014

Still, there are different aspects which are significant while designing and planning technology-enhanced learning systems. Some of the instructional design issues related to learning goals, like authentic vs. abstract problem solving [2], can influence the final success of the learning process. Furthermore, the real implementation of the technology in the education demands changes in teaching style, changes in learning approaches and access to information [3].

In this study, we explore the utilization of the information technology in primary education, by introducing traditional children games into the learning program, thus enhancing the learning curriculum. These games are implemented through different technological tools, making an interactive learning experience for the young population in their classroom activities. Through the research, we try to distinct the relevant factors which influence the positive students' experience and provide relationship among different variables while the game-based learning is introduced in the primary school education. We have evaluated this learning approach in primary schools in cities and villages, so we can provide relevant information when the technology is used in different environments, with students having different experience with the latest technology. In this study, we follow user-oriented approach, while exploring social element and students' subjective expectations, in terms of Quality of Experience (QoE) [4]. In the recent years QoE [1], [7] has emerged as a full scale evaluation of the technological implementations in terms of end-user expectations, so we believe the provided results can give relevant information for future development and successful integration of technology into the learning process.

This paper is organized as follows: section 2 provides related work and the novelty of our approach, research methodology is presented in section 3, section 4 demonstrates a case study on this topic, section 5 provides actual results and analyses, while section 6 concludes the paper.

2   Related Work

The educational process should be constantly improved to reach higher level of students' perceived knowledge according to their potential. Therefore different approaches, methodologies and teaching practices must be adopted and put into practice so positive results can be achieved at the end. The ICT offers different perspective, but some studies [5], [8], [11] already show that teachers are not integrating technology sufficiently as a teaching/learning tool. Teachers are generally very professional and are committed to provide quality education to their pupils, but these studies show several problems which usually occur when new technologies and changes have to be introduced in the educational process.

The computer simulated games and learning have increased their mutual connection in the recent years. Different games were developed to provide instructional, problem solving challenges and testing of specific skills. Graphical representation, models and modeling are very helpful to learning [6], because they allow certain aspects of students' experience to be incorporated in the problem solving, making the abstract problem more concrete and understandable to the students. Games posses interesting graphical representation and are usually built on certain models, which makes them appropriate for the learning process. Still, even though certain skills, such as problem solving ability increase within a game, the real challenge comes when these skills and learned content have to be used outside of the gaming environment [9]. Different studies [9], [10] have shown modest to low evidence, when gamed learning skills or content are transferred outside of the digital environment.

Our study researches implementation of specific technology enhanced games in the certain primary education schools in Macedonia, while evaluating different variables which can influence higher level of positive students' QoE and increase Quality of Learning (QoL). The novelty of our approach is that specific games are selected, which children, their parents or grandparents played without the use of technology, enhanced through proper ICT implementation and adopted accordingly in the everyday learning process. The research was conducted in several primary schools in different cities and villages in Macedonia. Teachers in these schools found proper way to embrace the new technology and enhance the standard state primary school curriculum with computer games, properly chosen for the specific content.

We believe that the idea of learning through games can improve the learning process, if the computer games are developed to incorporate adequate pedagogical components, based on didactic principals as highly organized and properly guided pupils' activities. We follow this idea in this study, while trying to present research finding that support it. The learned skills and content through such learning activities can be successfully transferred outside of the gaming environment, with increased students' motivation and QoE.

3   Integrating Computer Games in Primary Education

The simulations and digital games may deeply contribute in creation of a powerful learning environment in numerous ways. Traditional classroom education, without the utilization of the technology is no longer considered as adequate preparation for future development and successful life in general. Having in mind future challenges we have to analyze different approaches, priorities, obstacles and strategies for integration of computer games into the everyday learning process of primary school children and their teachers.

Our study tries to tackle these issues and provide positive example where digital games can be incorporated in the primary school educational program. During the process of planning, organizing and even programming of the games, we focused on proper connection of theory and practice, adaptation of the games according to students capabilities, their school activates, emotional interest and relation to their everyday life routine. Our efforts were especially aligned towards the goals and principles of the educational component of each involved school subject, in respect to the state educational program in Macedonia.

This approach which integrates the computer games in the learning environment offers exclusive possibility to transfer the learning process into maximum mental and physical activity of each student, while the learned skills and content through the games can be successful transferred outside of the digital environment. The traditional classroom learning offers little physical activities and interaction among students. These lessons tend to be static and monotones, which decrease students' focus on the subject's content and motivation to learn.

We have basically chosen to use games which can be adapted to incorporated technological advantages into the pedagogical practice and provide optimal involvement of the student, which can be mentally and physical challenging.

Besides focusing on proper pedagogical planning and implementation of the games in the school classes, this study also follows user-oriented approach while determining the relevant factors which influence the positive level of students' experience (QoE) during these innovative changes, comparing them to the traditional classroom education on the same subject with similar content.

During the study we have identified several primary education schools in different cities and villages in Macedonia. These schools have the capabilities to transition, from the traditional priorities and conventional learning goals, into an approach that changes the way of teaching, while engaging teachers to use digital games and fulfill the schools' highest ambitions for their students. They worked on similar school subject, while performing one activity enhanced with a computer game and similar activity with the traditional teaching method. The students involved in these activities were from 10 to 12 years of age, which mostly use computers and technology regularly in their everyday life for different purposes.

A survey was developed, which was distributed to students after class, both traditional and with computer games, to gather students' subjective opinion on different subjects. They were able to grade different questions on a scale from 1 to 5, where 1 is strongly disagree and 5 is strongly agree on each question. The surveys were conducted on-line after class, while the responses from different schools were collected in a central database for further analyses and results.

As a result of the students' responses we have identified several variables which illustrate their opinion regarding the traditional and non- traditional class. These variables are: simplified way of learning during the class, motivation to learn, effectiveness and quicker learning and students' QoE during the lecture. The feedback regarding these variables can show how they are connected among each other and results can be compared for differences in the traditional and computer game enhanced class.

Furthermore, in our study these observed variables regarding classes with computer games are used to construct a path analysis model, while exploring relationships between them, analyze regression weights and define how students' QoE is predicted as a combination of the other observed variables. These statistical analyses were conducted using Statistical Package for Social Sciences (SPSS) and Analysis of Moment Structures (AMOS) software. The quantitative numbers retrieved from the surveys' database can show the proper model fitting and testing of our path analysis model and explore different hypotheses, which are:

- Hypothesis 1: The simplified way of learning through computer games influences students' motivation to learn and increases effectiveness for quicker learning;

- Hypothesis 2: The motivation to learn during the class enhanced with computer games increases the level of positive students' QoE during the class;
- Hypothesis 3: The effectiveness and simplified way of learning also influence students' QoE from the non-traditional class.

We were also able to analyze if the computer games provided higher level of perceived knowledge, through quick evaluations at the end of each class and compare it in both environments.

In this article, we present a case study on a specific subject and content, which gives us possibility to test our methodology and proposed path analyses model, provide preliminary results on the relations among observed variables and justification of the proposed hypotheses.

4   Case Study: Game Zavor in a Primary Education Class

Following the methodology of this research study, we have organized a case study while incorporating a specific computer game, for a specific subject in several primary education schools in Macedonia. The schools that participated in this case study were in the capital city and two villages, more precisely OU "Ss. Cyril and Methodius" in Skopje, OU "Mancu Matak" in the village Krivogastani, Prilep and OU "Petar Pop Arsov" in the village Bogomila. The total number of evaluated students was 114, which included children at age 11-12, 46% male and 54% female. 92.1 % of these children use computers on daily basis, 7.02 % use computers two-three times a week, while only 0.88 % never used computers. Therefore, even though these participants live in different environments, this sample of students can be considered highly familiar with technology, representing the global situation in primary schools in Macedonia.

For the purpose of this case study we have chosen mathematics as a subject for evaluation. The content included lessons on the study of the metric system. During this part of the educational curriculum on this subject, students have to learn how to convert standard units of length, weight, volume, and temperature in the metric system.

We have chosen to use traditional class while teaching students weight metric conversion (kilograms, grams etc.) and a class with computer game for the length metric conversion (meters, centimeters, millimeters etc.). A traditional game named Zavor, in which participants throw a ball into a distance, measuring each participant's score for the length of the covered distance, was chosen for class for length metric conversion. Special game program according to traditional game was developed for Microsoft Kinetic, which is a motion sensing input device by Microsoft for the Xbox 360 video game console, present at all involved primary schools. With this computer game, students could play Zavor, by throwing a virtual ball into the distance, while the length of the covered distance is displayed on the screen in different metric, first in meters, centimeters etc. This game was properly aligned to the curriculum, while students were involved in mental and physical activity. During the computer game based class all of the students were completely involved, learning the metric conversion content, which can be totally abstract to children at that age, in an innovative and motivating environment.

Figure 1. illustrates one of the classes where Zavor was played with the Microsoft game console.

Fig. 1. Game Zavor played during a metric conversion class

The first student's impression after the class was their positive experience because new technology and physical activities were involved. Still deeper analyses regarding the learning benefits and the factors for possible increased level of students' QoE have to be conducted. Following our methodology, we have conducted the necessary surveys after the traditional and non-traditional classes in the involved primary schools and gathered students' evaluations data on previously described variables. Table 1 shows the variables and their connection to the surveys' questionnaire.

Table 1. Input variables from the students' responses

Variable Name	Description in the questionnaire
Simple	Students' think the learning is more simplified this way
Motivation	Students' were motivated to learning thought this approach
Quick	Students believe they can learn quicker and more effectively this way
QoE	Student's overall experience from the class

The gathered results on all of the variables were compared among the traditional and non-traditional class. Both types of classes were on the same subject and content, so we could provide relevant quantitative analysis and comparative numbers for the different learning methodologies. These results can point out if the computer based class provided better learning experience to the students and does this unique approach really makes a difference in the primary education.

Furthermore, the variables listed in Table 1 were subjected as observed input variables in our path analysis model, so we can elaborate the relationships and regression weights among them, test our hypotheses, while defining how students' QoE is predicted in the non-traditional learning environment.

5   Analysis and Results

It is globally accepted that education quality may improve through the promotion of technology integration. Still the technology use in classrooms does not necessary means that it is properly aligned to learning curriculum and teacher's believes as one of the factors for technology integration [12]. Our research study explores the idea of incorporating proper computer games into the primary education classes, while evaluating the students' experience. But this research also tries to generate constructive practices and positive examples where teachers play important role in proper technology integration which should improve the educational quality. The case study was performed according to our methodology and the obtained results are analyzed in the effort to support our claims.

The students' feedback from the questionnaires was analyzed and results are compared among the same children who attended the traditional class and the class where the game Zavor was utilized. While participating in both types of classes, on the same subject, these students could express their subjective experience, which gives relevant information for comparison purposes.

While evaluating the statistical data from the students' responses we can draw immediate conclusion if the introduction of the computer games provided positive change in the learning environment and higher level of students' Qoe.

The summary of the students' responses regarding the researched variables and their reliabilities during the traditional class is represented in Table 2.

Table 2. Statistical information regarding the variables and their reliabilities for the traditional class (n = 114)

Variable Name	Min / Max	Mean	Standard Deviation	Skewness	Kurtosis
Simple	1 / 5	3.25	1.443	-0.425	-1.155
Motivation	1 / 5	3.25	1.277	-0.289	-0.815
Quick	1 / 5	3.5	1.311	-0.455	-0.938
QoE	1 / 5	3.38	1.376	-0.409	-1.083

Students' responses covered all possible grades from 1 to 5 on different variables, but generally these responses were constructive with low standard deviation. From the statistical point of view, to evaluate the normality of the surveys' data, we need to look at the skewness and kurtosis, having in mind that absolute values of skew > 3.0 are described as "extremely" skewed and kurtosis absolute values > 8.0 suggest a problem [13]. Therefore the surveys' data presented in Table 2 provides relevant information from the traditional class.

Table 3 lists the statistical information regarding variables during the class with the game Zavor and the necessary parameters for the normality of the data.

Table 3. Statistical information regarding the variables and their reliabilities for the class which included the game Zavor (n = 114)

Variable Name	Min / Max	Mean	Standard Deviation	Skewness	Kurtosis
Simple	2 / 5	4.46	0.824	-1.488	1.480
Motivation	2 / 5	4.30	0.849	-1.034	0.286
Quick	1 / 5	4.39	0.862	-1.442	2.182
QoE	2 / 5	4.40	0.849	-1.669	3.182

The results show significantly higher students' grades on all observed variables when the computer game was introduced in the learning environment. The lower values for standard deviation in Table 3 represents lower errors from the mean scores and the skewness and kurtosis values indicate that surveys' data from this approach is relevant too.

The hypothesized model was also subjected to the students' responses and tested to see how closely the model matches the data. Figure 2 illustrates the proposed path analysis model with relationships and regression weights among the observed variables for the non-traditional class.

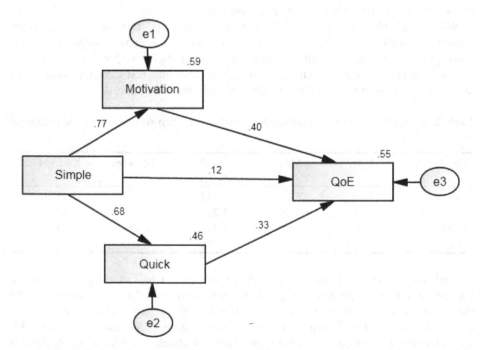

Fig. 2. Path analysis model with regression weights among observed variables

The produced values for the necessary indicators: CMIN/DF = 0.002 (relative chi-square where smaller values are preferable [14]); CMIN = 0.002; GFI = 1.00 (Goodness of Fit Index which should exceed 0.9 for good model [15]); RMSEA = 0.00 (Root Mean Square Error of Approximation where smaller values below 0.08 are preferable [16]); indicate proper model fit and alignment between the hypothesized model and the collected data. The path between Simple and QoE reported $p<0.05$, while the rest of them reported $p<0.001$ which means that all of them are significant (paths with values for $p>0.05$ are usually considered not relevant).

The results show close connection among Simple and Motivation (0.77) and Simple and Quick (0.68), which fully supports Hypothesis 1. The variable QoE is significantly predicted by Motivation (0.40), which leads to conclusion that the students' motivation for learning during the class with included computer game mostly influences positive students' QoE (Hypothesis 2). Furthermore, QoE is also predicted by Quick (0.33) and Simple (0.12) (Hypothesis 3). Even though the regression weight between Simple and QoE is low, the model reports high indirect connection among these variables (0.55), due to the connection between Simple and Quick/Motivation.

From the pedagogical point of view, the students showed better understanding of the content during the non-traditional class. This was evident through the quick evaluation of perceived knowledge at the end of each class, while the game-based class produced better results. Even more, the practical presentations of the material through the game, enabled students to deeper understand the abstract content and use it outside of the learning environment.

The study's analyses and results provide actual information about the students' subjective factors, which have to been taken in consideration for the improvement of the learning process and suggest incorporation of this new and innovative idea in the primary education practice.

6   Conclusion

In this paper we have presented a novel approach for integrating computer games in the primary educational program. Following the user-oriented approach we focus on positive students' experience, a key driver of technology acceptance, adoption and usage behavior. The results of the performed case study support our methodology and provide relevant information for successful implementation of computer games in the learning process, for increased students' motivation and QoE.

In our future work, we will continue to use the proposed methodology while enlarging the scope of work in different directions. We will evaluate benefits and students' experience when computer games are introduced in primary educational classes with subject like language, arts, science and mathematics, provide comparative analyses and test for the model behavior. Following the technology oriented approach, our future work will include involvement of distance education tools like videoconferencing and streaming media, integrated with properly selected computer games.

Acknowledgements. The authors would like to recognize the effort of the primary education teachers Slavica Karbeva, Viktor Gjorgonovski and Ubavka Butleska while participating in this research; and Petre Lameski for preparation of the case study.

References

1. Kilkki, K.: Quality of experience in communications ecosystem. Journal of Universal computer science 14(5), 615–624 (2008)
2. Collins, A.: Design issues for learning environments. International perspectives on the design of technology-supported learning environments, pp. 347–361 (1996)
3. Watson, D.M.: Pedagogy before Technology: Re-thinking the Relationship between ICT and Teaching. Education and Information Technologies 6(4), 251–266 (2001)
4. Wikipedia: Quality of experience (QoE), http://en.wikipedia.org/wiki/Quality_of_experience
5. Reynolds, D., Treharne, D., Tripp, H.: ICT - The Hopes and the Reality. British Journal of Educational Technology 34(2), 151–167 (2003)
6. Lehrer, R., Schauble, L.: Developing modeling and argument in the elementary grades. Understanding Mathematics and Science Matters, pp. 29–53 (2005)
7. Jarschel, M., Schlosser, D., Scheuring, S., Hossfeld, T.: An Evaluation of QoE in Cloud Gaming Based on Subjective Tests. In: 2011 Fifth International Conference on Innovative Mobile and Internet Services in Ubiquitous Computing (IMIS), pp. 330–335. IEEE (2011)
8. Bauer, J., Kenton, J.: Toward Technology Integration in the Schools: Why it isn't Happening. Journal of Technology and Teacher Education 13(4), 519–546 (2005)
9. Egenfeldt-Nielson, S.: Third generation educational use of computer games. Journal of Educational Multimedia and Hypermedia 16(3), 263–281 (2007)
10. Curtis, D., Lawson, M.: Computer adventure games as problem-solving environments. International Education Journal 3(4), 43–56 (2002)
11. Chen, C.H.: Why Do Teachers Not Practice What They Believe Regarding Technology Integration? The Journal of Educational Research 102(1), 65–75 (2008)
12. Levin, T., Wadmany, R.: Teachers' beliefs and practices in technology-based classrooms: A developmental view. Journal of Research on Technology in Education 39(2), 157–181 (2006)
13. Curran, P.J., West, S.G., Finch, J.F.: The robustness of test statistics to nonnormality and specification error in confirmatory factor analysis. Psychological Methods 1(1), 16–29 (1996)
14. Holmes-Smith, P.: Introduction to structural equation modeling using AMOS 4.0 & LISREL 8.30. In: School Research, Evaluation and Measurement Services, Melbourne, Australia (2000)
15. Joreskog, K.G., Sorbom, D.: LISREL-VI User's Guide. In: Scientific Software, 3rd edn. (1989)
16. Browne, M.W., Cudeck, R.: Alternative ways of assessing model fit. Sage Focus Editions 154, 136–136 (1993)

Modeling a Quality of Experience Aware Distance Educational System

Tatjana Vasileva-Stojanovska and Vladimir Trajkovik

Faculty of Computer Science and Engineering,
Ss. Cyril and Methodius University, Skopje, Republic of Macedonia
tatjanav@nbrm.mk, trvlado@finki.ukim.mk

Abstract. A successful distance educational system has to provide certain commonly accepted educational objectives in order to assure its high acceptability and successful implementation. One of the main objectives of any distance educational system is to bring the educational material and its presentation closer to the individual student's needs thus delivering high Quality of Experience (QoE). The overall QoE is a subjective perception impacted by factors from objective technical nature and subjective nature. In this paper we propose few educational scenarios that enable dynamical presentation of educational services and content delivery to the students depending on their personal affinities. Students can choose the preferred media presentation and after completion of the course an evaluation and comparison of the student's performances on the different educational scenarios is made. The aim of the study is to explore the impact of the Quality of Experience (QoE) on the performance and thus the overall Quality of Learning (QoL) of students. For comparative analysis of the educational scenarios, t-test statistic is proposed. In addition, we propose a neuro fuzzy inference system for QoE evaluation, utilizing an ANFIS controller to compare and predict the expected QoE based on the input parameters affecting the QoE. The model should provide an aid to design of educational courses and media presentation in a way that aims toward student satisfaction and educational benefits for the current and future users of distance educational systems. This paper is focused on the description of the proposed QoE aware educational system, leaving the discussion of the experimental results for future work.

Keywords: distance education, quality of experience, quality of learning.

1 Introduction

In the present world of web-based educational media, building a successful distance educational system (DES) faces many challenges from both technological and pedagogical aspect. The approach these challenges are met can lead to a successful acceptance and appliance of the DES. The overall acceptability of the system can be measured with the quality of experience factor (QoE) that is a subjective measure as perceived by the end users of the system [7].

V. Trajkovik and A. Mishev (eds.), *ICT Innovations 2013*,   45
Advances in Intelligent Systems and Computing 231,
DOI: 10.1007/978-3-319-01466-1_4, © Springer International Publishing Switzerland 2014

The interactive educational environment used in our research aims towards provision of the following educational objectives for the users of distance educational information systems [14], [12]:

- providing appropriate and high quality educational content
- delivery of learning quality (QoL) trough optimization of the subjective Quality of Experience (QoE) and objective Quality of Service (QoS) paradigms.
- identification and classification of the user preferences using user modeling
- delivery of dynamic services and presentation of the educational content according to the identified preferences for optimal QoE
- prediction of the future user behavior through development of intelligent neuro fuzzy algorithms
- dynamical measurement of student's knowledge and adapting content delivery to user learning and other higher-level needs

Our distance educational system uses the Moodle interactive interface for management of the student-content and teacher-content interaction. It provides appropriate access and presentation of the educational services for both students and lecturers. The lecturers create the learning environment for each subject, and the students are able to access the environment for every subject they are enrolled. In order to meet the objectives of user's classification and future behavior prediction, appropriate user models are being developed. A user model is considered as a set of information structures designed to represent the following elements [3]:

- representation of assumptions about the knowledge, goals, plans preferences, tasks and/or abilities about one or more types of users
- representation of relevant common characteristics of users pertaining to specific user subgroups (stereotypes)
- classification of a user in one or more of these subgroups
- recording of user behavior
- formation of assumptions about the user based on the interaction history; and/or
- generalization of the interaction histories of many users into stereotypes

The main goal of this paper is to present a user model of a DES that provides a learning environment offering appropriate educational context for different stereotypes of students, aiming towards optimization of the quality of experience and learning results. The experimental data is to be employed as an input in a neuro fuzzy system for QoE assessment and prediction and the results will be used to offer the most suitable educational scenarios for the current and further students depending on the preferences of the stereotype they are classified in.

The scope of this paper is to give a theoretical presentation of a model of QoE aware DES and development of the idea of personalization of the environment through different educational scenarios. The experimental results are not in the focus of this paper, and will be detailed in the further work.

Second section of the paper gives an overview of the related work. Third section describes the offered educational scenarios and deals with few pedagogical aspects that will arise from the results. The fourth section describes the proposed ANFIS system for QoE evaluation. The fifth section proposes a metric for assessment and interpretation of the obtained results from the educational scenarios. The last section concludes the paper.

2   Background

Distance education as concept emerged from the primal idea of giving the adult students an opportunity to continue their education, evolving to concept of modern learner-centered educational systems utilized and integrated into mainstream education. The growing trend of distance learning enrollments and on line services offered by universities is evident in yearly reports by institutions analyzing the DE trends and development [2].

The factors influencing the educational gains in distance educational systems researched in [11] are: quality of instruction and learning, cost of attendance, the needs of a typical DE student, student satisfaction from the DE and factors affecting the instructional efficacy [11].

The ability of the DE system to adapt to the learning preferences of every individual student, leads to improved experience with it. The level of acceptance of the educational system for the majority of the students impacts the overall success of the educational system. The acceptance of the DE systems should be lifted to a level of "educational appliance" [14]. The educational gains from this approach are measured with appropriate metric and compared to traditional classroom teaching concept in order to evaluate the results and determine the directions for further improvement and development of the distance educational systems [1].

The impact of the media presentation on the acceptance of the e-learning technology is given in [8]. Authors explored three different media types i.e. text, streamed audio and streamed video. The perceived usefulness and the user's attitude were used to predict the intention to use the system. The effect of the media on the student's satisfaction and engagement is given in [4]. The authors conclude that the asynchronous rich media presentations increase the student's satisfaction with the on-line courses.

The adaptation of the service performance on variable network conditions in order to satisfy the user's Quality of Experience is explored in [6]. The authors propose an ANFIS based neuro-fuzzy model using a combination of application and physical layer parameters to predict the delivered video quality to the users. A methodology for quantification of the correlation between QoS and QoE is given in [15].

In our work we are exploring the connection of the objective and subjective parameters influencing the individual student's QoE. Specifically, we are exploring few educational scenarios that offer adaptive presentation of the educational material in order to bring the content and educational environment closer to the user's expectations and thus obtain maximum QoE. The proposed neuro-fuzzy model should give an estimation of the QoE based on the given inputs, providing an aid to content presentation and adaptation for current and further users of the DE system.

3   Educational Scenarios of QoE Aware DE System

In our educational research approach, we propose a user model that classifies the students into few stereotypes based on assumptions about their learning preferences and evaluates the results from the final exams that would approve or reject the given assumptions.

For the purposes of this research, we conduct few experimental courses with two groups of students. The educational content of each course can be presented as:

1. **off line document content**; the lecturer creates a learning environment in which provides educational content in form of documents, presentations, url links, etc. Student is able to access the environment and learn at his own step and schedule.
2. **off line video content**; the lecturer records himself teaching the lecture, and uploads the video to a location accessible for students. Again the student is able to learn at his own step and schedule.
3. **on line video conferencing**; The lecture is scheduled at exact time and the participants are expected to enroll for the class. After the lecture, the participants are given a chance to interact with the lecturer and among themselves. Each student has to follow the predefined schedule for the class.

The learning sessions are conducted with two different groups of students (A and B) and two different courses (C1 and C2). Our goal is to experiment with the presentation of the educational content, the lecturing scenarios and to compare the results after the completion of the course.

The following four lecturing scenarios are proposed:

Course C1:

1. The group of students A, are asked to choose their preferred presentation of the educational content from the given choices. Following the students choice, they are divided into three stereotypes and to each stereotype the lectures are presented according to the preferred choice.
2. To the second group of students B, the educational content of each lecture is presented according to the lecturer's choice. The lecturer chooses one of the three options to present each lecture, without consulting the student's preferences. Every student enrolled for the course, follows the lesson according to the lecturers choice.

Course C2:

3. For the C2 course the students from the other group B can choose their preferred presentation of the educational content
4. The lecturer chooses the content presentation for the group A.

At the end of the courses C1 and C2, questionnaires are given to each participant to provide in the subjective estimation of the QoE as perceived by each student.

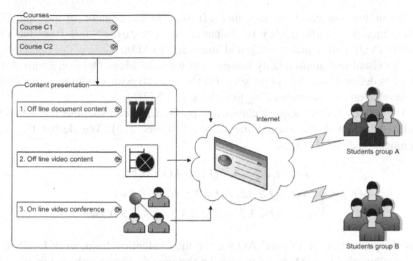

Fig. 1. Educational scenarios in interactive educational environment

The student's performances are examined upon completion of the courses and comparison of the results from each scenario is made. These comparisons should be able to provide conclusions on pedagogical issues regarding the correlation of the student's performance to the educational scenario, the relation between QoE and QoL, the comparative performance between on-line and off-line educational scenarios, the preferred media presentation etc.

In addition results will be employed in a neuro-fuzzy inference system for QoE evaluation. The inference system utilizes ANFIS (Adaptive Neuro Fuzzy Inference System) hybrid system described by R.Jang [5], for mathematical representation of the student's behavior in the distance educational environment. ANFIS integrates the ability of neural networks to learn from sample data with fuzzy logic techniques for human behavior representation.

4   The ANFIS Based Approach for QoE Assessment

4.1   Input and Output Variables

To evaluate the students QoE we consider both subjective and objective variables that influence the overall student's experience [6]:

1. **Educational content type.** The values for this variable are taken from the term set of educational content presentations types: {1=off line document content, 2=off line video content, 3=on line video conferencing }
2. **Media technical quality.** The values for the second variable are obtained as a composite value from few technical aspects typical for each of the three media content types. When the content type is off line document we assume a constant value for the technical quality variable.

When the content type is either off line video content or on line video conference, the media quality is evaluated as a composite value from the visual quality (VQ), audio quality (AQ) and audio delay (AD).

The visual and audio quality is measured with the Mean Opinion Score (MOS). MOS is defined as a five point scale (ITU-T P.800) with minimum threshold for acceptable quality corresponding to a MOS of 3.5 [7].

The values for the audio delay are expressed in seconds, and the acceptable audio delay must not exceed the value of 0.5sec [13]. We define the media technical quality (MTQ) as:

$$MTQ = \begin{cases} excelent, VQ > 4, AQ > 4, AD < 0.2\sec \\ good, VQ \geq 3.5, AQ \geq 3.5, AD < 0.5\sec \\ poor, VQ < 3.5 \ or \ AQ < 3.5 \ or \ AD > 0.5\sec \end{cases} \quad (1)$$

For the evaluation of VQ and AQ we use the evaluation framework EvalVid [6]. To achieve the best MTQ of utmost importance is to properly setup the video equipment i.e. the camera focus, the placement of one or more microphones close to the main presenter but away from sources of external distraction such as fans, speakers, etc.

3. **Quality of Service** (QoS). The values for the third variable refer to the quality of content delivery over the network. In our context we assume that the QoS is concerned primarily with the Network QoS (NQoS). The distance educational system has to provide a proper NQoS policy to enable reliable delivery of multimedia data over the transport infrastructure. The values for the NQoS are obtained from the EvalVid framework depending on the concrete readings for the latency, jitter and packet loss during the learning session.

4. **Content's educational quality.** The values for the fourth variable depend on the subjective opinion of the students regarding the educational content. To obtain the students opinion, after the learning session students are invited to answer short survey expressing their personal opinion of the quality of the presented educational content. The content's educational quality (CEQ) is defined as:

$$CEQ = \begin{cases} excelent, & if \ the \ answers \ of \ all \ the \ questions \ are \ positive \\ good, & if \ the \ answers \ of \ any \ two \ questions \ are \ positive \\ poor, & if \ less \ than \ two \ answers \ of \ any \ questions \ are \ positive \end{cases} \quad (2)$$

5. **Student's preferred stereotype.** The values for the fifth variable depend on the subjective choice of each student regarding the preferred educational content, i.e.

0=off line learning, either off line document or video content.
1=online video conference

6. **Lecturing scenario.** The values for the sixth variable are from the term set {1, 2, 3, 4}. Each value corresponds to the lecturing scenarios described in the section 2 of this document.

7. **Quality of Experience (QoE).** QoE is the single output variable. The values of this variable are measured with the MOS scale. Our objective is to keep MOS value above the value of 3.5

The first three variables are considered objective (directly measurable), and the next three variables are considered subjective, i.e. depending on the subjective affinities of each student and the lecturer.

4.2   Rule Base

Rule base definition means building a type III fuzzy controller, i.e. Takagi-Sugeno type, which uses a two-pass learning cycle, a forward pass and a backward pass.

A typical rule set is first-order Sugeno fuzzy model with six input variables, consists of rules of the following type:

Rule i: if x_1 is A_i and x_2 is B_i and x_3 is C_i and x_4 is D_i and x_5 is E_i and x_6 is F_i then $f_i=p_ix_1+q_ix_2+l_ix_3+m_ix_4+n_ix_5+s_ix_6+r_i$

4.3   The System Structure

The ANFIS based structure consists of five layers, each layer containing nodes of different structures and connections. The input signals for every node come from the output signals from the previous level. The output from the i-th node in k-th layer is noted as $O_{k,i}$. The following layers are identified:

Layer **0**: input variables layer

Layer **1**: Fuzzification layer. In this layer the membership functions and the term sets of each variable from the previous layer are defined. Each value from the term sets of the input variables represents a node in this layer.

Layer **2**: Every node in layer 2 represents the firing strength of each rule using the product (or soft-min) of all incoming signals as an output signal.

$$O_{2,i}=w_i=\mu_{Ai}(x_1)\cdot \mu_{Bi}(x_2)\cdot \mu_{Ci}(x_3)\cdot \mu_{Di}(x_4)\cdot \mu_{Ei}(x_5)\cdot \mu_{Fi}(x_6). \tag{3}$$

Layer **3**: This layer is called a *normalization layer*. Outputs of the nodes are called normalized firing strengths.

$$O_{3,i} = \overline{w_i} = w_i/(w_1 + w_2 + w_3 + w_4 + w_5 + w_6). \tag{4}$$

Layer **4**: The $O_{3,i}$ from the previous layer weighs the result of its linear regression $f_i=p_ix_1+q_ix_2+l_ix_3+m_ix_4+n_ix_5+s_ix_6+r_i$ in the fourth layer called the *function layer*, generating the rule output

$$O_{4,i} = \overline{w_i} f_i = \overline{w_i}(p_ix_1 + q_ix_2 + l_ix_3 + m_ix_4 + n_ix_5 + s_ix_6 + r_i). \tag{5}$$

Layer **5**. The output parameter is the overall output as sum of all incoming signals from layer 4, i.e.

$$O_{5,i} = \sum_i \overline{w}_i f_i = \sum_i w_i f_i / \sum_i w_i .$$ (6)

Graphical representation of proposed ANFIS model is given on Fig 2:

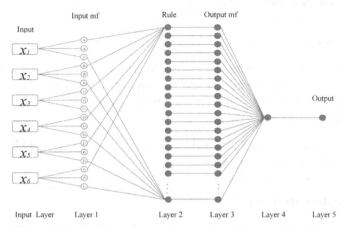

Fig. 2. The ANFIS system with six input and one output variable

5   Data Analysis and Learning Assessment

The assessment of the student performances in a distance educational system should provide the instructor with a set of qualified information about the cognitive and metacognitive level of the students, thus aiding to the proper choice of different pedagogical approaches in order to optimize the Quality of Learning (QoL) [10]. In this section we will propose a common statistical model for analysis and assessment of the performance of the students participating in each of the educational scenarios given in section 3. The aim of this analysis is to provide an objective estimation on the advantages or disadvantages of the pedagogical approach of each assessed educational scenario. The main assessment objectives that can be analyzed with the tools provided in the Moodle interactive learning environment are summarized in Table 1:

For each student participating in the educational scenarios given in section 3, the assessment process will result in eight metrics for each objective given in table 1. The instructor will give the student an averaged final score from the obtained eight metrics. The overall score is a crisp value from the interval [1-5]. At the end of the assessment, a data set containing the student's final scores will be produced for each educational scenario, thus obtaining a total of four data sets labeled as DS1, DS2, DS3 and DS4.

The data analysis on the resultant four data sets is focused around two main objectives. First objective is to analyze and compare the performances of the students regarding the educational scenarios given in section 3. Second objective is to compare

Table 1. Assessment objectives, instruments and metrics in DES

Assessment Objective	Assessment instruments	Metrics
Verify the participation level and degree of interaction among participants	Discussion Forum / Chat	Total number of interactions computed from the posted messages
Verify the capability of "asynchronous" elaboration of the participant	Off-Line Activities	Similarity Index between the Issued Answer and the Expected Standard Answer quantified
Verify the capability of "synchronous" elaboration of the participant	On-Line Activities (tools for collaborative edition)	Similarity Index between the Issued Answer and the Expected Standard Answer quantifie
Verify student knowledge level about a given topic with instant feedback	Objective Questions	Similarity Index between the Issued Answer and the Expected Standard Answer
Verify student knowledge level about one or various topics	Written Questions	Similarity Index between the Issued Answer and the Expected Standard Answer quantified, for example, by keywords in the solution
Identify learner confidence level about the learner own knowledge before studying a topic	Self-Assessment	Confidence Level
Verify student participation level	Access Statistics	Access index of the student in tools by executed actions
Indicate the Knowledge Acquisition Level of the Student	Grades Board	Metrics that summarizes the student performance indicating student cognitive and meta-cognitive profiles

the quality of experience of the participants in the presented distance educational system with the results obtained from the ANFIS.

To meet the first objective, we will perform a descriptive statistical analysis on the collected data sets. For that purpose data sets DS1, DS2, DS3 and DS4 will be paired, resulting in total of six pairs of data sets: (DS1, DS2), (DS1, DS3), (DS1, DS4), (DS2, DS3), (DS2, DS4) and (DS3, DS4). Then we perform a t-test on each of the paired data sets.

T-test is performed for each pair (DSi, DSj). We state research and null hypotheses, determine the significance level of 0.05 i.e. 95% confidence level of the obtained result.

Hypotheses:

- Null hypothesis: there is no difference of the performance of the students in the compared scenarios.
- Research hypothesis: there is difference of the performance of the students depending on the educational scenario.

Interpret the Results: The P-value is compared to the significance level. The null hypothesis is rejected if the P-value is less than the significance level.

The performed statistical analysis provides a metric for comparison of the student performances participating in the paired scenarios i and j. If the null hypothesis is accepted we will conclude that students do not experience significant difference in the QoL in scenario i and j. Otherwise, if the null hypothesis is rejected, the analysis should provide a clear answer of the degree of difference of student's performances in scenario i and j. The analysis of the obtained p-values for all the six data pairs will provide the conclusion on which of the four scenarios produces the best student's performance and thus help the instructor to adjust the pedagogical approach for optimal learning results.

The same t-test statistic can be employed to analyze the QoE perceived by the students after the completion of the experimental courses, and the predicted QoE from the proposed neuro fuzzy system.

6   Conclusion

A Quality of Experience aware distance educational system presented in this paper is an active educational environment providing a student content interaction via Moodle interface. Few learning scenarios utilizing different content presentations are proposed in order to make a comparison and adjust the learning environment as to provide the best performances for different stereotypes of users. The research is aiming towards development of a user model to manage the different stereotypes of users as well as prediction of the performances of current and future users of the system based on their past behavior and interaction history. A mathematical model utilizing a neuro fuzzy controller for QoE evaluation is proposed in the paper in order to predict the expected QoE based on the objective and subjective input parameters. An evaluation of student's performances comparing the results from the student's evaluation regarding the four proposed distance educational scenarios is given. A descriptive statistical analysis on the collected data sets from the scores of each participant should reveal whether the proposed educational scenarios can provide an aid to the current educational system for provision of optimal learning experience.

References

1. Allen, M., Mabry, E., Mattrey, M., Bourhis, J., Titsworth, S., Burrell, N.: Evaluating the effectiveness of distance learning: A comparison using meta-analysis. Journal of Communication 54(3), 402–420 (2004)

2. Allen, I.E., Seaman, J.: Going the Distance: Online Education in the United States. In: ERIC 2011 (2011)
3. Frías-Martínez, E., Magoulas, G.D., Chen, S., Macredie, R.: Recent soft computing approaches to user modeling in adaptive hypermedia. In: De Bra, P.M.E., Nejdl, W. (eds.) AH 2004. LNCS, vol. 3137, pp. 104–114. Springer, Heidelberg (2004)
4. Havice, P.A., Foxx, K.W., Davis, T.T., Havice, W.L.: The Impact of rich media presentations on a distributed learning environment. Quarterly Review of Distance Education 11(1), 53 (2010)
5. Jang, J.S.R., Sun, C.T., Mizutani, E.: Neuro-Fuzzy and Soft Computing-A Computational Approach to Learning and Machine Intelligence. Prentice Hall inc (1997)
6. Khan, A., Sun, L., Fajardoi, J.-O., Liberal, F., Ifeachor, E.: An ANFIS-based Hybrid Quality Prediction Model for H.264 video over UMTS Networks. In: 2010 IEEE International Workshop Technical Committee on Communications Quality and Reliability (CQR), pp. 1–6. IEEE (2010)
7. Kuipers, F., Kooij, R., De Vleeschauwer, D., Brunnström, K.: Techniques for measuring quality of experience. In: Osipov, E., Kassler, A., Bohnert, T.M., Masip-Bruin, X. (eds.) WWIC 2010. LNCS, vol. 6074, pp. 216–227. Springer, Heidelberg (2010)
8. Liu, S.H., Liao, H.L., Pratt, J.A.: Impact of media richness and flow on e-learning technology acceptance. Computers & Education 52(3), 599–607 (2009)
9. Moore, J.C.: The Sloan consortium quality framework and the five pillars. The Sloan Consortium (2005)
10. Paechter, M., Maier, B., Macher, D.: Students' expectations of, and experiences in e-learning: Their relation to learning achievements and course satisfaction. Journal of Computers & Education 54(1), 222–229 (2010)
11. Shachar, M.: Meta-Analysis: The preferred method of choice for the assessment of distance learning quality factors. The International Review of Research in Open and Distance Learning 9(3) (2008)
12. Sun, P.-C., Tsai, R.J., Finger, G., Chen, Y.-Y., Yeh, D.: What drives a successful e-Learning? An empirical investigation of the critical factors influencing learner satisfaction. Journal of Computers & Education 50(4), 1183–1202 (2008)
13. Tang, J.C., Isaacs, E.A.: Why do users like video? Studies of multimedia-supported collaboration. Computer Supported Cooperative Work (CSCW) 1(3), 163–196 (1993)
14. Vouk, M.A., Bitzer, D.L., Klevans, R.L.: Workflow and end-user quality of service issues in Web-based education. IEEE Transactions Knowledge and Data Engineering 11(4), 673–687 (1999)
15. Wu, W., Arefin, A., Rivas, R., Nahrstedt, K., Sheppard, R., Yang, Z.: Quality of experience in distributed interactive multimedia environments: toward a theoretical framework. In: Proceedings of the 17th ACM international conference on Multimedia, pp. 481–490. ACM (2009)

Adaptive Multimedia Delivery in M-Learning Systems Using Profiling

Aleksandar Karadimce and Danco Davcev

Faculty of Computer Science and Engineering,
Ss. Cyril and Methodius University, Skopje, 1000, R. Macedonia
akaradimce@ieee.org, danco.davcev@finki.ukim.mk

Abstract. The importance of mobile services in our everyday life is growing and there is an increased necessity for adapting the multimedia content according to the user's requirements. Therefore, first we need to determine user cognitive preference in order to be able to make adaptation of the multimedia content to mobile user desires. This process of estimation of user profile characteristics and determining which services can be offered for the end user is profiling. This research will inspect the influence of profiling regarding the visualizer-verbalizer dimension of cognitive style in m-learning systems. We have conducted research experiments for different scenarios by using discrete event simulation in OPNET simulator. This paper considers the significance of high QoS requirements that are essential to achieve higher continuity of real-time delivery of multimedia contents.

Keywords: M-learning, adaptive multimedia, profiling, OPNET.

1   Introduction

The existing perception for mobile learning (M-learning) consists of teaching a lecture or e-books with lot of pages of text and graphics [1], delivered on a very small screen. However, with the establishment of new mobile information technology [2] and ubiquitous expansion of wireless technology, these conventional techniques will be replaced by the brand-new mobile learning [3]. Recent years, M-learning is a field which chains mobile computing and electronic learning (e-learning) [3], and provides more interactive and personalized content based on the learner's context [4] and learner cognitive profile [5]. Content adaptation [6] offers the most suitable applications according to students' computing context [7], referred to devices, network, location, and time, which have influence on students' mobile access of multimedia learning content [3].

In particular, network-aware Quality of Service (QoS) parameters that have huge influence on the bandwidth allocation in the process of multimedia content delivery [8] are lower startup delay and reduced end-to-end delay. M-learning systems deliver

V. Trajkovik and A. Mishev (eds.), *ICT Innovations 2013*,
Advances in Intelligent Systems and Computing 231,
DOI: 10.1007/978-3-319-01466-1_5, © Springer International Publishing Switzerland 2014

an interactive environment [9], using the right tools and support, as presented in [10] and [11], have shown that students can gain significantly more and achieve a greater level of skill and performance. In order to reach appropriate adaptation of multimedia content we need to consider the user cognitive perception of multimedia content during the m-learning process. Another research [12] has revealed the need of reduced file size and bandwidth demand only by using the multiple view perspective approach. It also has also shown that multiple view perspective is less expensive and more inclusive, scalable, flexible and easier to be deployed especially for mobile devices [12].

This paper is organized as follows: Section II presents the process of estimating user profiles in M-learning system. Section III presents the simulation results of delivery of multimedia content in M-learning environment. Finally, Section IV concludes the paper.

2   Estimating User Profiles in M-learning System

The introduction of mobile learning systems have provided a potential for anytime and anywhere educational environments that support the learning process of each individual student and keep the continuity of life-long learning. Existing m-learning environments still experience diverse technological and Quality of Service (QoS) problems, such as delivery of different kind of multimedia materials; adaptation of the learning material to individual student needs.

According to Mayer [13] the user experience is divided into four parts: perception measures, rendering quality, physiological measures, and psychological measures. His research is limited for cognitive style, learning preference, and cognitive ability to individual differences along the visualizer–verbalizer dimension within a multimedia learning environment [14]. In research the content adaptability will be performed by determining the user profile first and then delivering the preferable multimedia content i.e. any combination of text, graphics, audio, video or animation that best suites individual's cognitive perception. We have adopted Learning Scenario Questionnaire [14] to be used for estimating the preferences in five learning situations from the research area Database systems. The Learning Scenario Questionnaire, which we originally designed as a measure of learning preference, is used to provide the cognitive style factor (visualizer, verbalizer or bimodal dimension). Visualizers prefer to receive multimedia information via graphics, animation, video and images, whereas, mainly because their visual memory is much stronger than their verbal. On the other hand, verbalizers would prefer to process information in the form of words, expressed by audio or text based form. Certainly, bimodal users are equally comfortable using either modality of perception of multimedia content.

Learning Scenario Questinaire

Q1) Which format do you prefer a scientific description of a Database System?
  1) a paragraph describing each part
  2) a label diagram showing each part

Q2) Which format do you prefer in learning scietific explanation of how executing SQL query works?
  1) an essay describing what happens with each command of SQL query statement
  2) a series of labeled diagram showing the status of each part of execution scenario of SQL query

Q3) Which format do you prefer for following directions for how to draw an ER diagram?
  1) verbal direction including telling you that you have to draw first entities, attributes and relations, respectivly
  2) a map showing how to draw the entities, attributes, and how are they connected with relations

Q4) Which format do you prefer for following instructions for how to aggregate data using SQL statement?
  1) a list of steps in words
  2) a labeled diagram showing the steps

Q5) Which format do you prefer for describing the results of the executed SQL queries?
  1) a list of results in table
  2) a graphical report with chart, diagram or pie.

Fig. 1. Learning Scenario Questionnaire

We have conducted survey on a group of 30 students, which have been given Learning Scenario Questionnaire (LSQ) for estimating preferences in five learning situations based on text descriptions form the LSQ questionnaire.

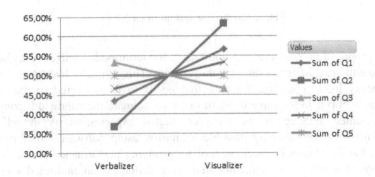

Fig. 2. Results from Learning Scenario Questionnaire

Results, given on Figure 2, confirm that there is clear distinction of students that are visualizers from students that prefer to process information in text based form – verbalizers. These results from LSQ questionnaire have been used as input parameters for profile definition in the OPNET network simulator.

3   Simulation Results of Delivery of M-learning Multimedia Content

We have proposed architecture of M-learning system as presented in Figure 3. The simulation scenario is consisted of two similar laboratories for M-learning that have wireless access point router and mobile clients. This way all the students using variety of mobile devices (smart phones, tablets and mobile phones) can easily connect, through the wireless router, to the multimedia streaming server.

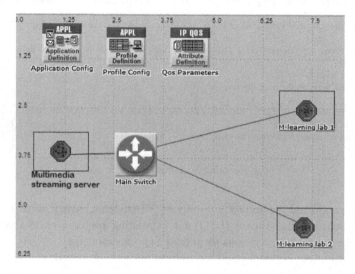

Fig. 3. M-learning system in OPNET

In our proposed architecture, the bottleneck for multimedia delivery lies between the multimedia streaming server and the switch. This happens because of the increased heavy load of multimedia data that needs to be delivered to the two M-learning laboratories in the same time. In order to estimate the bandwidth congestion in the presented architecture we have conducted experiments by modeling the proposed network and defining possible scenarios using software simulation tool. Therefore, the simulation is excellent tool for studying performance and identifying the Quality of Service (QoS) factors that have influence on multimedia contents delivery.

We have used Discrete Event Simulation (DES) [15] because it enables modeling in a more accurate and realistic way. It creates an extremely detailed, packet-by-packet model for predicting the activities of the network. Multimedia streaming server (application configuration) is configured for streaming real-time audio, video and multimedia data (text or images), similar like in the real M-learning systems, see Figure 4.

Fig. 4. Application config in OPNET

In order to simulate the M-learning laboratory we have configured two subnets that each contains wireless access router and 5 to 9 mobile clients. Requests for multimedia content are streamed from the Multimedia streaming server to the mobile clients in the M-learning laboratories 1 or 2. Mobile clients using the profile configuration have been settled to three different cognitive learning skills: visualizer, verbalizer and bimodal users (FTP profile), see Figure 5.

This way every mobile user depending of his cognitive learning style can receive appropriate multimedia content that completes the process of profiling. Depending of the context-aware conditions that are present the configured QoS parameter is analyzed for three different simulation scenarios. The first OPNET scenario is using FIFO packet delivery, represented with blue line, for the bottleneck link between the multimedia streaming server and the switch. The second OPNET scenario is using Priority Queuing (PQ) algorithm to determine the delivery of multimedia packets. Finally, the third OPNET scenario is using the Weighted-Fair Queuing (WFQ) algorithm for transferring of multimedia packets.

The network simulator was configured to run one hour of multimedia content in the established m-learning system for the three different OPNET scenarios (FIFO, PQ and WFQ algorithms). The blue line represents the results from the FIFO packet delivery algorithm simulation, the red line represents the results from the PQ packet delivery algorithm simulation and the green line gives the results from WFQ packet delivery algorithm simulation.

Analyzing the wireless LAN delay from the M-learning system in the three different OPNET scenarios, see Figure 6, we conclude that the FIFO algorithm is generating the

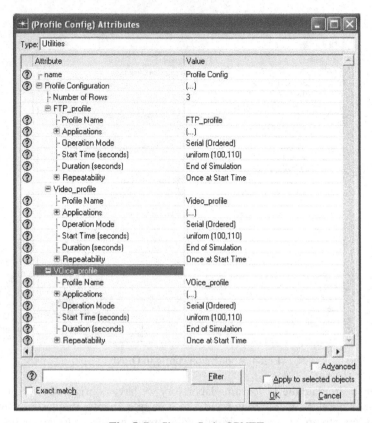

Fig. 5. Profile config in OPNET

biggest delay. On the other side the scenarios with PQ and WFQ algorithms have decreased delay, which can be described with the existence of CDN nodes that provide improved delivery of multimedia traffic in the M-learning systems.

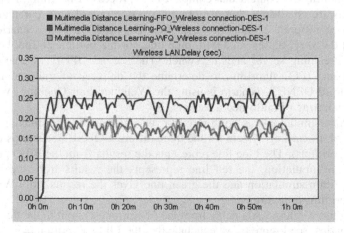

Fig. 6. Results from wireless LAN delay in OPNET simulator

The results from the Voice traffic QoS parameter, average packet delay, for the three different OPNET scenarios are shown in Figure 7. From the verbalizer cognitive perception the process of voice delivery have revealed lowest average delay in the OPNET scenario that uses WFQ packet delivery algorithm.

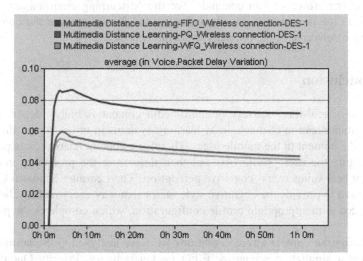

Fig. 7. Results from Voice average packet delay in OPNET simulator

The results from the Video traffic QoS parameter, end-to-end average packet delay, for the three different OPNET scenarios are shown in Figure 8. The visualizer cognitive perception given by the video delivery has shown lowest average delay in the both of OPNET scenarios the PQ and the WFQ packet delivery algorithm.

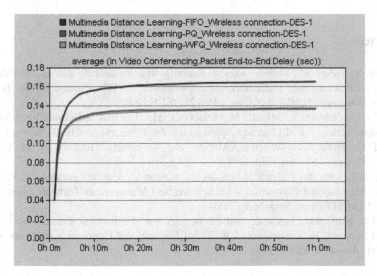

Fig. 8. Results from Video end-to-end average packet delay in OPNET simulator

Considering the received results we can see that almost similar results are achieved for the both of scenarios that are using the PQ and the WFQ packet delivery algorithm. Although, the FIFO algorithm for packet delivery that represents standard communication network has shown the poorest results for all of the monitored QoS parameters. Therefore, we can conclude that the M-learning environment that uses profiling with the presence of priority and weighted-fair queuing in the communication link have delivered improved delivery of multimedia traffic.

4   Conclusion

The process of real-time delivery of multimedia content is highly dependent from efficient communication channels in M-learning systems in order to ease the transfer of multimedia content to the mobile users. Therefore, content adaptation is performed by determining the user profile first and then delivering the preferable multimedia content that best suites users' cognitive perception. The Learning Scenario Questionnaire was used to provide the cognitive style factor that way every mobile clients have been provided with appropriate profile configuration, which completes the process of profiling.

Considering the context-aware conditions we have analyzed QoS parameters into three different simulation scenarios: FIFO packet delivery, Priority Queuing (PQ) algorithm and Weighted-Fair Queuing (WFQ) algorithm for transferring of multimedia packets. In conclusion the PQ and the WFQ packet delivery algorithm OPNET scenarios with decreased delay have led to improved collaborative response to the mobile clients. Eventually, this means efficient and increased learner perception and satisfaction from the M-learning system that is in direction for better quality of learning.

References

1. Joseph, S., Uther, M.: Mobile language learning with multimedia and multi-modal interfaces. In: Proceeding of IEEE International Workshop on Wireless and Mobile Technologies in Education (ICHIT 2006), pp. 124–128. IEEE (2006)
2. Gil, D., Andersson, J., Milrad, M., Sollervall, H.: Towards a Decentralized and Self-Adaptive System for M-Learning Applications. In: Proceeding of IEEE Seventh International Conference on Wireless, Mobile and Ubiquitous Technology in Education (WMUTE), pp. 162–166. IEEE (2012)
3. Yang, S.R., Lim, Y.P.: Towards Content Adaptation for Mobile Learning. In: Proceeding of 2nd International Conference on Education and Management Technology, pp. 77–81. IPEDR IACSIT Press, Singapore (2011)
4. La, H.J., Kim, S.D.: A Conceptual Framework for Provisioning Context-aware Mobile Cloud Services. In: Proceeding of IEEE 3rd International Conference on Cloud Computing, pp. 466–473. IEEE (2010)

5. Muntean, V.H., Muntean, G.M.: A novel adaptive multimedia delivery algorithm for increasing user quality of experience during wireless and mobile e-learning. In: Proceeding of IEEE International Symposium on Broadband Multimedia Systems and Broadcasting (BMSB), pp. 1–6. IEEE (2009)
6. Pettersson, O., Gil, D.: On the Issue of Reusability and Adaptability in M-learning Systems. In: Proceeding of 6th IEEE International Conference on Wireless, Mobile, and Ubiquitous Technologies in Education, pp. 161–165. IEEE (2010)
7. Wang, J., Kourik, J.L.: Delivering database knowledge with web-based labs. In: Proceeding of American Society of Business and Behavioral Sciences, ASBBS in Las Vegas, pp. 923–931 (2012)
8. Laghari, K.R., Pham, T.T., Nguyen, H., Crespi, N.: QoM: A new quality of experience framework for multimedia services. In: Proceeding of IEEE Symposium on Computers and Communications (ISCC), pp. 851–856. IEEE (2012)
9. Jung, I.: The Dimensions of E-Learning Quality: From the Learner's Perspective. Educational Technology Research and Development 59, 445–464 (2011)
10. Shariffudin, R.S., Julia-Guan, C.H., Dayang, T., Mislan, N., Lee, M.F.: Mobile Learning Environments for Diverse Learners in Higher Education. In: IJFCC 2012, pp. 32–35 (2012)
11. Saranya, M., Vijayalakshmi, M.: Interactive Mobile Live Video Learning System in Cloud Environment. In: Proceeding of IEEE-International Conference on Recent Trends in Information Technology, ICRTIT, pp. 673–677. IEEE (2011)
12. Kushalnagar, R.S., Cavender, A.C., Pâris, J.F.: Multiple view perspectives: improving inclusiveness and video compression in mainstream classroom recordings. In: Proceedings of the 12th international ACM SIGACCESS conference on Computers and accessibility (ASSETS 2010), pp. 123–130. ACM, New York (2010)
13. Mayer, R.E.: Multimedia learning. Cambridge University Press, New York (2001)
14. Mayer, R.E., Massa, L.J.: Three facets of visual and verbal learners: Cognitive ability, cognitive style and learning preference. Journal of Educational Psychology 95, 833–846 (2003)
15. Zhang, J.: Discrete Event Simulation Enabled High Level Emulation of a Distribution Centre. In: Proceeding of 14th International Conference on Computer Modelling and Simulation (UKSim), pp. 470–475 (2012)

Influence of Stop-Words Removal on Sequence Patterns Identification within Comparable Corpora

Daša Munková, Michal Munk, and Martin Vozár

Constantine the Philosopher University in Nitra,
Tr. A. Hlinku 1, 949 74 Nitra, Slovakia
{dmunkova,mmunk,mvozar}@ukf.sk

Abstract. Short texts like advertisements are characterised by a number of slogans, phrases, words, symbols etc. To improve the quality of textual data, it is necessary to filter out noise textual data from important data. The aim of this work is to determine to what extent it is necessary to carry out the time consuming data pre-processing in the process of discovering sequential patterns in English and Slovak advertisement corpora. For this purpose, an experiment was conducted focusing on data pre-processing in these two comparable corpora. We try to find out to what extent removing the stop words has an influence on a quantity and quality of extracted rules. Stop words removal has no impact on the quantity and quality of extracted rules in English as well as in Slovak advertisement corpora. Only language has a significant impact on the quantity and quality of extracted rules.

Keywords: natural language processing, comparable corpora, text mining, data pre-processing, stop words, sequence rule analysis.

1   Introduction

The today's world is characteristic of enormous amount of available textual information on one hand, but often a lack of knowledge on the other hand. A huge amount of textual data has a weak predictive value. A concept of knowledge discovery was created for this purpose. Knowledge discovery is characterised by a wide range of variables and data sources.

The biggest differences in areas of knowledge discovery occur during the phase of data pre-processing in the process of managing CRISP-DM methodology (Cross Industry Standard Process for Data Mining). Data pre-processing represents the most time consuming phase in the whole process of knowledge discovery. It converts a document transformation from an original textual data source into a form which is suitable for applying various methods of extraction, in order to transform unstructured form into structured representation, i.e. to create a new collection of texts fully represented by concepts [1]. According to Feldman and Sanger, two steps of textual data pre-processing are inevitable.

V. Trajkovik and A. Mishev (eds.), *ICT Innovations 2013*,
Advances in Intelligent Systems and Computing 231,
DOI: 10.1007/978-3-319-01466-1_6, © Springer International Publishing Switzerland 2014

Firstly, it is an identification of features (keywords) in a way that is computationally most efficient and practical for pattern discovery. Secondly, it is an accurate capture of the meaning of an individual text (on the semantic level) [1].

Based on a huge amount of texts collected and the nature and assumptions of the techniques, textual data have to be of a very good quality in order to be effective [2, 3, 4, 5]. To improve the quality of textual data, many authors have proposed different techniques to extract an effective stop word list for a particular corpus [2], [6, 7]. Stop words lists have not been examined in great detail, which has resulted in the use of pre-existing stop word lists. These might not be suitable in each context of the textual sources as evidenced in our experiment. Research in the area has identified weaknesses of standardized stop words list [2], [8, 9, 10].

Short texts like advertisements are characterised by a number of slogans, phrases, words, symbols etc. To improve the quality of textual data it is necessary to filter out noise textual data from important data. Noise textual data are data (text) not relevant to the task at hand [11, 12]. Stop words are good examples of noise data.

The aim of this paper is to determine to what extent it is necessary to carry out the time consuming data pre-processing in the process of discovering sequential patterns in English and Slovak advertisement corpora. For this purpose, an experiment was conducted by focusing on data pre-processing in these two comparable corpora. Due to influence of works [13, 14] during the realisation of an experimental plan we used our own model for text representation, which is similar to bag-of-words model [15, 16].

The paper is further divided into several sections which are as follows: in section 2 we focus on content vs. function words. We define stop words and summarize related works dealing with stop words issues. We summarize the transaction/sequence model in section 3. Subsequently, we particularize research methodology in section 4. This section describes how we prepared texts on different levels of data pre-processing. Section 5 provides a summary of the experiment results in detail. Finally, the discussion of the results and a conclusion follows in section 6.

2   Content vs. Function Words

Words differ in the role they perform. We can divide words into two groups. One group is content words referring to objects, actions and properties, the second group is function words telling us how the words from the first group are mutually related [17]. Linguists define two categories of words: open-class words and closed-class words. Open-class words represent content words and closed-class words represent function words.

In terms of parts of speech (syntactical categories of words) content words (open-class words) include nouns (objects), verbs (actions) and adjectives and adverbs (properties that quantify nouns and verbs). On the other hand, function words (closed-class words) consist of determiners, pronouns, prepositions, conjunctions, numbers etc. [17].

Text is made up of a sequence of words, which are separated by a tokenization process [17]. Some frequent words make up most of the text. The words which are frequent occurred in the most of the texts in given corpus are called Stop words. Stop words carry less important meaning than other words occur in document.

Stop words are functional, general, and common words of the language that usually do not contribute to the semantics of the documents and have no read added value [12]. Myerson [18], stated two conditions for a stop word. It should have a high document frequency (DF) and the statistical correlations with all the classification categories should be small. Zou et al. [19] define a stop word as a word with stable and high frequency in documents. According to M. Khosrow [20] stop words are words having no significant semantic relation to the context in which they exist.

For example, in English language, articles "the, a, an", prepositions "on, up etc." conjunctions "and, or etc.", pronouns "it, us etc." are usually defined as stop words. Stop words may also be document-collection specific words [20], [12], e.g. the word "to help" would probably be a stop word in a collection of advertisements but certainly not in a collection of News articles. Several authors [21, 22, 23] have argued for the removal of stop words which make the selection of the useful words more efficient and reduce the complexity of the structure of the document.

The most common approach how to create a stop words list is to manually assemble it from a list of words or terms having no natural useful information [2]. This approach is used by several authors [24], [23] and others.

3   Transaction/Sequence Model

Text mining is analogous to KDD. Sometimes it is enough to slightly adapt the existing methods and procedures from other areas of knowledge discovery. In our case we chose a representation of short texts, and we found the inspiration in area of KDD and web usage mining. We used transaction/sequence model for text representation, similar to bag-of-words model, which allows us to examine the relationships between the examined attributes and search for associations among the identified words in corpus. The structure and data character predetermine the use of specific methods for analysis - data modelling. In case of the use of transaction/sequence model for text representation, it is mainly association rule analysis and sequence rule analysis. Association/sequence rule analysis has its application in area of quantitative syntax analysis [25].

Examined variables: *Language, Text ID, Sentence ID, Transaction/Sequence ID* - it consists of previous two/three variables, *Sequence* - an order of words in text/sentence, *Word, Part of speech* - words classification (nouns, verbs, adjectives, adverbs, articles, pronouns, prepositions, conjunctions and others), and *Stop words* - words which do not contain important significant information or occur so often that in text that they lose their usefulness (Snowball list of stop words was used).

4   Experiment Research Methodology

We aimed at specifying the inevitable steps to improve the quality of textual data represented by transaction/sequence model. We focused on a sequence identification and stop words removal. We tried to find out to what extent has the stop words elimination an influence on a quantity and quality of extracted rules. Especially we assessed the impact of these techniques on the quantity and quality of the extracted rules representing sequential patterns in comparable advertisement corpora (EN, SK).In our experiment we used a pre-existing list of stop words for English (Snowball stop words for English) and similar for Slovak [26].
   Experiment was conducted in following steps:

1. Text collection (Data collection-comparable advertisement corpora).
2. Format removal.
3. Data pre-processing on different levels:
   (a) a sentence sequence identification without stop words removal for English corpus (File EN1),
   (b) a sentence sequence identification with stop words removal for English corpus (File EN2),
   (c) a sentence sequence identification without stop words removal for Slovak corpus (File SK1),
   (d) a sentence sequence identification with stop words removal for Slovak corpus (File SK2).
4. Data analysis - searching for sequential patterns in individual files. We used *STATISTICA Sequence, Association and Link Analysis* for sequence rules extraction. It is an implementation of algorithm using the powerful a-priori algorithm [27, 28, 29, 30] together with a tree structured procedure that only requires one pass through data.
5. Understanding of the output data - a production of data matrices from the analysis outcomes, defining assumptions.
6. Comparison of results of data analysis elaborated on various levels of data pre-processing from the point of view of quantity and quality of the found rules - sequential patterns.

We articulated the following two assumptions:

1. we expect that the stop words elimination will have a significant impact on the quantity of extracted rules, and
2. we expect that the stop words elimination will have a significant impact on the quality of extracted rules in the terms of their basic measures of the quality in examined comparable advertisement corpora.

5   Results

In this section we describe the results of comparisons of the quality and the quantity of extracted rules in examined files.

5.1   Data Understanding

Text is rarely translated sentence by sentence or word by word. Long sentences may be split into short sentences or vice versa. Therefore our analysed texts represent collections of short texts - advertisements from the comparable corpora (Slovak advertisement corpus and English advertisement corpus) i.e. we created corpora in two different languages (Slovak and English) with the same subject matter. They write about the same topics (products), but they are not translations of each other (no direct translations).

The experiment used two different corpora. A corpus of English written advertisements contains over 31390 words. The second, Slovak corpus of written advertisements consists of 28070 words. We used our own analyser for determining the parts of speech. Among the most frequent parts of speech in English advertisement corpus are nouns with portion higher than 26 %, verbs and adjectives with portion higher than 14 %, then others and pronouns, each with approximately 10 % of the total number of words. For Slovak advertisement corpus, there is a difference: nouns with portion higher than 36 %, adjectives with portion higher than 18 %, verbs with portion higher than 14 % and then conjunctions with approximately 10 % of the total number of words.

Based on Snowball list of stop words, 42.59 % of stop words were determined in English advertisement corpus. Pronouns and prepositions are the parts of speech most frequently used as stop words, with portion higher than 21 %, followed by others and verbs with portion higher than 15 % of stop words. A similar stop words list was used for the Slovak advertisement corpus where 26.75 % of stop words were identified. From the point of view of parts of speech, pronouns and prepositions are the parts of speech most frequently used as stop words, with portion higher than 33 %, then verbs and pronouns with higher than 11 % of the total number of words used. In English advertisement corpus nouns, verbs, adjectives, pronouns and others (articles, interjection and symbol) belong to the most frequently occurring parts of speech. On the contrary, in Slovak advertisement corpus nouns, adjectives, adverbs and conjunctions belong to the most frequently used. The differences are mainly in the verb incidence, pronoun, conjunction and others. Based on the cross-tabulation analysis there is a low dependency between the incidence of parts of speech and language in case of Slovak vs. English advertisement corpus, the contingency coefficient (V = 0.27) is statistically significant (Chi-square = 416.7343; df = 8; p = 0.0000), i.e. the incidence (use) of parts of speech depends only on the language of corpus (Slovak or English).

Furthemore we examined whether there is also a difference in the incidence of parts of speech in stop words in Slovak and English advertisement corpus.

The results of cross-tabulation analysis showed that there is a medium dependency between the incidence of parts of speech in stop words and language in case of Slovak vs. English advertisement corpus, the contingency coefficient (V = 0.37) is statistically significant (Chi-square = 280.1117; df = 8; p = 0.0000), i.e. the incidence of parts of speech in stop words depends on the language (Slovak or English).

5.2 Comparison of the Quantity of Extracted Rules in Examined Files

The analysis (Table 1) resulted in sequence rules, which we obtained from frequented sequences fulfilling their minimum support (in our case min s = 0.1). Frequented sequences were obtained from identified sequences based on the length of sentence.

Table 1. Incidence of discovered sequence rules in particular files

Body	⟹ Head	SK1	SK2	EN1	EN2
(verb)	⟹ (preposition), (noun)	1	0	0	0
...	...				
(adjective)	⟹ (verb)	1	1	1	1
Count of derived rules		65	73	45	50
Percent 1's		57.02	64.04	39.47	56.14
Percent 0's		42.98	35.96	60.53	56.14
Cochran Q Test	Q = 20.20266; df = 3; p < 0.000154				

Most rules were extracted from file with sentence sequence identification without stop words in Slovak corpus; concretely 73 were extracted from the file (File SK2), which represents over 64 % of the total number of found rules. Based on the results of Q test (Table 1), the zero hypothesis, which reasons that the incidence of rules does not depend on individual levels of text pre-processing or language is rejected at the 1 % significance level.

Kendall's coefficient of concordance represents the degree of concordance in the number of the found rules among examined files. The value of coefficient (Table 2) is 0.059, while 1 means a perfect concordance and 0 represents discordance. Low value of coefficient confirms Q test results.

From the multiple comparison (Tukey HSD test) three homogenous groups (Table 2) consisting of files (File EN1, File EN2), (File EN2, File SK1) and (File SK1, File SK2) were identified in terms of the average incidence of the found rules. Statistically significant differences on the level of significance 0.05 in the average incidence of found rules were proved between files (File EN1, File SK1), (File EN1, File SK2) and (File SK2, File EN2).

Statistically significant differences were proven only in language in terms of the average incidence of found rules. That means that only language has an important impact on the quantity of extracted rules. Naturally, the Slovak language belongs to a morphologically richer language family than the English language, so the morphological differences between them are axiomatic. The same fact was proven in the average incidence of found rules. On the contrary, removing the stop words has no significant impact on the quantity of extracted rules in particular languages.

Table 2. Homogeneous groups for incidence of derived rules in examined files

File	Mean	1	2	3
EN1	0.39474	****		
EN2	0.43860	****	****	
SK1	0.57018		****	****
SK2	0.64035			****

Kendall Coeff. of Concordance 0.05907

5.3   Comparison of the Quality of Extracted Rules in Examined Files

Quality of sequence rules is assessed by means of two indicators [27]: support and confidence. Results of the sequence rule analysis showed differences not only in the quantity of the found rules, but also in the quality. Kendalls coefficient of concordance represents the degree of concordance in the support of the found rules among examined files. The value of coefficient (Table 3a) is 0.21, while 1 means a perfect concordance and 0 represents discordancy.

Table 3. Homogeneous groups for (a) support of derived rules; (b) confidence of derived rules

Support	Mean	1	2		Confidence	Mean	1	2
EN1	30.8393		****		File EN1	47.2998	****	
EN2	34.2194	****	****		File EN2	50.8743	****	****
SK1	35.9736	****			File SK1	52.3577	****	****
SK2	36.8069	****			File SK2	52.9699		****

Kendall Coeff. of Concordance 0.2100	Kendall Coeff. of Concordance 0.2500

From the multiple comparison (Tukey HSD test) two homogenous groups (Table 3a), one consisting of files File EN2, File SK1 and File SK2; and other consisting of files File EN1 and File EN2 were identified in terms of the average support of found rules. Statistically significant differences on the level of significance 0.05 in the average support of found rules were only proved among File EN1 and files File SK1, File SK2, i.e. again only between languages. There were demonstrated differences in the quality in terms of confidence characteristics values of the discovered rules among individual files. The coefficient of concordance values (Table 3b) is 0.25, while 1 means a perfect concordance and 0 represents discordancy.

From the multiple comparison (Tukey HSD test) two homogenous groups (Table 3b), first consisting of files File EN1, File EN2 and File SK1, second consisting of files File EN2, File SK1 and File SK2 were identified in terms of the average confidence of found rules. Statistically significant difference on the level of significance 0.05 in the average confidence of found rules was proved between File EN1 and File SK2.

Results (Table 3a, Table 3b) show that the largest degree of concordance in the support and confidence is among the rules found in the files without stop words removal and files with stop words removal. On the contrary, discordance is between languages. Again it was proven that stop words removal has no impact on the quality of extracted rules and only language has a significant impact on the quality of extracted rules.

6   Discussion and Conclusions

As a result, the in-depth analysis of comparable corpora of advertisement texts proved that discrepancies in principles of syntactic sentence structures, relating to both languages, influence the frequency of parts of speech combinations as well. It is the reason why the fixed word order of English sentences reflects the fact that the most frequent combination of parts of speech in English advertisements consisted of a noun or a pronoun functioning as a subject and a verb which are considered to be the most important sentence elements necessary for creating sentences. Moreover, the combination of a verb followed by an adverb is a consequence of the fixed unmarked word order of English declarative sentences.

On the contrary, due to the fact that Slovak advertisements have a rather loose word order (which does not necessarily require the presence of verbs in the sentences) [31], the most frequent combination of the parts of speech is the one of a noun and an adjective. Since a lot of the advertisements are based on a description of a promoted product in general the adjectives modifying the nouns can be observed. Similarly, the combination of a verb and an adjective can be explained by the fact that simple verb phrases consisting of intensive verbs are often used for a description of the products in the advertisements in both languages.

The first assumption, removing stop words has no significant impact on the quantity of extracted rules in both comparable corpora (SK, EN), was not proved. Only language has a statistically significant impact on the quantity of extracted rules. Removing stop words has influence on increasing a number of extracted rules from Slovak as well as from English advertisement corpus but this increase is not statistically significant.

The second assumption was also not proved. Stop words removal has no significant impact on the quality of extracted rules in both examined corpora (SK, EN). Again it demonstrated that only language has a significant impact on the quality of extracted rules.

It is important what list of stop words is used (pre-existing or generated). In this study a Snowball list of English stop words (and its equivalents for Slovak) was used. It was turned out that these lists are likely ineffective for advertisement corpus. It was shown that stop words list depends on corpus. The question remains whether removing stop words has an impact on the quantity and quality of extracted rules. Therefore, in further research we will attempt to propose an effective stop words list for advertisement corpora (EN, SK) and focus on identifying an impact of proposed list of stop words in extraction of knowledge.

Acknowledgements. This paper is published with the financial support of the projects of Slovak Research and Development Agency (SRDA), project number APVV-0451-10 and Scientific Grant Agency (VEGA), project number VEGA 1/0392/13.

References

1. Feldman, R., Sanger, J.: The text mining handbook. Cambridge University Press (2007)
2. Choy, M.: Effective Listings of Function Stop words for Twitter. International Jurnal of Advanced Computer Science and Application 3(6), 8–11 (2012)
3. Cooley, R., Mobasher, B., Srivastava, J.: Data Preparation for Mining World Wide Web Browsing Patterns. Knowledge Information Systems 1(1), 1–27 (1999)
4. Tayi, G.K., Ballou, D.P.: Examining Data Quality. Communications of the ACM 41(2), 54–57 (1998)
5. Jung, W.: An Investigation of the Impact of Data Quality on Decision Performance. In: Proceedings of the 2004 International Symposium on Information and Communication Technology (ISICT 2004), pp. 166–171 (2004)
6. Salton, G.: The SMART Retrieval System-Experiments in Automatic Document Processing. Prentice-Hall, Inc., Upper Saddle River (1971)
7. Rose, S., Engel, D., Cramer, N., Cowley, W.: Automatic keyword extraction from individual documents. In: Berry, M.W., Kogan, J. (eds.) Text Mining: Applications and Theory. John Wiley and Sons, Ltd. (2010)
8. Chakrabarti, S., Dom, B., Agrawal, R., Raghavan, P.: Using Taxonomy, Discriminants, and Signatures for Navigating in Text Databases. In: Proceedings of the 23rd International Conference on Very Large Databases, pp. 446–455 (1997)
9. Chakrabarti, S., Dom, B., Agrawal, R., Raghavan, P.: Scalabe Feature Selection, Classification and Signature Generation for Organizing Large Text Databases into Hierarchical Topic Taxonomies. The VLDB Journal 7, 163–178 (1998)
10. Silva, C., Ribeiro, B.: The Importance of Stop Word Removal on Recall Values in Text Categorization. In: Proceedings of the International Joint Conference on Neural Networks, vol. 3, pp. 1661–1666. IEEE (2003)
11. Nisbet, R., Elder, J., Miner, G.: Handbook of statistical analysis and data mining applications. Academic Press, Elsevier (2009)
12. Alajmi, A., Saad, E.M., Darwish, R.R.: Toward an ARABIC Stop-Words List Generation. International Journal of Computer Applications 46(8), 8–13 (2012)
13. Munk, M., Kapusta, J., Švec, P.: Data Preprocessing Evaluation for Web Log Mining: Reconstruction of Activities of a Web Visitor. In: International Conference on Computational Science, ICCS 2010, Procedia Computer Science, vol. 1, pp. 2273–2280 (2010)

14. Munk, M., Drlík, M.: Impact of Different Pre-Processing Tasks on Effective Identification of Users' Behavioral Patterns in Web-based Educational System. In: International Conference on Computational Science, ICCS 2011, Procedia Computer Science, vol. 4, pp. 1640–1649 (2011)
15. Munková, et al.: Analysis of Social and Expressive Factors of Requests by Methods of Text Mining. In: Pacific Asia Conference on Language, Information and Computation, PACLIC 26, pp. 515–524 (2012)
16. Munková, D., Munk, M., Vozár, M.: Data Pre-Processing Evaluation for Text Mining: Transaction/Sequence Model. In: International Conference on Computational Science, ICCS 2013, Procedia Computer Science, vol. 18, pp. 1198–1207 (2013)
17. Koehn, P.: Statistical Machine Translation. Cambridge University Press (2010)
18. Myerson, R.B.: Fundamentals of social choice theory. Discussion Paper No. 1162 (1996)
19. Zou, F., Wang, F.L., Deng, X., Han, S., Wang, L.S.: Automatic Construction of Chinese Stop Word List. In: Proceedings of the 5th WSEAS International Conference on Applied Computer Science, pp. 1010–1015 (2006)
20. Khosrow, M.: Encyclopedia of Information Science and Technology. Information Sci. 2 edn. (2009)
21. Sinka, M.P., Come, D.W.: Evolving Better Stoplists for Document Clustering and Web Intelligence. In: Proceedings of the 3rd Hybrid Intelligent Systems Conference. IOS Press, Australia (2003)
22. El-Khair, I.A.: Effect of Stop Words Elimination for Arabic Information Retrieval: A comparative Study. International Journal of Computing & Information Sciences 4(3), 119–133 (2006)
23. Yao, Z., Ze-wen, C.: Research on the construction and filter method of stop-word list in text Preprocessing. In: Fourth International Conference on Intelligent Computation Technology and Automation (2011)
24. Fox, C.: Lexical analysis and stoplists. Information Retrieval - Data Structures & Algorithms 7, 102–130 (1992)
25. Khler, R.: Quantitative Syntax Analysis. De Gruyter, Berlin (2012)
26. Snowball, http://snowball.tartarus.org/algorithms/english/stop.txt
27. Agrawal, R., Imielinski, T., Swami, A.N.: Mining association rules between sets of items in large databases. In: Proceedings of the 1993 ACM SIGMOD International Conference on Management of Data (1993)
28. Agrawal, R., Srikant, R.: Fast Algorithms for Mining Association Rules in Large Databases. In: Proceedings of the 20th International Conference on Very Large Data Bases (1994)
29. Han, J., Lakshmanan, L.V.S., Pei, J.: Scalable frequent-pattern mining methods: an overview. In: Tutorial notes of the seventh ACM SIGKDD International Conference on Knowledge Discovery and Data Mining (2001)
30. Witten, I.H., Frank, E.: Data Mining: Practical Machine Learning Tools and Techniques. Morgan Kaufmann, New York (2000)
31. Gadušová, Z., Gromová, E.: Discourse Analysis in Translation. In: 1st Nitra Conference on Discourse Studies. Trends and Perspectives, pp. 59–64 (2006)

Modelling of Language Processing Dependence on Morphological Features

Daša Munková, Michal Munk, and Ľudmila Adamová

Constantine the Philosopher University in Nitra
Tr. A. Hlinku 1, 949 74 Nitra, Slovakia
{dmunkova,mmunk,ladamova}@ukf.sk

Abstract. The order, association and variability of the advertising language is different in every language and culture, because it is based on different rules in the given culture. Therefore, the study is focused on comparative linguistic data analysis of advertisements written in Slovak and English randomly collected from online sources. The transaction/sequence model for text representation was used and an association rules analysis was applied as the research method. The results are significant mainly in terms of the differences in the incidence of parts of speech in English and Slovak written advertisements. Based on the morphological features of the examined languages, different models of language of advertising were being created.

Keywords: Natural language processing, text mining, association rules analysis, advertisements.

1 Introduction

The present era is characterised by the amount of available electronic data on one hand, but often a lack of knowledge on the other hand [1], [2], [3], [4]. The gist of text mining is processing of unstructured (textual) information and extraction of meaningful variables from a text document, so that the information from the text can be used for various statistical methods and methods of machine learning. It builds on theoretical and computational linguistics by data pre-processing [5-7], [8], [9], [10], [11]. It allows us, for instance, to analyse the words used in a given text, their association or order, or to analyse whole texts in terms of determining similarities among them, relations among variables, or how the incidence of one variable depends on others and so on.

The order, association and variability of the advertising language is different in every language and culture, because it is based on different rules in the given culture – based on a general but also on an individual level.

In our paper we focus on the analysis of the language of advertising, especially on morphological characteristics through a description of association rules found in the Slovak and English advertisements. Within the morphological structure of advertisements, we will try to find similarities and differences in the use or transaction of parts of speech in both languages. The advertisements which we used as a data source for our research were randomly obtained from websites. Before the analysis they were classified according to advertising products into several categories such as food, cars, detergents, etc. Regarding the created corpus it is necessary to point out that since our research was focused on comparative analysis of morphological features of both languages as present in written advertisements, only those English advertisements were included in the corpus which had their equivalents in the Slovak language. As a result, the corpora created in the both languages can be considered to be equal and thus suitable for mutual comparison.

Due to influence of works by [12-13] during the realization of a research the transaction/sequence model for text representation was used [14], and an association rules analysis was applied as the research method.

The paper is further divided into several sections which are as follows: in section 2 the morphological features of the advertising language together with the references to the related linguistic works are summarized. Furthermore, data pre-processing and the linguistic data analysis of the advertisements are particularized in section 3. In addition, two analyses, the cross-tabulation analysis and the association analysis, together with a summary of the results are included. Finally, the discussion of the results and a conclusion follows in section 4.

2 Morphological Characteristics of Advertising Language

The English word *advertisement* has its root in the Latin verb *advertere* meaning *to turn towards*. As a result, the purpose of this text type is to get attention of the recipients with the intention of the originator to promote branded products and thus to benefit materially or to enhance status or image [15].

From a linguistic point of view an advertisement thus can be seen as a sophisticated system consisting of elaborated linguistic and stylistic devices present at all language levels. In general, the stylistic features of advertising language are common for all languages; however, some differences can be observed resulting from national and thus cultural differences of the recipients as well as from the nature of the language used. Therefore, the study is focused on comparative linguistic data analysis of advertisements written in Slovak and English randomly collected from online sources (websites) with the aim to compare the morphological features of both languages.

Regarding the parts of speech, full words play the most important role. In advertising, texts contain catchy slogans as well as simple and comprehensible text, also simple verb forms in the present tense and active voice are usually expected. As Leech states, verbal groups are mostly of maximum simplicity, consisting of only one word [16]. Compared with verbs, nouns are used more complexly in advertising texts. From a semantic point of view the lexemes which modify the nouns are important, because with the aim to draw the recipients' attention a lot of semantically and linguistically unusual and original modifiers are used to describe the product in as interesting and attractive way as possible [17-18]. Since the advertisements are often based on description of a product, noun phrases are frequently used while verb phrases are sometimes completely omitted. In advertising texts numerals are also important [19-20], [21].

Within the lexis of the advertisements the morphological structure of the weasel claims used in the Slovak and the English advertisements was analyzed as well. They can be defined as the claims suggesting the particular meaning without actually being specific. They usually negate a positive claim that follows in a way that the readers do not notice it. It is popular strategy frequently exploited by the copywriters causing that the recipients consider the promoted products to be better than they actually are. The term comes from the egg-eating habits of weasels and thus the words or claims that appear substantial upon first look, but disintegrate into hollow meaninglessness when analyzed are called weasels, for example, to help, virtual, can be, up to, tackles, fights, as much as, to fortify, etc. [22].

3   Linguistic Data Analysis of Advertisements

The structure and data character predetermine the use of specific methods for analysis – data modelling. In the case of the use of transaction/sequence model for text representation, it is mainly association rule analysis and sequence rule analysis. The difference between association and sequence rule analysis is that we do not analyse the sequences but the transactions in association rule analysis, which means, we do not include the sequence variable representing the order of the words in text into the analysis. The transaction represents a set of words occurring in text, whereby the order of incidence of the identified words (content words) in the given text is not taken into account.

Association/sequence rule analysis has its application also in areas of quantitative morphology or syntax analysis. Specifically, in our case, we focused on morphological aspect of a language of advertising in the Slovak and English advertisements.

3.1   Examined Variables

Advertisement ID; *Paragraph ID*; *Sentence ID* - within a paragraph; *Language* - language of advertising; *Transaction/Sequence ID* - a set ID of tokens in text, it consists of previous three/four variables; *Content word* - word that refers to object, action or property; *Parts of speech* - (POS) words classification (nouns, verbs, adjectives, adverbs, pronouns, prepositions, conjunctions and the others- articles, interjection and abbreviation); *Sequence* - an order of parts of speech in text/paragraph/sentence.

3.2   Cross-Tabulation Analysis

In our case, a cross-tabulation analysis consists of an analysis of advertisements written in Slovak and English languages. These advertisements were randomly collected from online sources. With the help of the cross-tabulation analysis we investigated whether there is a difference in the incidence of parts of speech in Slovak and English advertisements.

The only requirement (a validity assumption) of the use of chi-square test is a large amount of expected frequencies. The requirement is not violated; the expected frequencies are large enough. The contingency coefficient represents the degree of dependence between two nominal variables.

Table 1. Results of cross-tabulation analysis - Slovak vs. English advertisements

	Chi-square	df	p
Pearson	416.7343	8	0.0000
Cont. coeff. C	0.2559		
Cramér's V	0.2647		

The value of coefficient (Table 1) is approximately 0.26, where 1 means perfect dependency and 0 means independency. There is a medium dependency between the incidence of part of speech and the language in case of Slovak vs. English advertisements, the contingency coefficient is statistically significant. The zero hypotheses (Table 1) are rejected, i.e. the incidence (use) of parts of speech depends on the language (Slovak or English).

The *nouns, verbs, adjectives, pronouns* and *others* (articles, interjection and abbreviations) belong to the most frequently used POS in English. On the contrary, in Slovak the *nouns, adjectives, adverbs* and *conjunctions* were the most used. The differences are mainly in the *verb* incidence, *pronoun, conjunction* and *others*. In English, as an analytic language, the word order is an important grammatical indicator of sentence function, because it has lost most of its inflection over centuries. Therefore, in declarative sentences a verb follows a subject and the change of their position is not possible due to missing morphological inflection. On the other hand, in Slovak there is the grammatical function of the word order secondary, because synthetic relations among sentence elements are indicated by morphological means, i.e. it has a grammatical system based on modifications in the form of the words by means of inflections (endings and vowel changes) to indicate grammatical functions such as case, number, tense or aspect. As a result, the Slovak language has more grammatical endings and thus is less dependent on word order and function words than English [23].

As we mentioned previously, the Slovak language belongs to a morphologically richer language family than the English language, so the morphological differences between them are axiomatic. The same fact was proven in the language of advertising, i.e. there is a difference in the use (incidence) of POS in Slovak and English advertising language. Furthermore, we wanted to know whether it is the same in weasel claims, which are typical for advertisements. We focused on weasel claims in both languages. We investigated whether there is a difference in the incidence of POS in weasel claims in Slovak and English written advertisements. We articulated the following assumption: we expect that there is significant difference in terms of incidence of parts of speech in weasel claims of both languages.

The results of cross-tabulation analysis showed that there is a medium dependency between the incidence of POS and the weasel claims in case of Slovak vs. English advertisements, the contingency coefficient is statistically significant.

Table 2. Results of cross-tabulation analysis - Slovak vs. English weasel claims

	Chi-square	df	p
Pearson	33.0338	8	0.0001
Cont. coeff. C	0.2795		
Cramér's V	0.2691		

The zero hypotheses (Table 2) are rejected, i.e. the incidence of parts of speech in weasel claims depends on the language (Slovak or English). The *nouns, adjectives* and *verbs* belong to the most frequently used POS in Slovak weasel claims. On the contrary, the *nouns, verbs* and *prepositions* were the most used in English weasels claims. The differences are mainly in the *verb* incidence, *preposition, conjunction, adverb* and *others*. Our assumption was correct; there is significant difference in Slovak and English weasel claims in term of incidence POS.

3.3 Association Rule Analysis

Based on different rules in every language, the sentence structure, word order or associations of languages are different. How it is in Slovak and English written advertisements we will particularize in this section. The association rule analysis represents a non-sequential approach to the data being analyzed. We will not analyse the sequences but transactions, so we will not include the order in which parts of speech occurred in the analysis. In our case, a transaction represents the set of POS observed in the texts of advertisements separately for English and Slovak.

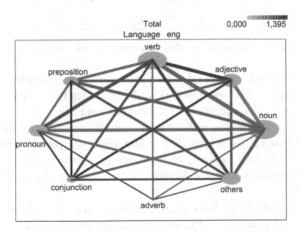

Fig. 1. Web graph – a visualization of the discovered rules – English advertisements

The web graph (Fig. 1) depicts the discovered association rules for the texts of advertisements written in English, specifically the size of node represents the support of incidence of POS, the thickness of the line represents the support of rule – pairs of POS (probability of incidence in the pair) and the darkness of the line colour presents a lift of the rule – the measure of a pair incidence in transaction. We can see from the graph (Fig. 1) that POS: *noun, verb, adjective, others* and *pronoun* (support > 59%) belong to the most frequently used. Similarly, like the combination of these parts of speech` pairs *(noun, verb)*, *(noun, adj.)*, *(verb, pronoun)*, *(adj., verb)* (support > 54%), the parts of speech *conj.==>prep., conj.==>adj., conj.==>others, adverb==>pronoun, adverb==>others, prep.==>others, pronoun==>conj., pronoun==>prep.* and *verb==>adverb* occur in sets of parts of speech more often together than as separate items (lift > 1.19). In these cases the highest degree of interestingness was achieved – the lift, which defines how many times the chosen POS occur more often together as if they were statistically independent. If the lift is more than 1, the selected pairs occur more often jointly than separately in the set of incidence of POS. It is necessary to take into account that in characterising the degree of interestingness – the lift, the orientation of the rule does not matter.

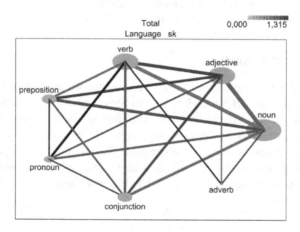

Fig. 2. Web graph – a visualization of the discovered rules – Slovak advertisements

We found different association rules for advertisements written in Slovak than for those written in English. The web graph (Fig. 2) illustrates the discovered association rules. The most frequently occurred POS are *noun, adjective, verb, conjunction* and *preposition* (support > 49%), as well as their pairs *(noun, adj.)*, *(noun, verb)*, *(adj., verb)* and *(noun, prep.)* (support > 47%). The parts of speech *verb==>pronoun, verb==>adverb, conj.==>pronoun, conj.==>prep., verb==>conj., adj.==>prep.* and *adj.==>conj.* occur more often together in transactions of used POS than separately (lift > 1.15).

As a result, the in-depth analysis of advertising texts proved that discrepancies in principles of syntactic sentence structures, valid for both languages, influence the frequency of POS combinations as well. It is the reason why the fixed word order of English sentences reflects the fact that the most frequent combination of POS in English advertisements consisted of a noun or a pronoun functioning as a subject and a verb which are considered to be the most important sentence elements necessary for creating sentences. Moreover, the combination of a verb followed by an adverb is the consequence of the fixed unmarked word order of English declarative sentences. On the contrary, due to the fact that texts written in the Slovak language have a rather loose word order, the most frequent combination of POS is the one of a noun and an adjective. Since in general a lot of the advertisements are based on a description of a promoted product the adjectives modifying the nouns can be observed. Similarly, the combination of a verb and an adjective can be explained by the fact that simple verb phrases consisting of intensive verbs are often used for description of the products in the advertisements in both languages.

In the second case, a transaction represents the set of POS observed in weasel claims separately for English and Slovak written advertisements.

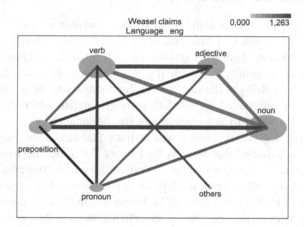

Fig. 3. Web graph – a visualization of the discovered rules of weasel claims – English

The graph (Fig. 3) depicts that POS: *noun, verb, adjective* and *preposition* (support > 47%) belong to the most frequently used. Similarly, like the combination of these POS` pairs (*noun, verb*), (*noun, prep.*), (*adj., verb*) and (*noun, adj.*) (support > 35%) are used the most. The parts of speech *prep.==>pronoun, verb==>others, verb==>pronoun, prep.==>noun, verb==>adj.* and *pronoun==>noun* occur in sets of POS more often together than as separate items (lift > 1.04).

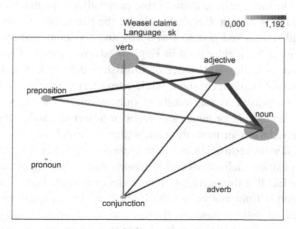

Fig. 4. Web graph – a visualization of the discovered rules of weasel claims – Slovak

We also found different association rules for weasel claims in Slovak than for English written advertisements. The most frequently occurred POS are *noun, adjective, verb, preposition* and *conjunction* (support > 25%), as well as their pairs (*noun, adj.*), (*noun, verb*), (*adj., verb*), (*noun, prep.*) and (*adj., prep.*) (support > 29%). The POS *adj.==>prep., adj.==>noun, adj.==>conj., verb==>conj.* and *noun==>prep.* occur more often together in transactions of used POS than separately (lift > 1.08).

4   Discussion and Conclusion

In our study we aimed to analyse the advertisements written in Slovak and English. We focused on free (online) advertisements on the internet. Advertisements can be defined as the promotion of goods or services for sale through impersonal media [24]. We classified them, according to advertising products, into several categories (food, cars, detergents, etc.). Being of the same or similar content was the main condition for our analysis, i.e. in our created corpus if there is an English advert for Persil detergent, there must also be a Slovak advertisement for the same product .

Based on morphological, syntactical and stylistic features of the English and Slovak languages, we assumed that there is significant difference between Slovak and English advertisements. Our assumption, confirmed by the cross-tabulation results, is that there is a statistically significant dependence between a language of advertising and an incidence of POS (Cont. coeff. 0.26). In the research we focused only on morphological differences of the language of advertising, but for the future we will consider analysing also other aspects of advertising language.

From the point of view of morphology, nouns (26.79%), verbs (17.11%), adjectives (14.88%), others (10.58%) and pronouns (10.16%) occurred more frequently in English advertisements. For Slovak advertisements, there is a difference: nouns (36.23%), adjectives (18.13%), verbs (14.89%) and conjunctions (9.9%) were more frequently used.

From the point of view of different proportion of POS between languages, in Slovak advertisements nouns, adjectives, conjunctions and numbers occurred more often than in English. On the contrary, more verbs, prepositions, pronouns and others are used in English adverts than in Slovak. Similarly, the pairs, the combination of nouns and verbs or of adjectives and nouns or of verbs and pronouns and also of verbs and adjectives, were more frequently used in English advertisements. These findings are caused by differences in the word order of English and Slovak. English word order and sentence structure is more constant, while Slovak language has a loose word order which was also proved by the results of our analysis of association rules. Pairs such as adjectives and nouns or nouns and verbs or adjectives and verbs or nouns and prepositions occurred together more frequently than separately.

What we consider interesting is the fact that conjunctions occurred more frequently together with prepositions, adjectives and pronouns than occurring alone in either language, despite the fact that they have different sentence constructions. The further interesting finding resulted from analysis of the English advertisements, where there were fewer verbs combined with prepositions than in common English language, where they are often used. Another interesting paradox is that the Slovak advertisements address or offer the products to someone and very often the verb is combined with a pronoun, whereas the English advertisements are more frequently impersonal. We consider it to be caused by using the passive voice, which is more typical for the English language. The frequent incidence of a pair of a verb and an adverb is common in Slovak as well as in English advertisements.

Our analysis was also focused on weasel claims, which belong to characteristic features of advertising texts. The cross-tabulation results showed that there was also a statistically significant dependence between the weasel claims and an incidence of POS in Slovak and English advertisements (Cont. coeff. 0.27) i.e. it was proven the incidence POS in weasel claims depends on language.

In English weasel claims, nouns (23.26%), verbs (21.40%) and prepositions (15.81%) occurred the most frequently and adverbs (1.40%) and numbers (3.26%) the least. In Slovak weasel claims nouns (31.73%), adjectives (21.15%) and verbs (20.67%) occurred the most frequently, and the others (0.0%) and numbers (3.37%) the least.

From the point of view of the different proportion of POS between languages, adjectives, adverbs, nouns and conjunctions are dominant in Slovak weasel claims, whereas in English weasel claims pronouns, prepositions and verbs occurred more frequently. In English weasel claims, nouns are combined with verbs or prepositions or adjectives, whilst in Slovak weasel claims nouns and adjectives are often joined with conjunctions. The construction of Slovak weasel claims does not need to involve a verb. This is the biggest difference between Slovak and English weasel claims in written advertisements.

Moreover, we consider as a noteworthy finding the fact that adjectives are more frequently joined with prepositions and conjunctions in Slovak weasel claims than in English, and if they do occur in English, then they are usually combined with verbs and nouns. The results are significant mainly in terms of the differences in the incidence of POS in English and Slovak written advertisements.

We consider these findings remarkable, because we examined the same advertisements, but in different languages. Based on the morphological features of the examined languages, different models of language of advertising were being created.

As a result, the transaction/sequence model for text representation has proved to be suitable for short texts, like advertisements, because it allows us to examine the relationships among the examined attributes and search for associations among the identified parts of speech not only in advertisements, but also in their weasel claims.

Acknowledgements. This paper is published with the financial support of the projects of Slovak Research and Development Agency (SRDA), project number APVV-0451-10 and Scientific Grant Agency (VEGA), project number VEGA 1/0392/13.

References

1. Chapman, P., Clinton, J., Kerber, R., Khabaza, T., Reihartz, T., Shearer, C., Wirth, R.: CRISP-DM 1.0 Step-by-step Data Mining Guide (2000)
2. Fayyad, U.M., Piatesky-Shapiro, G., Smyth, P., Uthurusamy, R.: Advances in Knowledge Discovery and Data Mining (1996)

3. Paralič, J.: Objavovanie znalostí v databázach, 80 p. Elfa, Košice (2003)
4. Sullivan, D.: Document Warehousing and Text Mining: Techniques for Improving Business Operations, Marketing and Sales. John Willey & Sons, Inc. (2001)
5. Hajičová, E., Panevová, J., Sgall, P.: Úvod do teoretické a počítačové lingvistiky. Karolinum, Praha (2003)
6. Hearst, M.A.: Untangling Text Data Mining. In: Proceedings of the 37th annual meeting of the Association for Computational Linguistics on Computational Linguistics, pp. 3–10 (1999)
7. Houškova Beranková, M., Houška, M.: Data, Information and Knowledge in Agricultural Decision-making. Agris On-line Papers in Economics and Informatics 3(2), 4–82 (2011)
8. Neuendorf, K.A.: The Content Analysis Guidebook, 320 p. Sage Publications (2002)
9. Paralič, J., et al.: Dolovanie znalostí z textov, 184 p. Equilibria, Košice (2010)
10. Titscher, S., et al.: Methods of Text and Discourse Analysis. Sage, London (2002)
11. Weiss, S.M., Indurkhya, N., Zhang, T.: Text Mining: Predictive Methods for Analyzing Unstructured Information. Springer (2005)
12. Munk, M., Drlík, M.: Impact of Different Pre-Processing Tasks on Effective Identification of Users' Behavioral Patterns in Web-based Educational System. In: International Conference on Computational Science, ICCS 2011, Procedia Computer Science, vol. 4, pp. 1640–1649 (2011)
13. Munk, M., Kapusta, J., Švec, P.: Data Preprocessing Evaluation for Web Log Mining: Reconstruction of Activities of a Web Visitor. In: International Conference on Computational Science, ICCS 2010, Procedia Computer Science, vol. 1(1), pp. 2273–2280 (2010)
14. Munková, D., Munk, M., Fraterova, Z., Durackova, B.: Analysis of Social and Expressive Factors of Requests by Methods of Text Mining. In: Pacific Asia Conference on Language, Information and Computation, PACLIC 26, pp. 515–524 (2012)
15. Goddard, A.: The Language of Advertising. Written Texts, 131 p. Routledge, London (1998)
16. Leech, G.N.: English in Advertising: A Linguistic Study of Advertising in Great Britain. English Language Series, 210 p. Longman, London (1966)
17. Myers, G.: Words in Ads, 222 p. Hodder Arnold, London (1994)
18. Vestergaard, T., Schroder, K.: The Language of Advertising, 182 p. Vasil Blackwell Inc., New York (1985)
19. Čmejrková, S.: Reklama v češtině, 258 p. Leda, Praha (2000)
20. Cook, G.: The Discourse of Advertising, 272 p. Routledge, London (2011)
21. Mistrík, J.: Štylistika, 598 p. Slovenské pedagogické nakladateľstvo, Bratislava (1997)
22. Schrank, J.: The Language of Advertising Claims,
    http://home.olemiss.edu/~egjbp/comp/ad-claims.html
23. Gadušová, Z., Gromová, E.: Discourse Analysis in Translation. In: 1st Nitra Conference on Discourse Studies. Trends and Perspectives, pp. 59–64 (2006)
24. Cook, G.: The Discourse of Advertising, 256 p. Routledge, London (2001)

Pagerank-Like Algorithm for Ranking News Stories and News Portals

Igor Trajkovski

Faculty of Computer Science and Engineering,
"Ss. Cyril and Methodius" University in Skopje,
Rugjer Boshkovikj 16, P.O. Box 393, 1000 Skopje, Macedonia
trajkovski@finki.ukim.mk
http://www.finki.ukim.mk/en/staff/igor-trajkovski

Abstract. News websites are one of the most visited destinations on the web. As there are many news portals created on a daily basis, each having its own preference for which news are important, detecting unbiased important news might be useful for users to keep up to date with what is happening in the world. In this work we present a method for identifying top news in the web environment that consists of diversified news portals. It is commonly know that important news generally occupies visually significant place on a home page of a news site and that many news portals will cover important news events. We used these two properties to model the relationship between homepages, news articles and events in the world, and present an algorithm to identify important events and automatically calculate the significance, or authority, of the news portals.

Keywords: news, ranking, pagerank, portals.

1 Introduction

According to a recent survey [1] made by Nielsen/NetRatings for Newspaper Association of America, news browsing and searching is one of the most important Internet activities. Specifically, 60% of the users rank news portals as one of the top 3 places they visit when surfing the Internet. The creation of many independent online news portals has created a large increase in news information sources available to the users. What we are confronted with is the huge amount of news information coming at us from different news sources, and we as users are interested for latest important news events. Thus key problem is to identify those news that report important events. However, not all important news have similar importance, and how to rank news articles according to their importance becomes a key issue in this field. In this work, we mainly discuss the problem of finding and ranking these publicly important news, with no respect to user's personal interests.

It is hard to distinguish important news, since each news portal has its own preference in reporting events. But generally speaking, the following five properties can be used for detecting and ranking important news:

V. Trajkovik and A. Mishev (eds.), *ICT Innovations 2013*,
Advances in Intelligent Systems and Computing 231,
DOI: 10.1007/978-3-319-01466-1_8, © Springer International Publishing Switzerland 2014

1. *Time awareness.* The importance of a piece of news changes over the time. We are dealing with a stream of information where a fresh news story should be considered more important than an old one.
2. *Important news articles are clustered.* An important event is probably (partially) covered by many sources with many news articles. This means that the (weighted) size of the cluster where it belongs is a measure of its importance.
3. *Authority of the sources.* The algorithm should be able to assign different importance to different news sources according to the importance of the news articles they produce. So that, a piece of news coming from BBC can be more authoritative (important) than a similar article coming from gossiping portal, since BBC is known for producing good stories.
4. *Visually significant.* Important news usually occupies a visually significant place in the homepage (such as headline news).
5. *Diversity.* Events reported by big number of sources should be more important (and all the news reporting about it) than events covered by small number of sources.

In this paper, we present a method to detect those news that posses these five properties. The visual significance of the news in a homepage can be seen as the recommendation strength to the news by the homepage. The strength of the recommendation can be modeled by the size (in number of pixels) of the box where the title/picture of the news or by the position on the homepage where the link of the news is places.

We denote the term credibility/authority to describe the extend to which we can believe a homepage's recommendation. Credibility/authority of the homepages and importance of the news pages exhibit a mutual reinforcement relationship, which is similar to that between hub pages and authoritative pages in a hyperlinked environment [2]. Similarly, importance of news articles and importance of events also exhibit such a mutual reinforcement relationship. By event we mean implicate event, defined by the cluster of related news. We model the relationship between homepages, news articles and events into a tripartite graph and present an algorithm to determine important news by seeking the equilibrium of those two mutual reinforcement relationship in this graph.

Related work on this topic is 'important story detection' that is mainly studied within the topic detection and tracking (TDT) community [3] [4] [5] . TDT tries to detect important stories from broadcast news. Our work differs from them in the way that we consider this problem in the web environment where more independent information sources are available.

The organization of this paper is as follows. In Section 2, we present the relationship between homepages, news articles and events and model it by a tripartite graph. Then we present the algorithm for calculation of importance of news articles by exploiting the five mentioned principles and the presented model in Section 1. In Section 3 we give an overview of the system that implements our algorithm, we describe the experiments that evaluate the proposed models, and experimental results are discussed. We summarize our contributions and conclude in Section 4.

2   Model for Calculating Event Importance and Portals Credibility

To identify important news that follows mentioned properties, we investigate two kind of information from homepages and news pages respectively.

News homepages not only provide a set of links to news pages, they also work as visual portals for users to read news. They are delicately designed to help user acquire information quickly. Examples include headline news, top story recommendation, etc. One of the most general forms is that all pieces of news are presented by homepages with different visual strength, either by the different size of the graphics, or by the placement on a different position. The most important pieces od news is often put in the top place, accompanied by some image or snippet, while each of those less important ones is just a short sentence with a hyperlink. From another point of views, the visual layout of each homepage reflects its editor's viewpoint on importance news at that time. Such kind of information is quite helpful to identify important news.

Each news article generally has a title, an abstract and the content. Thus we can compare the content of two news pages and estimate whether they are reporting the same event. Furthermore, from the corpus of multiple news sources, we may estimate how many pieces of news are reporting the same event.

2.1   Reinforced Importance

News portals vary in credibility. Each portal generally contains two kinds of homepages: portal page and category pages. A portal page often summarizes important news from different classes, while each category page focuses on one kind of news, such as world, business, sports, entertainment, etc. The headline news in category page is possibly important only within the corresponding class. So generally speaking, portal pages are likely to be more creditable within a site. Besides, homepages of prestigious sites are averagely more creditable than those of non-famous sites. Credibility/authority of news portals and importance of news exhibit a mutually reinforcing relationship as follows:

Observation 1: *News portals vs. News articles*

- News presented by more creditable portals with stronger visual strength is more likely to be important.
- More creditable portals are expected to recommend important news more reliably.

All news articles are driven by the similar sets of events taking place in the world. Here we take the definition of event from detection and tracking community [3]. *Event* is something that happens at specific time and place. E.g. a specific election, sport, accident, crime, or natural disaster.

The importance of news articles and importance of events also exhibit a mutually reinforcement relationship.

Observation 2: *News articles vs. Events*

- Important events are likely to be reported by more news articles.
- A news article that reports an important event is important.

2.2   Tripartite Graph Model

We take a tripartite graph to model the relationships between three mentioned objects: homepages, news articles and events. The graph is a five-tuple $G = \{S, N, E, Q, P\}$, where:

$S = \{S_1, S_2, \ldots, S_n\}, N = \{N_1, N_2, \ldots, N_n\}, E = \{E_1, E_2, \ldots, E_l\}$ are three sets of vertices corresponding to homepages(sources), news articles and events respectively. Q is defined as an $n \times m$ matrix such that Q_{ij} represents the recommendation strength of N_j by S_i. We assume that the maximum recommendation strength equals for all homepages. Therefore, Q is normalized along rows so that $\sum_{j=1}^{m} Q_{ij} = 1$. P is an $m \times k$ matrix such that P_{jk} is the probability that N_j is reporting about E_k. Here it also holds that $\sum_{k=1}^{l} P_{jk} = 1$. Here P and E are unobservable directly, but can be obtained by clustering the news articles according to their semantic similarity. Each obtained cluster will be one event, and belongingness of a news article in a cluster will determine the strength between a news article and event. This belongingness can be discrete, 0/1, or can be probabilistic, which depends from the clustering algorithm.

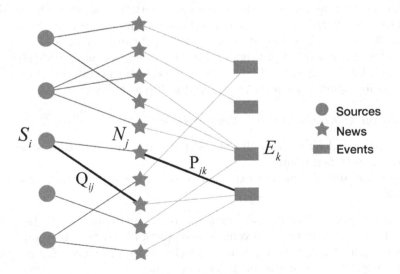

Fig. 1. Tripartite graph model of sources, news articles and events

Beside mentioned properties, we associate credibility weight w_i^S for each source S_i, importance weight w_j^N for each news article N_j and importance weight w_k^E for each event E_k. We maintain the invariant that the weights of each type are normalized:

$$\sum_{i=1}^{n} w_i^S = 1, \quad \sum_{j=1}^{m} w_j^N = 1, \quad \sum_{k=1}^{l} w_k^E = 1. \quad (1)$$

2.3 Importance Propagation

Based on the described reinforced importance model, we identify the credibility of sources w_i^S, importance of the news articles w_j^N, and importance of events w_k^E, by finding equilibriums in those relationships. It can be done by iterative algorithm.

Corresponding to observations 1 and 2, we can define the following **forward** operations:

$$w_j^N = \sum_{i=1}^{n} (w_i^S * Q_{ij}), \quad (2)$$

$$w_k^E = \sum_{j=1}^{m} (w_j^N * P_{jk}). \quad (3)$$

Similarly we define the following **backward** operations:

$$w_j^N = \sum_{k=1}^{l} (w_k^E * P_{jk}), \quad (4)$$

$$w_i^S = \sum_{j=1}^{m} (w_j^N * Q_{ij}). \quad (5)$$

In each iteration, after computing the new weight values, they are normalized using the equations (1). The last four equations (2), (3), (4) and (5) are the basic means by wich w^S, w^N and w^E reinforce one another. The equilibrium values for the weights can be reached by repeating (2), (3), (4) and (5) consecutively. Thus weight values converge to w^{S^*}, w^{N^*} and w^{E^*}.

2.4 Inclusion of Time Awareness and Diversity

As we mentioned in the introductory section every model for ranking news relevance should include *time awareness* and *news source diversity*. In our model we can include these two properties by manipulating the weights w^N and w^E.

For inclusion of time awareness, each time when new version of w^N is calculated, before normalization, we multiply news articles weight w_j^N by $e^{-\alpha(t-t_j)}$ where t is the current time, t_j is the publication time of news article N_j. The value α, which accounts for the decay of "freshness" of the news article, is obtained from the half-life decay time θ, that is the time required its weight to halfe its value, with the relation $e^{-\alpha\theta} = \frac{1}{2}$.

For inclusion of news source diversity, each time when new version of w^E is calculated, before normalization, we multiply events weight w_k^E by $(0.5+ENT)$, where ENT is entropy of the set of news sources that are reporting the event. ENT has a maximum value of 1, when all sources are different, and minimum value of 0, when there is only one source reporting the event.

2.5   Recommendation Strength by Visual Importance

The visual importance of a news article is decided by its block size, position and whether it contains an image.

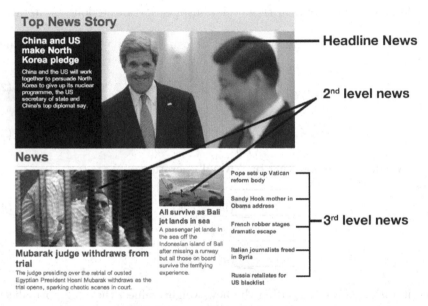

Fig. 2. A snapshot of a home page

In our model all news articles from a news source are classified into one of a four categories: Headline news, 2nd level, 3rd level and 4th level news. In the experiments for each source we wrote regular expressions that identifies which news articles belong to which category. If a news portal does not have four levels of categorization, several categories were merged, for example if there are several headline news and all remaining news belonged in one category by visual importance, that news portal will have 2 categories: Headline news and 4th level news.

We set the recommendation strengths of each category in the following way: Headline news had 5 times, 2nd level news had 3 times and 3rd level news had 2 times bigger recommendation strength than 4th level news. When relative recommendation strength for the news articles is known, it is easy the computation of absolute values of the elements of matrix Q.

3 Experiments and Results

We implemented a system to verify the advantages of our algorithm over the standard baseline model where authority of the sources is constant and visual recommendation strength is not considered.

The system monitors a set of news portals, crawls their home pages and linked news articles in a certain frequency. By our algorithm, each news source/portal, news article and event gets a ranking score. For detecting events/clusters, we used standard hierarchical agglomerative clustering (HAC) algorithm, and we used cosine similarity as distance measure and traditional TF-IDF document representation [6]. Events were ranked by their weight score w_k^E.

We first explain how we collect data and set up the ground truth. Then sets of experiments were conducted to investigate our algorithm and evaluate the implemented system.

3.1 Dataset Description

We monitor 10 news portals for one working day. They are one of the most visited news web sites in Macedonia. All homepages of the portals were crawled in the frequency of ten minutes. In each iteration, there were 144 iterations, we calculated ranking scores w^{S^*}, w^{N^*} and w^{E^*}. We were especially interested in local/Macedonian news because they are popular and comparable among all portals. Statistics of crawled data are shown in Table 1.

Table 1. Statistics of Experimental Data

News portal	#news	#macedonia	News portal	#news	#macedonia
DNEVNIK	111	20	ALFA	60	14
UTRINSKI	97	18	KANAL5	81	16
KURIR	110	34	NETPRESS	59	12
PLUSINFO	99	25	PRESS24	71	18
SITEL	115	29	MKD	55	13

It is quite difficult to give an importance value for each news article. The key problem is that users can only evaluate importance of each event instead of each news article. So it is necessary to associate each news article into some event for comparing our method to the ground truth from the users. We used fixed clustering algorithm (HAC) to deal with this task. We defined three importance levels and their corresponding weight values. (See Table 2).

We asked 10 users to label 20 events/clusters. These 20 events were from the same iteration and were composed of 10 highest ranked events and 10 other random events. Calculated ranking of the 20 presented events was not show to the users. They were presented in random permutated order. For each event, the average value is taken as its importance value.

Table 2. Importance level

News portal	Weight
Very important	10
Important	5
Normal	0

We evaluated three models. The first model is a baseline model (Baseline-BASE) where authority of the sources is constant and visual recommendation strength is not considered. The second model (PageRank-PR) is the case where authority of the sources is calculated, and the third model (PageRank Visual-PRV) where authority of the sources and visual recommendation strength is included in the model.

3.2   Scope - Average Importance

We took a strategy like scope-precision, to evaluate the performance of the three models. Here the scope is the number of top important events returned. Precision is the average importance value of these top events. We also define the ideal case for comparison. It represents the best performance we can expect from the user labeling. Fig. 3 illustrates the results. The PRV model outperforms both the BASE and PR model remarkably.

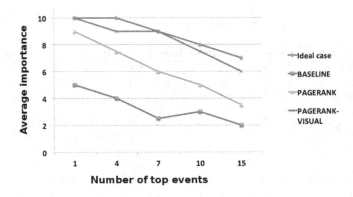

Fig. 3. Scope - Average importance

3.3   Time Delay

Another criterion to evaluate the performance of the models is *time delay* for reporting events. Given one event, its time delay is defined as the period from the earliest time when a news article reporting about the event appears, to the time we can identify it as member of the three most important events. We hope the time delay should be as short as possible so that the system can report important news to users as soon as possible.

We randomly select a set of events from those that can be identified as important by all the three models. These events are listed in Table 3. The second column is the earliest time that our crawler found a news article reporting about the event. We used the archives of the news portals to determine the publication timestamps of news articles, and we used the crawling html archives of TIME.mk, the biggest Macedonian news aggregator, for determining the visual recommendations strengths of the news portals at those moments.

Table 3. Importance level

Event	First news	Description
1	2013-03-27 13:16	Nikola Mladenov has died in a car crash
2	2013-02-24 11:06	Actor Branko Gjorcev dies
3	2011-03-11 08:53	Tsunami in Japan
4	2013-02-15 16:49	Chelyabinsk meteor in Russia
5	2012-06-20 07:33	Assange seeks asylum at Ecuador embassy

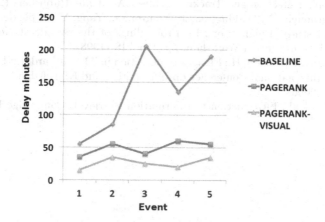

Fig. 4. Time delay

The delays are illustrated in Fig. 4. The average delays for three models are 26, 49 and 132 minutes respectively for PRV, PR and BASE. The time delay by the PRV model is very small because a piece of important news occupies significant importance as soon as it appears. While the BASE model requires a significant time delay and can hardly identify important events in the first time. It is because in this model an event can be identified as important only after many sites have reported it. The PR model is quite close to the PRV for most cases.

4 Conclusions

In this work, we propose a method to detect and to rank important news stories in web environment. We use visual layout information in homepages and content similarity information in news articles. The relationship between news sources, news articles and events is modeled by a tripartite graph. Then we present an iterative algorithm to find the importance equilibrium in this graph. Based on this algorithm, we implemented a system for automatically finding important news. Experiments show the whole framework is effective. We evaluated three models (BASE, PR and PRV) with two criteria, for accuracy and time delay. The PRV can identify important news fastest and most accurately.

References

1. Newspaper Association of America (2009),
   http://www.naa.org/News-and-Media/Press-Center/Archives/2009/
   Newspaper-websites-attract-more-than-70-million-visitors.aspx
2. Kleinberg, J.M.: Authoritative Sources in a Hyperlinked Environment. Journal of the ACM 46, 604–622 (1999)
3. Allan, J.: Topic Detection and Tracking. Kluwer Academic Publishers (2002)
4. Allan, J., Carbonell, G., Doddington, J., Yamron, J., Yang, Y.: Topic detection and tracking pilot study: Final report. In: Proceedings of the Broadcast News Understanding and Transcription Workshop, pp. 194–218 (1998)
5. Allan, J., Lavrenko, V., Jin, H.: First story detection in TDT is hard. In: Proceedings of the Ninth International Conference on Information and Knowledge Management, pp. 374–381 (2000)
6. Manning, C., et al.: Introduction to Information Retrieval. Cambridge Press, New York (2008)

A System for Suggestion and Execution of Semantically Annotated Actions Based on Service Composition

Milos Jovanovik, Petar Ristoski, and Dimitar Trajanov

Faculty of Computer Science and Engineering,
Ss. Cyril and Methodius in Skopje, Republic of Macedonia
{milos.jovanovik,dimitar.trajanov}@finki.ukim.mk,
petar.ristoski88@gmail.com

Abstract. With the growing popularity of the service oriented architecture concept, many enterprises have large amounts of granular web services which they use as part of their internal business processes. However, these services can also be used for ad-hoc actions, which are not predefined and can be more complex and composite. Here, the classic approach of creating a business process by manual composition of web services, a task which is time consuming, is not applicable. By introducing the semantic web technologies in the domain of this problem, we can automate some of the processes included in the develop-and-consume flow of web services. In this paper, we present a solution for suggestion and invocation of actions, based on the user data and context. Whenever the user works with given resources, the system offers him a list of appropriate actions, preexisting or ad-hoc, which can be invoked automatically.

Keywords: Semantic web services, automatic composition, semantic web technologies, service oriented architecture.

1   Introduction

The growing trend in software architecture design is to build platform-independent software components, such as web services, which will then be available in a distributed environment. Many businesses and enterprises are tending to transform their information systems into linked services, or repeatable business tasks which can be accessed over the network. This leads to the point where they have a large amount of services which they use as part of predefined business processes. However, they face the problem of connecting these services in an ad-hoc manner.

The information an employee works with every day, can be obtained from different sources – local documents, documents from enterprise systems or other departments, emails, memos, etc. Depending on the information, the employee usually takes one or more actions, such as adding a task from an email into a To-Do list, uploading attachments to another company subsystem for further action or analysis, or sending the attachments to the printer.

V. Trajkovik and A. Mishev (eds.), *ICT Innovations 2013*,
Advances in Intelligent Systems and Computing 231,
DOI: 10.1007/978-3-319-01466-1_9, © Springer International Publishing Switzerland 2014

Additionally, with the increasing number of cloud services with specialized functionalities in the last years, the common Internet user comes across the need to routinely perform manual actions to interchange data among various cloud services – email, social networks, online collaboration systems, documents in the cloud, etc. – in order to achieve more complex and composite actions. These actions always require a certain amount of dedicated time from the user, who has to manually change the context in which he or she works, in order to take the appropriate actions and transfer data from one system to another.

In this paper we present a way of using the technologies of the Semantic Web [1], to automate the processes included in the develop-and-consume flow of web services. The automatic discovery, automatic composition, and automatic invocation of web services provide a solution for easier, faster and ad-hoc use of specialized enterprise services for an employee in the company, and of public services for the common Internet user.

The paper is structured as follows: In Section 2 we provide an overview of existing related solutions and approaches. In Section 3 we give a detailed explanation of the system architecture and its components. In Section 4 we describe the algorithm for detection and selection of the most suitable action for returning the requested output from the set of provided inputs. In Section 5 we discuss the advantages and applications of the system. We conclude in Section 6 with a short summary and an outlook on future work.

2   Related Work

As the semantic web technologies proved their usability in a large number of IT systems [2], [3], and as most of the applications and systems are now being built upon the Service Oriented Architecture (SOA) model [4], many solutions combining the two fields have been developed. These solutions apply semantic web technologies into SOA systems, in order to automate various complex processes within them [5], [6].

There are many tools and solutions for designing and running standard BPEL processes, such as Oracle Fusion Middleware[1] and IBM Websphere[2] [7]. However, they usually don't provide the ability to describe and characterize the services with semantics. Without information about the service capabilities and behavior, it is hard to compose collaborative business processes.

One of the solutions for this problem is the OntoMat-Service [8], a framework for discovery, composition and invocation of semantic web services. OntoMat-Service does not aim at intelligent and completely automatic web service discovery, composition and invocation. Rather, it provides an interface, the OntoMat-Service-Browser, which supports the intelligence of the user and guides him or her in the

---

[1] http://www.oracle.com/technology/products/middleware/index.html

[2] http://www.ibm.com/software/websphere/

process of adding semantic information, in a way that only a few logically valid paths remain to be chosen.

The system described in [9] can deal with preexisting services of standard enterprise systems in a semantically enriched environment. By transforming the classic web services into semantic web services, the services are prepared to be invoked within a prebuilt business process. The system described in [10] presents a web service description framework, which is layered on top of the WSDL standard, and provides semantic annotations for web services. It allows ad-hoc invocation of a service, without prior knowledge of the API. However, this solution does not support the ability of creating a composition of atomic semantic web services.

The authors in [11] propose a planning technique for automated composition of web services described in OWL-S process models, which can be translated into executable processes, like BPEL programs. The system focuses on the automatic composition of services, disregarding the user's context and provided inputs to suggest the most reliable and relevant composition.

Another approach [12] describes an interface-matching automatic composition technique that aims to generate complex web services automatically by capturing user's expected outcomes when a set of inputs are provided; the result is a sequence of services whose combined execution achieves the user goals. However, the system always requests the user's desired output, which means that the system is unable to suggest new actions. Additionally, it is not guaranteed that the system would always choose the most reliable compositions of services, as the compositions are built based only on two factors: the execution time and the similarity value between the services in the composition, expecting only one user input.

Similar approaches have been further studied in [13], [14] and [15]. However, none of the related systems fully automate the workflow of discovery, ranking and invocation of web services and web service compositions, but they only automate a certain part of it. In our solution, we fully automate the workflow of web service and web service composition invocation, which includes automatic fetching of possible actions for a given context, automatic ranking and composition, and automatic invocation.

3 Solution Description

Our approach is based on web service invocation. We refer to the invocation as *taking an action*. As an *action* we consider a single RESTful service, a single SOAP web service, or a composition of more than one SOAP web services. The system tries to discover all of the possible actions that can be taken over the given resources in a given context, and provides the user with a list of available actions to execute. The user can then quickly execute complex actions by a single click. These actions can be discovered in an ad-hoc manner, i.e. they do not have to be predefined and pre-modeled.

The solution is developed in the Java programming language, using the Play MVC framework[3]. The system architecture, shown in Fig. 1, consists of several components.

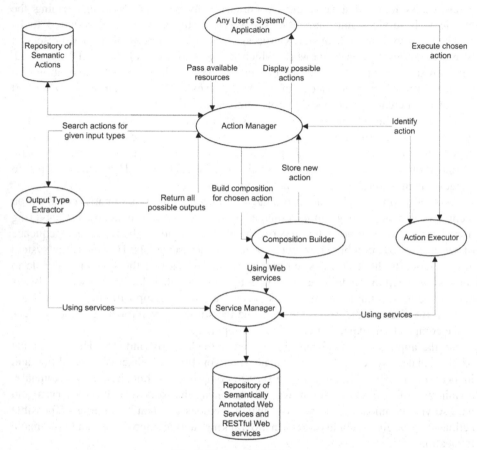

Fig. 1. System Architecture

The *repository of semantically annotated web services and RESTful web services* (SAWSRWS) holds information about all of the semantically annotated web services. There is no restriction on the technology used to develop the web services. After a web service is developed, in order for it to be uploaded onto the repository it has to be semantically annotated. The system provides a simple form for annotation and saving the information for the new web services into the repository. For SOAP web services, the semantically annotated WSDL file for the service is stored. The SOAP web services are annotated using the Semantic Annotations for WSDL and XML Schema (SAWSDL) framework[4]. For RESTful web services, we store an XML file with

---

[3] http://www.playframework.org/
[4] http://www.w3.org/TR/sawsdl/

details about the service, such as the base-URL, the method type, the names of the input parameters and the output parameter, along with their semantic annotations. This information is stored within an XML file in the repository.

The *service manager* is responsible for handling the requests to the SAWSRWS repository: adding, removing, updating and loading services. On system startup, the manager is indexing the services, and this index is updated only when a service is added or removed from the SAWSRWS repository. With this index, the number of accesses to the SAWSRWS repository is decreased and the system performance is improved. Newly created actions are stored in the *repository of semantic action* in three strictly defined storage forms. The first storage form includes a unique ID for the action, the inputs that are needed to invoke the action, and the output of the action. The second storage form is an upgrade of the first storage form, which includes a list of all the web services that compose the action, which are described with their name, the function, and the inputs and output of the function. The third storage form is used only for RESTful web services, which cannot be part of a composition in our system. This storage form includes the base-URL of the RESTful web service, the input parameters, the output parameter and the method type.

The *action manager* is intended to improve the performance of the system. On system startup, if the repository is not empty, the action manager creates an index of the actions. The action manager is handling the requests from the user applications and systems. When a request arrives, the resource types are aligned as inputs. The action manager is iterating the index to find if previously created actions for these inputs exist. If not, the action manager sends a request with the list of inputs to the *output type extractor*, and receives a list of all possible outputs that can be obtained from the available services. From the list of outputs, the action manager creates actions in the first storage form, for SOAP web services, and creates actions in the third storage form for RESTful web services. Then it stores them in the semantic action repository, updates the action index and returns the list of actions, in XML form, to the user application or system. When the user wants to invoke an action, a request to the *action executor* is sent, which identifies the action in the action manager, based on the action unique ID. The action manager checks in the index of actions for the storage form of the action with the given ID. If the action is in the second or third storage form, the actions' details are sent to the action executor. If the action is in first storage form, the action manager sends the action details to the *composition builder*, and receives a composition of functions from the web services with their name, list of inputs and the output. Then the action storage form is upgraded to the second storage form, and the action details are sent to the action executor.

The *output type extractor* receives the list of inputs from the action manager and iterates the index of services in the service manager, in order to find all of the possible outputs from the services for the given list of inputs. In the list of outputs we add only the outputs of the service functions which can be invoked with the given list of inputs, or a subset of the list of inputs. When a new output is detected, it is added both to the list of outputs and to the list of inputs. Then the extractor iterates the index of services again, with the new list of inputs. When there are no more new outputs, the iteration stops. Then the extractor sends the list of detected outputs to the action manager.

The *composition builder* receives a list of inputs and one output from the action executor. The composition builder uses an intelligent algorithm, described further, for building an optimal composition of semantic web services, in order to provide the needed output for the list of given inputs. The RESTful web services are not included in compositions.

The *action executor* receives requests from the user applications and systems. The request contains the action ID and a list of values, which represent the input values for the action. The action executor identifies the action in the action manager, and receives a RESTful web service, a single function from a SOAP web service, or a composition of functions from SOAP web services, ordered for invocation. In the former two cases, the invocation is done in a single step. But for the latter case, the action executor invokes the first SOAP web service function, for which every input value is provided by the user. If the function is successfully executed, the result value is added in the initial list of input values, and the values used for invocation are removed from the list. The same steps are repeated for the rest of the functions. When the last function is executed, if the function has an output, it is displayed to the user; otherwise, a message for successful invocation is displayed to the user. If any function fails to execute, the algorithm stops, and an error message is displayed to the user. For the invocation of the actions we use the Apache Axis2 engine[5].

4   Service Composition

The composition builder uses a specially created algorithm for building an optimal composition of semantic web services, in order to provide the needed output for a list of given inputs. The algorithm works with a set of inputs and an output parameter, provided to the composition builder by the action manager. The algorithm tries to identify if the service repository contains a single service which returns an output of the same semantic type, as the requested output value. If there are one or more such services, it checks to see if their input parameters match the inputs provided to the composition builder.

If o_r is the semantic type of the requested output parameter, and o_i is the semantic type of the output from the ith web service from the repository, what the algorithm tries to find are services for which

$$o_r = o_i \tag{1}$$

is true. These semantic web services become candidate web services for providing the requested output.

For each of the semantic web services which satisfy the equation (1), the algorithm has to compare the input types set $I_r = \{i_{r1}, i_{r2}, ..., i_{rn}\}$, provided to the composition

---

[5] http://axis.apache.org/axis2/java/core

builder, and the set of input types of the ith web service, $I_i = \{i_{i1}, i_{i2}, ..., i_{im}\}$. If the two sets satisfy that

$$|I_i| < |I_r|,\tag{2}$$

the algorithm eliminates the ith semantic web service from the list of potential candidate web services, because the number of input parameters of the services is less than the number of input parameters provided to the composition builder. This way, the potential lack of precision in the output, caused by lesser constraints, is eliminated.

If the input sets satisfy that

$$I_i = I_r,\tag{3}$$

it means that the number and the semantic types of the inputs provided to the composition builder match the number and the semantic types of the ith web service. The algorithm assigns this semantic web service with a *fitting coefficient*:

$$F = 1.$$

The fitting coefficient – F, represents the suitability of a given semantic web service, or a composition of semantic web services, to provide the requested output.

When the ith web service satisfies (3), the service is considered to be the most suitable – the requested output can be returned in just one step. Therefore its fitting coefficient is equal to the highest value. When the algorithm discovers at least one semantic web service with $F = 1$, the discovery of candidate web services ends.

If the algorithm does not find a suitable semantic web service for the received request, i.e. does not find a service which satisfies (1), the algorithm ends without success and does not return a suitable semantic web service or a composition of semantic web services.

If the sets of input satisfy that

$$|I_i| \geq |I_r|,\tag{4}$$

but the types of all of the input parameters of the ith semantic web service do not match the types of the input parameters provided to the composition builder, the fitting coefficient of the ith semantic web service is calculated as

$$F_i = \frac{y_{fi}}{y_i},\tag{5}$$

where $y_{fi} = |I_{fi}|$ is the number of parameters from the ith semantic web service which have a matching semantic type with the input parameters from I_r, $I_{fi} \subseteq I_i$, and $y_i = |I_i|$.

In this case, the algorithm continues to search for the other input parameters which do not belong to I_{fi}. The algorithm starts again, but now the requested output o_r is the input parameter which does not belong to I_{fi}. If we have more than one such parameter, this secondary search is performed for each of them. This way, we search for outputs from other services which can be used as inputs for the discovered service.

If the services discovered in the secondary search have the same types of input parameters as the inputs provided to the composition builder, they can be invoked and their outputs can be used as inputs for the semantic web service discovered in the first iteration. If they too have input parameters with types which do not match those provided to the composition builder, the algorithm performs a tertiary search for services which can provide them. These iterations last until the algorithm does not come to the state in which all of the discovered semantic web services have the same types of input parameters as the input parameters provided to the composition builder and as the outputs provided from other services in the composition, or the state in which a suitable service or composition cannot be discovered.

By creating a composition of semantic web services in this manner, we raise the cost for getting the required output. Depending on the number of services and levels in the composition, the time necessary to get the output from the list of given inputs increases. Additionally, as the composition grows larger, so does the possibility of an error occurring during a call to a web service from the composition. Therefore, we must somehow take this into account in our calculations for the fitting coefficient.

We add a *coefficient for fitness degradation*:

$$K_i = \sum_{j=1}^{p} (k_1^{-1} + y_j k_2^{-1}), \tag{6}$$

where p is the total number of services in the composition, without the initially discovered service, y_j is the number of input parameters from the jth service, and k_1 and k_2 are *factors for fitness degradation*. k_1 is a factor of influence of the number of services from the composition. k_2 is a factor of the influence of the number of parameters of services from the composition. Generally, the values of these factors should always be $k_1 < k_2$, because the number of services has a bigger impact on the total call time of the composition, compared to the number of parameters of the services. The default values for the factors are chosen to be $k_1 = 10$ and $k_2 = 100$, and can be modified within the composition builder.

From (6) we can see that the algorithm does not take into account the level of composition at which the jth service is positioned. This is because the calls to web services from the same level are performed sequentially, just as the calls to web services from different levels. Therefore, the cost for getting the requested output depends only on the number of services in the composition, and not their level distribution.

From (6) we can also see that the coefficient depends on the number of parameters used for each of the services, disregarding whether they are provided to the composition builder, or returned from another service. This is because the number of parameters represents the amount of data which has to be transferred for the calls to the services, so the nature of the parameters is irrelevant.

We add the coefficient for fitness degradation to (5):

$$F_i = \frac{y_{fi}}{y_i} - K$$

$$F_i = \frac{y_{fi}}{y_i} - \sum_{j=1}^{p} (k_1^{-1} + y_j k_2^{-1}). \tag{7}$$

The algorithm uses (7) to calculate the fitness of all candidate semantic web services which satisfy (4). The fitting coefficient is larger when a candidate service uses more of the input parameters provided to the composition builder. The coefficient drops with the number of additional services and the number of their input parameters.

Once the algorithm calculates F for each of the candidate web services, they are ranked and the atomic service or a composition of services with the highest value of F is selected as most suitable for providing the requested output.

5   Advantages and System Usability

The flexible architecture of the solution allows it to be used from within various systems. In order for it to be prepared for use in a new domain, it requires semantically annotated services from the domain, which can be taken from different enterprise systems and cloud infrastructures. After this step is completed, the users can receive a list of possible actions for any resources and data they are working with, within their own environment. These ad-hoc actions can then be executed by a single click, which is time-saving; an advantage towards which all modern tools aim. Additionally, because of its modularity, the solution can be easily updated and extended.

5.1   Use-Case

In this scenario, the user application works with geographic data, and it uses several web services which have the functions given in Table 1.

In this use-case, a user uses an application which works with the web services from Table 1, and has a name of a certain municipality as the only useful information in the context of the working environment, e.g. an email message. In this case, the application has only one service which can be invoked for the context of the user – WSF6 from Table 1 – so in a standard SOA architecture with service discovery the system will only offer this action to the user.

Table 1. List of web service functions from the use-case example

WSF #	Web Service Function
1	DialingCode *getDialingCode* (Country country, City city);
2	DialingCode *getDialingCode* (Continent continent, Country country, City city, Municipality municipality);
3	Continent *getContinent* (Country country);
4	Country *getCountry* (Municipality municipality, City city);
5	Country *getCountry* (City city);
6	City *getCity* (Municipality municipality);
7	City *getCity* (City city);

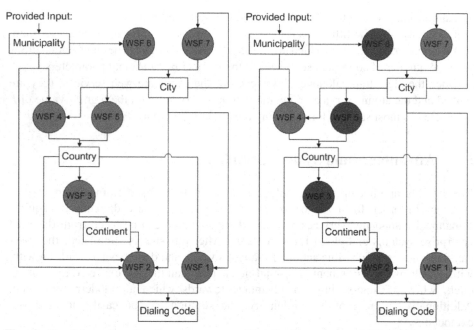

Fig. 2. Possible action for the given user input, i.e. the name of the municipality.

Fig. 3. The most suitable semantic web service composition for returning a dialing code, based on the name of the municipality.

However, if the application is connected with the system presented in this paper, the number of possible actions will grow. The system, besides WSF6 as an atomic web service, will automatically detect the possible semantic web service compositions, as shown on Fig. 2. This means that four different actions: *getCity*, *getCountry*, *getContinent* and *getDialingCode* can be executed over the information extracted from the context of the user, i.e. the name of the municipality, either as atomic web services or as web service compositions. When an action represents a composition of SOAP web services, its name is derived from the name of the last web service function in the composition. When the action consists of a single SOAP or RESTful web service, it has the same name as the atomic web service.

These actions are built from the most optimal and most reliable web services or compositions of web services, according to the algorithm from Section 4. One of these actions is the action which can return the dialing code, based only on the municipality name. This action can be derived from four different web service compositions, shown in Table 2, all of which end with either WSF1 or WSF2. Therefore, the name of this action is *getDialingCode*.

For each of the compositions which return the dialing code, we calculate the fitting coefficient F (for $k_1 = 10$ and $k_2 = 100$), using (7), and the results are shown in Table 2. The most suitable composition, the one with the highest value of F, is chosen for representing the action of deriving the dialing code based on the municipality name. In this scenario, it is the first composition.

Table 2. List of possible compositions for providing the dialing code based on the municipality name

Action	Fitting Coefficient
WSF6 → WSF5 → WSF3 → WSF2	$F = 0.19$
WSF6 → WSF4 → WSF3 → WSF2	$F = 0.18$
WSF6 → WSF4 → WSF1	$F = -0.17$
WSF6 → WSF5 → WSF1	$F = -0.18$

The user can choose to execute any of the actions from Fig. 2, just by a single click. The list contains all of the possible actions for the given input, by creating compositions of web services. Thus, introducing new actions, which were previously not part of the user system, or the user was unaware that they existed, is done automatically. Even if the user was aware that these actions were available, he or she would have had to pre-connect the services manually into a business process, which takes much longer and is not an easy task to do. This is essential in systems where the services are constantly changing, and new services are added regularly.

5.2   Application

The solution has been implemented as part of Semantic Sky, a platform for cloud service integration [16]. Semantic Sky enables connectivity and integration of different cloud services and of local data placed on the users machines, in order to create a simple flow of information from one infrastructure to another. It is able to automatically discover the context in which the users are working, and based on it and by using the solution described in this paper, provide them with a list of actions which can be executed over their data. In this way, the users can completely focus on their tasks in their work environment, and get relevant information and executable actions in their current context. By automating the discovery and execution of relevant tasks, the system improves the productivity, information exchange and efficiency of the users.

6   Conclusion and Future Work

This paper presents a solution for automatic discovery and invocation of atomic web services and web service compositions, by employing semantic web technologies. The solution provides a list of all possible actions which exist within the user system or in distributed environments and which can be executed over the data and information the user is currently working with. This approach offers the user a broader perspective and can introduce action for which he or she was previously unaware.

Additionally, the solution offers actions in an ad-hoc manner; the service compositions are created on-the-fly, overriding the need for pre-connecting the services into fixed compositions, i.e. creating pre-built business processes. This is essential in systems where the services are continually changing, and new services are added regularly. In such dynamic environments, the fast and automatic detection of new possible actions is of high importance.

Currently, the solution does not support building compositions of RESTful web services or combining RESTful web services with SOAP web services in a same composition. This is because we use primitive data types for semantic annotation of the inputs and the output of the services, and most of the RESTful services return a more complex value. This can be solved by adding more complex classes for annotation into the ontologies. These drawbacks will be our main focus in the future development of the solution.

References

1. Berners-Lee, T., Hendler, J., Lassila, O., et al.: The Semantic Web. Scientific American 284(5), 28–37 (2001)
2. Hitzler, P., Krotzsch, M., Rudolph, S.: Foundations of Semantic Web Technologies. Chapman, Hall/CRC (2011)
3. Sugumaran, V., Gulla, J. A.: Applied Semantic Web Technologies. Auerbach Pub. (2012)
4. He, H.: What is Service-Oriented Architecture? Publicação eletrônica em 30 (2003)
5. Hu, Y., Yang, Q., Sun, X., Wei, P.: Applying Semantic Web Services to Enterprise Web. International Journal of Manufacturing Research 7(1), 1–8 (2012)
6. Wang, D., Wu, H., Yang, X., Guo, W., Cui, W.: Study on automatic composition of semantic geospatial web service. In: Li, D., Chen, Y. (eds.) CCTA 2011, Part II. IFIP AICT, vol. 369, pp. 484–495. Springer, Heidelberg (2012)
7. Kloppmann, M., Konig, D., Leymann, F., Pfau, G., Roller, D.: Business process choreography in websphere: Combining the power of bpel and j2ee. IBM Systems Journal 43(2), 270–296 (2004)
8. Agarwal, S., Handschuh, S., Staab, S.: Annotation, Composition and Invocation of Semantic Web Services. Web Semantics: Science, Services and Agents on the World Wide Web 2(1), 31–48 (2004)
9. Martinek, P., Tothfalussy, B., Szikora, B.: Execution of Semantic Services in Enterprise Application Integration. In: Proc. of the 12th WSEAS International Conference on Computers, pp. 128–134 (2008)
10. Eberhart, A.: Ad-hoc Invocation of Semantic Web Services. In: IEEE International Conference on Web Services, pp. 116–123. IEEE (2004)

11. Traverso, P., Pistore, M.: Automated Composition of Semantic Web Services into Executable Processes. In: McIlraith, S.A., Plexousakis, D., van Harmelen, F. (eds.) ISWC 2004. LNCS, vol. 3298, pp. 380–394. Springer, Heidelberg (2004)
12. Zhang, R., Arpinar, B., Aleman-Meza, B.: Automatic Composition of Semantic Web Services. In: 1st International Conference on Web Services, pp. 38–41 (2003)
13. He, T., Miao, H., Li, L.: A Web Service Composition Method Based on Interface Matching. In: The eight IEEE/ACIS International Conference on Computer and Information Science, pp. 1150–1154. IEEE (2009)
14. Talantikite, H., Aissani, D., Boudjilda, N.: Semantic Annotations for Web Services Discovery and Composition. Computer Standards & Interfaces archive 31(6), 1108–1117 (2009)
15. Hamadi, R., Benatallah, B.: A Petri Net-Based Model for Web Service Composition. In: The 14th Australasian Database Conference, pp. 191–200. Australian Computer Society, Inc. (2003)
16. Trajanov, D., Stojanov, R., Jovanovik, M., Zdraveski, V., Ristoski, P., Georgiev, M., Filiposka, S.: Semantic Sky: A Platform for Cloud Service Integration based on Semantic Web Technologies. In: Proceedings of the 8th International Conference on Semantic Systems (I-SEMANTICS 2012), pp. 109–116. ACM (2012)

Optimization Techniques
for Robot Path Planning

Aleksandar Shurbevski, Noriaki Hirosue*, and Hiroshi Nagamochi

Department of Applied Mathematics and Physics, Kyoto University,
Yoshida-Honmachi, Sakyo-ku, Kyoto 606-8501, Japan
{shurbevski,n_hirosue,nag}@amp.i.kyoto-u.ac.jp

Abstract. We present a method for robot path planning in the robot's configuration space, in the presence of fixed obstacles. Our method employs both combinatorial and gradient-based optimization techniques, but most distinguishably, it employs a Multi-sphere Scheme purposefully developed for two and three-dimensional packing problems. This is a singular feature which not only enables us to use a particularly high-grade implementation of a packing-problem solver, but can also be utilized as a model to reduce computational effort with other path-planning or obstacle avoidance methods.

Keywords: Robot, path planning, combinatorial optimization, multi-sphere scheme, packing problems.

1 Introduction

Robotics is already a well-established area vastly adopted in the domains of industry, entertainment, and steadily moving on to be incorporated in daily life. Yet still, there are many challenges left to be solved or improved upon. Among them is certainly the most fundamental robot motion planning, which becomes ever more challenging as the autonomy and our expectations of robots increase; "The minimum one would expect from an autonomous robot is the ability to plan its own motions" [1].

Motion planning in its essence is defined as the problem of finding a collision-free continuous motion of one or more rigid objects from a given start to a known goal configuration. Commonly also referred to as the *Piano Movers' Problem* [2], the basic problem investigates the movement of a single body, assumes complete and accurate knowledge of the environment, and neglects dynamic and kinematic issues, or the possibility of interaction and compliance of objects, however, many extensions have been built upon this simplified model. The one closest to this investigation is the *Generalized Mover's Problem* [3], which deals with the motion of an object comprised of multiple polyhedra, linked at various vertices (akin to a robotic arm with multiple joints).

---

* Present address: 1-21-10 Matsumicho, Kanagawa-ku, Yokohama 221-0005, Japan.

V. Trajkovik and A. Mishev (eds.), *ICT Innovations 2013*,
Advances in Intelligent Systems and Computing 231,
DOI: 10.1007/978-3-319-01466-1_10, © Springer International Publishing Switzerland 2014

Furthermore, motion planning as conceived for the purpose of robotics has recently found its way into such diverse areas as video animation, or even biomolecular studies, where it has been used to study protein folding [4,5].

For the reasons above, motion planning has gathered a lot of attention, and extensive research has been done. Notably, it has been proven to be computationally intractable [3,6], with complexity increasing exponentially, not only with the degrees of freedom (DOF) of a robot, but also with the number of non-static obstacles [7]. Exact algorithms for motion planning do exist, however they adopt overly simplistic models which have little applicability in practice [8]. To alleviate the intrinsic computational complexity of robot path planning, several heuristic approaches have been developed, largely classified as Roadmap and Potential Field methods [1,9,10,11]. It is notable that sampling based methods, such as the roadmap methods [12,13] tend to incorporate a sort of potential field as well, and utilize the information which the intensity and gradients of this potential field can render with regard to the robot's configuration space. A comparative overview of probabilistic roadmap planners can be found in [14].

Our approach falls under the umbrella of roadmap methods. In particular, it is aimed to extend ideas presented in [1,15]. As a distinguishing feature, we use ideas purposefully developed for packing problems, such as the Multi-sphere Scheme [16], which is particularly useful in determining and penalizing collisions between objects [17], but also as a basis to implement object separation algorithms [18]. Moreover, using methods developed in two and three-dimensional packing solvers can greatly benefit our attempts to find collision-free configurations, with a very clear relation to interactions between physical objects.

2   Problem Description

Before proceeding we introduce formal description and notation used in the remainder of the paper.

An object S is a closed continuous subspace in $\mathbb{R}^{d_e}$, where $\mathbb{R}$ is the set of real numbers, and $d_e \in \{2, 3\}$ is its dimension. A robot, $\mathcal{R} = \{R_1, R_2, \ldots, R_n\}$, is a collection of rigid objects connected at certain defined points, commonly called *links* and *joints*, respectively. To emphasize that individual objects in $\mathcal{R}$ are not entirely independent but have a high degree of mutual interaction, we also describe the robot as a *multi-object system*. The Euclidean space in which the robot operates is commonly called a *workspace*, denoted by $\mathcal{W}$. (Note that the term workspace in some other publications is used to refer to the portion of Euclidean space reachable by the end-effector of a robotic arm [9].) We represent the workspace as $\mathbb{R}^{d_e}$. A point $r \in \mathcal{W}$ is a d_e dimensional vector. We always denote a vector by a lower-case bold letter and assume it is of proper dimension.

The workspace is populated by a collection $\mathcal{O} = \{O_1, O_2, \ldots, O_m\}$ of rigid and immovable objects, called *obstacles*. The obstacles can be of arbitrary shape, and we assume that we have complete and accurate knowledge of the obstacles' geometry and spatial distribution.

A key prerequisite of our problem is to be able to accurately determine the location of every point of the robot $\mathcal{R}$ in the workspace $\mathcal{W}$. For that purpose, we use a *configuration space*, denoted by $\mathcal{C}$ [19]. The configuration space $\mathcal{C}$ is an abstract d_f-dimensional space (Fig. 1(c)), where d_f stands for the *degrees of freedom* of the robot. The role of configuration spaces and objects' interactions in physical space are further discussed in Subsection 3.2. A point $q \in \mathcal{C}$ is called a *configuration*, and it uniquely defines the location of every point of $\mathcal{R}$ in $\mathcal{W}$, that is, the robot's position and orientation. However, the opposite is not true in general, for there may be (infinitely) many points in $\mathcal{C}$ corresponding to a single position and orientation of the robot. We have resolved this issue by adopting a convention for transforming a configuration $q \in \mathcal{C}$ into a canonical one giving the same position and orientation in the workspace. Therefore, it is justified to equate the terms "position and orientation" (in $\mathcal{W}$-space) with "configuration" (in $\mathcal{C}$-space). We write $\mathcal{R}(q)$ for the robot assuming configuration $q \in \mathcal{C}$.

The robot $\mathcal{R}$ has either natural or imposed restrictions on its range of motions, reflected as a restricted domain in $\mathcal{C}$-space. In general, there are restrictions on each degree of freedom of the robot. Let $\mathcal{D} = \{(r_{min}^j, r_{max}^j) \mid j = 1, 2, \ldots, d_f\}$ be a set of ordered pairs, where (r_{min}^j, r_{max}^j) gives respectively the minimum and maximum values for the jth degree of freedom. Thus, an ordered pair $(r_{min}^j, r_{max}^j) \in \mathcal{D}$ defines the admissible range of values of q^j, $j = 1, 2, \ldots, d_f$, for which $\mathcal{R}(q)$ is still valid.

We call the obstacles from $\mathcal{W}$ projected onto the configuration space, that is, areas $\mathcal{B} = \{q \in \mathcal{C} \mid \exists O_k \in \mathcal{O} \text{ s.t. } \mathcal{R}(q) \cap O_k \neq \emptyset\}$, $\mathcal{C}$-obstacles, and we call configurations $q \in \mathcal{B}$ *collision* configurations. Furthermore, there might be such configurations $q \in \mathcal{C}$ for which if the robot assumes the position and orientation $\mathcal{R}(q)$ in $\mathcal{W}$-space, individual links R_i and R_y, $i, y = 1, 2, \ldots, n$, $i \neq y$ come into contact with each other at points other than the predefined joints, called *self-collision* configurations. Points $q \in \mathcal{C}$, with $r_{min}^j \leq q^j \leq r_{max}^j$, $j = 1, 2, \ldots, d_f$, that are neither collision nor self-collision configurations are called *free $\mathcal{C}$-space*, and denoted by $\mathcal{C}_{\text{free}}$. For a configuration $q \in \mathcal{C}$ that is also in $\mathcal{C}_{\text{free}}$ we say to be a *free configuration* (Fig. 1(a)), otherwise we say it is *colliding* (Fig. 1(b)).

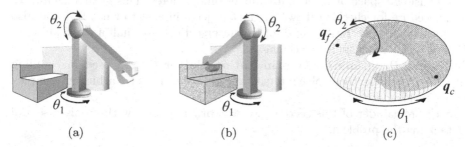

(a)   (b)   (c)

Fig. 1. A 2-DOF robot $\mathcal{R}$ in a workspace with obstacles along with its configuration space: (a) In a free configuration, $\mathcal{R}(q_f)$; (b) In a colliding configuration, $\mathcal{R}(q_c)$; (c) The configuration space $\mathcal{C}$, represented as a torus, and non-free $\mathcal{C}$-space shaded darker.

It is now possible to state more accurately the Motion Planning Problem:
Instance: A robot $\mathcal{R}$ with d_f degrees of freedom and a set $\mathcal{D}$ of restrictions on
the degrees of freedom, a set $\mathcal{O}$ of stationary and rigid obstacles with accurately
determined geometry and locations in the workspace $\mathcal{W}$.
Query: Given an initial configuration $\boldsymbol{q}_s \in \mathcal{C}$ and an objective configuration
$\boldsymbol{q}_t \in \mathcal{C}$, both of which are in $\mathcal{C}_{\text{free}}$, find a continuous motion of $\mathcal{R}$ through the
workspace $\mathcal{W}$ from $\mathcal{R}(\boldsymbol{q}_s)$ to $\mathcal{R}(\boldsymbol{q}_t)$ obeying DOF constraints, $\mathcal{D}$, without any
part of the robot, $R_i, i = 1, 2, \ldots, n$ coming into contact with any of the obstacles
in $\mathcal{O}$, nor with another part $R_y, y = 1, 2, \ldots, n, y \neq i$ of itself.

From the perspective of $\mathcal{C}$-space, the aim is to find a continuous curve (path)
from $\boldsymbol{q}_s$ to $\boldsymbol{q}_t$ lying entirely in $\mathcal{C}_{\text{free}}$ and not breaching any DOF constraint at
any time. Such a path connecting $\boldsymbol{q}_s$ to $\boldsymbol{q}_t$ ($\in \mathcal{C}$) is said to be *feasible*. It is not
always certain that a feasible path exists.

3 Solution Framework

As stated in the Introduction, the Robot Motion Planning problem has attracted
a lot of interest since its inception, and the literature boasts an abundance of
ideas and approaches. Even though the approach we present can clearly be identified as belonging to the class of Probabilistic Roadmap Methods [1,9,14,15,20],
what primarily sets it apart from existing approaches, is that to the best of
our knowledge, it is a first one to adopt methodologies specifically developed
with packing problems in mind. We expect this to greatly aid the search for a
feasible path, for both problems share some common features, namely, we need
to efficiently tell if two objects (of arbitrary shape, position and orientation) in
d_e-dimensional space overlap or not, even more, calculate the level of overlap, or
penetration depth (to be more precisely defined in Subsection 3.3). Further still,
once we have established that there does exist an overlap, we need to resolve it
in an efficient manner, both with regards to computational effort, and fitness to
contribute towards finding a feasible solution.

The two main tools we have adopted are:
The Multi-sphere Scheme [16], used to represent (approximate) an object in
d_e-dimensional space by a set of d_e-dimensional spheres. This greatly facilitates
the procedure for checking if two distinct objects intersect or not [17] and also
calculating the penetration depth of this intersection. We shall use the abbreviation MSS for Multi-sphere Scheme.
Nonlinear Programming Optimization Solver [18,21], used in coupling
with MSS to efficiently resolve intersections of two objects (of arbitrary complex
shape) in $\mathcal{W}$-space.

In the remainder of this section, we will briefly overview the tools essential
for tackling the problem.

3.1 The Probabilistic Roadmap Method, PRM

Following is but a short review of certain notions from PRM (for Probabilistic
Roadmap Method) [9,14,15] important to the presentation of our approach. It is

not aimed at describing the method itself, but just as a glossary of terminology associated with it.

The planning procedure with PRM consists of two phases: A *learning phase*, and a *query phase*. We carry out computationally and time demanding operations as a preprocessing in the learning phase, so that we will be able to almost instantaneously answer arbitrary queries.

In the learning phase, we begin with a set $\mathcal{O}$ of rigid and immovable obstacles in $\mathcal{W}$-space, as well as a description of a robot $\mathcal{R}$ together with the set $\mathcal{D}$ of DOF-constraints. The learning phase itself is a two-step process, having a *construction step* and an *expansion step*. In the construction step we perform random sampling of the d_f dimensional configuration space of the robot in an attempt to "learn" about the features of $\mathcal{C}_{\text{free}}$. Let V be a set of sampled points of $\mathcal{C}_{\text{free}}$.

While sampling, we also try to connect pairs of sampled points. The choice of a candidate pair is done by a heuristic function estimating their distance in $\mathcal{C}$-space. The procedure by which we try to make this connection is referred to as a *local planner*. We store the information of pairs connected by the local planer in a set E. Thus we have obtained a graph structure $G = (V, E)$, the set V is the graph's vertices, and E - the graph's edges. In the following, we will refer to the graph G as a *roadmap*.

The construction step terminates after a predefined time limit has been reached, or a certain population size of sample points in $\mathcal{C}_{\text{free}}$ has been obtained. However, it is very likely that the resulting roadmap consists of several isolated components and does not accurately reflect the topology of $\mathcal{C}$-space. The *expansion* step, which is the second step of the learning phase, aims to increase the connectivity of the roadmap. Regions where $\mathcal{C}_{\text{free}}$ is connected while there is a gap between components in the roadmap are considered "difficult". They might correspond to such regions in $\mathcal{W}$-space as narrow passages or regions cluttered with obstacles. We aim to populate these regions of $\mathcal{C}$-space with configuration samples so as to make a connection between different components of the roadmap. One of the features our approach boasts is that it sometimes relies on a sophisticated packing-problem solver to aid a local planer when trying to produce new candidate configurations.

After the learning phase is done, we can use the obtained roadmap to perform multiple queries for a given robot and a set of obstacles. Queries (q_s, q_t) are answered by trying to connect q_s and q_t to some vertices in V and then use a graph search algorithm to find a path which connects them in G.

3.2 Objects and their Interactions in Space

In this subsection we only touch upon notation essential for the exhibition of our proposed approach. We have followed standard geometric notions and their implementation as can be found in [1,9,19,22].

Following the robot definition, we say that an object S is given in the workspace, $\mathcal{W}$. A translation of an object in the workspace by a vector $x \in \mathcal{W}$ can be achieved by a Minkowski sum:

$$S \oplus x = \{s + x \mid s \in S\} . \tag{1}$$

We will be mainly interested in overlapping objects, as in Fig. 2(a). Given two overlapping objects, S and T, their *penetration depth* is judged according to:

$$\delta\left(S, T\right) = \min\left\{\|\boldsymbol{x}\| \mid S \cap \left(T \oplus \boldsymbol{x}\right) = \emptyset, \, \boldsymbol{x} \in \mathcal{W}\right\}. \tag{2}$$

The calculation of $\delta\left(S, T\right)$ is a highly non-trivial task. There do exist clever methodologies proposed for determining if two polygons in the plane intersect or not, but in the interest of space we refrain from further discussion. The interested reader is referred to [18] and references therein.

Working with movable objects in $\mathcal{W}$ has been simplified with the introduction of the configuration space, $\mathcal{C}$. Let $\Lambda(\boldsymbol{s}, \boldsymbol{q}) : \mathcal{W} \times \mathcal{C} \to \mathcal{W}$ be a *motion function* [18], taking as argument a point $\boldsymbol{s} \in \mathcal{W}$ and a configuration $\boldsymbol{q} \in \mathcal{C}$, and moves the point $\boldsymbol{s}$ by $\boldsymbol{q}$-variables. The choice of a motion function for a given robot may not be unique, but we adopt conventions by which $\mathcal{C}$ is always a differentiable manifold [9]. Thus, an object S in configuration $\boldsymbol{q}$ is given as:

$$S(\boldsymbol{q}) = \bigcup_{\boldsymbol{s} \in S} \Lambda(\boldsymbol{s}, \boldsymbol{q}). \tag{3}$$

Configurations of robots as articulated multi-object systems are a composition of configurations for each individual object, and certain relations between them exist, depending on the robot's representation as a serial or a parallel mechanism.

3.3 The Multi-sphere Scheme, MSS

The Multi-sphere Scheme [16] has been devised as a framework aimed to simplify certain geometrical issues commonly arising in packing and cutting stock problems [18]. It has gone beyond a mere approximation of geometrical objects for the purpose of fast collision detection [23]. In addition to the original application [18], MSS has continued to evolve [24,25], and has since been applied in fields such as layout planning and optimization [26,27], and the present, robot motion planning [28,29]. Without further ado, we state that we have a means of approximating any solid object, say O_k, in $\mathbb{R}^{d_e}$ by a set $\{O_1^k, O_2^k, \ldots, O_{n_k}^k\}$ of n_k d_e-dimensional spheres. Details on how to obtain such an approximation from other types of object representations can be found in [25]. In the context of MSS, we shall use O_k to denote both an object and the set of spheres used to approximate it. Now the object O_k can be though of as a multi-object system consisting of n_k spheres, where the sphere O_i^k has radius r_i^k, and the coordinates of its center are given by $\boldsymbol{c}_i^k \in \mathbb{R}^{d_e}$. The motion of the object O_k retains the form of Eq. (3), only the union is over all spheres in O_k, a finite number.

As already mentioned, MSS greatly facilitates the procedures we need to recognize if two distinct objects ever come into contact with little computational effort. Further still, we would like to be able to also quickly compute the amount of overlap between two colliding objects. Let O_k and O_l be two objects in $\mathbb{R}^{d_e}$, each approximated by a set of spheres. For brevity, we ommit stating their configurations explicitly. For two spheres $O_i^k \in O_k$ and $O_j^l \in O_l$, the penetration depth function from Eq. (2) can be simplified to:

$$\delta\left(O_i^k, O_j^l\right) = \max\left\{r_i^k + r_j^l - \left\|\boldsymbol{c}_i^k - \boldsymbol{c}_j^l\right\|, 0\right\}. \tag{4}$$

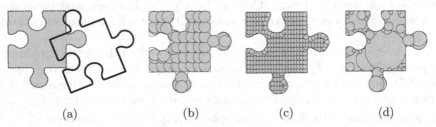

(a)   (b)   (c)   (d)

Fig. 2. Objects in $d_e = 2$-dimensions: (a) Two overlapping objects; (b) A coarse MSS approximation; (c) A finer MSS approximation; (d) MSS approximation with spheres of variable radii.

In addition to the penetration depth, for two spheres of different objects, $O_i^k \in O_k$ and $O_j^l \in O_l$, we introduce the *penetration penalty* function:

$$f_{ijkl}^{\text{pen}}\left(O_i^k, O_j^l\right) = \left(\delta\left(O_i^k, O_j^l\right)\right)^2 \tag{5}$$

and the overall penetration penalty function for two objects, O_k and O_l, becomes:

$$f_{\text{pen}}\left(O_k, O_l\right) = \sum_{i=1}^{n_k}\sum_{j=1}^{n_l} f_{ijkl}^{\text{pen}}\left(O_i^k, O_j^l\right). \tag{6}$$

Since the robot $\mathcal{R}$ was initially given as a multi-object system (Section 2), we write $R_i(q)$ for the link $R_i \in \mathcal{R}$ as it is when the configuration of the robot is given by q, for notational convenience.

As the robot $\mathcal{R}$ is assumed to be the only movable object in a given instance of $\mathcal{R}, \mathcal{O}$ in $\mathcal{W}$, it is the only one for which explicitly defining a configuration makes sense, therefore the notation $O_j(q)$, $j = 1, 2, \ldots, m$ is omitted and obstacles are simply referenced as $O_j \in \mathcal{O}$. We introduce a penalty function $F_{\text{pen}}(q)$, $q \in \mathcal{C}$ depicting intersection penalties in the entire scene $(\mathcal{R}(q), \mathcal{O})$:

$$F_{\text{pen}}(q) = \sum_{i=1}^{n}\sum_{j=1}^{m} f_{\text{pen}}(R_i(q), O_j) + \sum_{1 \leq k < l \leq n} f_{\text{pen}}\left(R_k(q), R_l(q)\right). \tag{7}$$

3.4 Nonlinear Optimization Solver

Eq. (7) enables us to utilize MSS beyond fast collision checking. It will bring us to efficient means to seek out a collision-free configuration, $q_f \in \mathcal{C}_{\text{free}}$, given a collision configuration q_c. We can get such a free configuration q_f as a solution to the following optimization problem:

$$\text{minimize } F_{\text{pen}}(q), \tag{8a}$$

$$\text{subject to } q \in \mathcal{C}. \tag{8b}$$

There are in fact infinitely many configurations which solve the problem given with Eq. (8), however our expectations to find a "nearby" collision free configuration are justified by using a gradient-based iterative method, such as the

Quasi-Newton method [18,21]. This can be a pitfall however, and methods to guide successive iterations towards a reasonably close solution are being considered [29].

In order to use a gradient based method we need to define the gradient of the penalty function, $F_{\mathrm{pen}}(q)$. This should be simple enough retracing through Eqs. (4-7), and as a base, we have that C is always a differentiable manifold. Details can be found in [9,18]. The only point of caution is that f_{ijkl}^{pen} from Eq. (5) is not differentiable if the two spheres (Eq. (4)) are concentric ($c_i^k = c_j^l$) and in such a case we need to use subgradients.

As a part of PRM, the optimization method meticulously built upon MSS is mainly applied in the expansion step of the learning phase. It can be also used in queries when trying to connect given q_s and q_t to an existing roadmap.

Let x and y be two configurations from the roadmap which are judged sufficiently close by a heuristic function, and yet a local planner fails to connect them. We invoke the iterative gradient-based solver starting from a collision configuration somewhere on the straight line between x and y, hoping that it will glide us to a free configuration q_f, which in turn will either connect to one (or both!) of x and y. If that fails, we will try starting from collision configurations on the straight lines between q_f and x, and q_f and y.

4   Results and Conclusion

We have implemented both the PRM as described in [15], adopting a *random walk* as a method to connect individual components in the expansion step, and a version incorporating the MSS as described above. In order to test the effectiveness of using our proposed method in the expansion step of PRM, over the instances we have limited the construction step to a small population size (only 10 candidate configurations in C_{free}). In this way, we aimed to have a roadmap of several largely separated components on which we would run the expansion step. Experiments were done for a 12-DOF and 15-DOF serial mechanism (an arm model of "free flying" base and 3, respectively 4, freely rotating joints). Both models had the same maximal length in a canonical configuration. All instances required finding a way through a narrow passage with width 1/3 of the robot's canonical length, and the level of difficulty was varied with the length of the passage. We set a time limit of $1800\,s$ for the expansion step.

Over the presented instances, employing our MSS in the expansion step largely increased the success ratio of obtaining a single connected component in the roadmap as compared to PRM without using MSS. A singular striking result is that the expansion step using MSS required far less newly generated configurations (nodes in the roadmap) to obtain a connected roadmap. As given in Fig. 3(c), we have observed an order of magnitude difference over the tested instances.

However, there were occasions in which MSS did not prove as favorable, and further research will be aimed in achieving performance benefit even in broader variation of scenarios and workspace topologies. From technical point of view, we

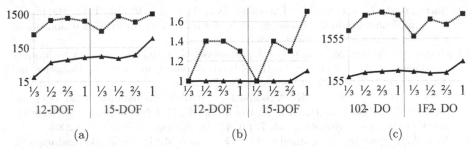

Fig. 3. Comparative results, averaged over 10 random runs. Original PRM is drawn by ■, and our MSS procedure by ▲. x-axis labels are ratios of the passage length over the robot's canonical length: (a) Running time $[s]$; (b) Number of components; (c) Number of nodes in the final roadmap.

would like to integrate our solver in a framework enabling us to experiment with different robot models, and include additional constraints, such as kinematic constraints and nonholonomicity.

Acknowledgement. The authors would like to express their sincere gratitude to Dr. Takashi Imamichi of IBM Research-Tokyo for his unreserved help in this project, and many valuable discussions.

References

1. Latombe, J.C.: Robot Motion Planning. Kluwer Academic Publishers, Norwell (1991)
2. Schwartz, J.T., Sharir, M.: On the piano movers' problem. II. General techniques for computing topological troperties of real algebraic manifolds. Advances in Applied Mathematics 4, 298–351 (1983)
3. Reif, J.R.: Complexity of the generalized movers' problem. In: Schwartz, J.T., Sharir, M., Hopcroft, J. (eds.) Planning, Geometry and Complexity of Robot Motion. Ablex Publishing Corporation (1987)
4. Amato, N.M., Song, G.: Using motion planning to study protein folding pathways. Journal of Computational Biology 9, 149–168 (2002)
5. Song, G.: A Motion Planning Approach to Protein Folding. PhD thesis, Texas A&M University (2003)
6. Canny, J.F.: The Complexity of Robot Motion Planning. The MIT Press (1988)
7. Reif, J.R., Sharir, M.: Motion planning in the presence of moving obstacles. Journal of the ACM 41, 764–790 (1994)
8. Ahrikencheikh, C., Seireg, A.A.: Optimized-Motion Planning: Theory and Implementation, 1st edn. John Wiley & Sons, Inc., New York (1994)
9. Choset, H., Lynch, K.M., Hutchinson, S., Kantor, G.A., Burgardand, W., Kavraki, L.E., Thrun, S.: Principles of Robot Motion. The MIT Press (2005)
10. Russell, S., Norvig, P.: Artificial Intelligence: A Modern Approach. Prentice Hall (2010)
11. Schwartz, J.T., Sharir, M., Hopcroft, J.: Planning, Geometry and Complexity of Robot Motion. Ablex Publishing Corporation (1987)

12. Barraquand, J., Kavraki, L., Latombe, J.C., Motwani, R., Li, T.Y., Raghavan, P.: A random sampling scheme for path planning. The International Journal of Robotics Research 16, 759–774 (1997)
13. Ratliff, N., Zucker, M., Bagnell, J.A., Srinivasa, S.: CHOMP: gradient techniques for efficient motion planning. In: IEEE International Conference on Robotics and Automation, ICRA 2009, pp. 489–494. IEEE (2009)
14. Geraerts, R., Overmars, M.H.: A comparative study of probabilistic roadmap planners. In: Boissonnat, J., Burdick, J., Goldberg, K., Hutchinson, S. (eds.) Algorithmic Foundations of Robotics, vol. 7, pp. 43–58. Springer, Heidelberg (2004)
15. Kavraki, L., Švestka, P., Latombe, J.C., Overmars, M.: Probabilistic roadmaps for path planning in high-dimensional configuration spaces. In: IEEE International Conference on Robotics and Automation, pp. 566–580. IEEE (1996)
16. Imamichi, T., Nagamochi, H.: A multi-sphere scheme for 2D and 3D packing problems. In: Stützle, T., Birattari, M., Hoos, H.H. (eds.) SLS 2007. LNCS, vol. 4638, pp. 207–211. Springer, Heidelberg (2007)
17. Imamichi, T., Nagamochi, H.: Performance analysis of a collision detection algorithm of spheres based on slab partitioning. IEICE Fundamentals of Electronics, Communications and Computer Sciences E91-A, 2308–2313 (2008)
18. Imamichi, T.: Nonlinear Programming Based Algorithms to Cutting and Packing Problems. Doctoral Dissertation, Kyoto University (2009)
19. Lozano-Perez, T.: Spatial planning: A configuration space approach. IEEE Transactions on Computers 100, 108–120 (1983)
20. Kavraki, L., Kolountzakis, M.N., Latombe, J.C.: Analysis of probabilistic roadmaps for path planning. IEEE Transactions on Robotics and Automation 14, 166–171 (1998)
21. Nocedal, J., Wright, S.J.: Numerical Optimization, 2nd edn. Springer (2006)
22. O'Rourke, J.: Computational Geometry in C, 2nd edn. Cambridge University Press (1998)
23. Hubbard, P.M.: Approximating polyhedra with spheres for time-critical collision detection. ACM Transactions on Graphics 15, 179–219 (1996)
24. Imamichi, T., Nagamochi, H.: Designing algorithms with multi-sphere scheme. In: Informatics Education and Research for Knowledge-Circulating Society, ICKS 2008, pp. 125–130. IEEE (2008)
25. Hiramatsu, M.: Approximating objects with spheres in multi-sphere scheme. Master's thesis, Kyoto Univeristy (2010)
26. Jacquenot, G., Bennis, F., Maisonneuve, J.J., Wenger, P.: 2D multi-objective placement algorithm for free-form components. ArXiv e-prints (November 2009)
27. Bénabès, J., Bennis, F., Poirson, E., Ravaut, Y.: Interactive optimization strategies for layout problems. International Journal on Interactive Design and Manufacturing (IJIDeM) 4, 181–190 (2010)
28. Beppu, K.: An application of multi-sphere scheme to robot path planning. Master's thesis, Kyoto University (2012)
29. Hirosue, N.: An application of multi-sphere scheme to robot path planning with 3D-motion. Master's thesis, Kyoto University (2013)

Global Path Planning in Grid-Based Environments Using Novel Metaheuristic Algorithm

Stojanche Panov and Natasa Koceska

Faculty of Computer Science, University "Goce Delcev" - Stip,
bul. Krste Misirkov bb. 2000 Stip, Macedonia
{stojance.panov,natasa.koceska}@ugd.edu.mk

Abstract. The global path planning problem is very challenging NP-complete problem in the domain of robotics. Many metaheuristic approaches have been developed up to date, to provide an optimal solution to this problem. In this work we present a novel Quad-Harmony Search (QHS) algorithm based on Quad-tree free space decomposition methodology and Harmony Search optimization. The developed algorithm has been evaluated on various grid based environments with different percentage of obstacle coverage. The results have demonstrated that it is superior in terms of time and optimality of the solution compared to other known metaheuristic algorithms.

Keywords: Artificial Intelligence, Free Space Decomposition, Global Path-Planning, Heuristic Algorithm, Robotics.

1   Introduction

The global path planning problem can be easily formulated as establishing a feasible route from a starting point to a destination point. On this route, a search agent might face obstacles in the searching environment, which need to be avoided, which leads to the fact that a collision-free path is crucial and necessary. This route, however, ought to be determined before the search agent has started it's travelling through this path. This means that one needs a certain effective algorithm to perform this task as fast as possible and it's recommended that the found solution is optimal. Since the nature of this problem is its NP-completeness, there have been several metaheuristic techniques applied to the global path planning. Although there were approaches that used completely metaheuristic algorithms, there were also research studies which utilized hybrid approaches, i.e. combination of deterministic and metaheuristic algorithms in order to obtain a technique which would provide desirable or near-optimal solutions.

The agent's navigation through the defined route can be divided into two sub-tasks, namely local and global path planning. This means that there's a part of the path planning approach, i.e. the local path planning, which deals with avoiding obstacles, acceleration, processing input data and other environmental dependent problems, whereas the global path planning issues the finding of an optimal route. The representation of the environment of interest for the search agent can be given as a

V. Trajkovik and A. Mishev (eds.), *ICT Innovations 2013*,
Advances in Intelligent Systems and Computing 231,
DOI: 10.1007/978-3-319-01466-1_11, © Springer International Publishing Switzerland 2014

real-world interpretation, or it can be discretized. There have been many researches that applied grid-based path planning, i.e. techniques that represented the search space as a two dimensional grid. This type of discretization is popular in path planning algorithms, as it can easily be manipulated by the algorithms utilized for its solving. Though this approach in real-world applications might not be satisfying, it is a crucial foundation for obtaining a path based on a priory provided data to the algorithm.

This paper presents a novel sophisticated approach to solving the path planning problem, namely the Quad Harmony Search algorithm. This metaheuristic algorithm has been accelerated by the popular quad-tree free space decomposition technique in order to divide the search space into smaller free space regions, which proved by our experimental results, gave great speed to the process of obtaining an optimal solution. In the end, we provide a detailed study of the effectiveness of the Quad Harmony Search algorithm.

2   Related Work

The global path planning problem is a popular topic that has been actively researched over the last several years. Most of these research studies involve using metaheuristic approaches, given they were purely metaheuristic or combined with other techniques as well. Also, one should emphasize that grid-based representations of the mobile robot's environment are explored thoroughly, so grid-based path planning plays a crucial role in solving today's real-world applications. Particular studies even use discretization of complex obstacles to a grid-based representation, which provided collision-free near optimal solutions as well [1].

Simulated annealing (SA) had been proposed as an algorithm in 1983. It is a well utilized metaheuristic algorithm which mimics the slow cooling appearing in the annealing process [2], analogous to the slow convergence and accepting more solutions. SA has been vastly used for path planning problems over the past years. Some research studies, referring to this algorithm, represented the robot's path by using a Voronoi diagram. A particular research used the Voronoi diagram to find a collision-free path by using Dijkstra's algorithm, thus utilizing this data to compute the best path by applying SA, satisfying the kinematic constraints [3]. Results have proved that this approach gives better results than the traditional SA algorithm [3]. SA has also been combined with other metaheuristic techniques in order to solve the path-planning problem for mobile robots. A concrete study used the genetic algorithm (GA) as an acceleration to the SA algorithm [4, 5]. Some researches presented that this hybrid approach of the GA and SA algorithms avoids premature convergence and gives better results [6]. In the end, there's a research paper stating that SA algorithm, used solely and compared to other metaheuristic techniques, was able to always find a solution and proved to be practical and effective, but compared to tabu search couldn't always find the shortest possible path [7].

Particle Swarm Optimization (PSO) [8] is an algorithm that is also one of the well utilized metaheuristic approaches, and it's based on imitating the social behaviors, where individual's capabilities are less valued than the global social interactions [9]. Global path planning has been previously solved by utilizing a quantum-behaved PSO algorithm (QPSO). Here, the robot's map between the initial and the end point is represented with coordinate system transferring. The algorithm has performed with accelerated convergence having no restrictions regarding the shapes of the obstacles [10]. In the past several years, there have been also many hybrid approaches which combined the PSO algorithm with other metaheuristic algorithms. These approaches have proven their efficiency through several experimental studies. One of them is the hybrid GA-PSO algorithm, which used the genetic algorithm as an acceleration to the existing PSO algorithm. This algorithm applied the mutations and crossover operators on the generated results from the particles. After finding a feasible route, a cubic B-spline technique is utilized to produce a smoother and better solution. This method has avoided the drawbacks of an early convergence, which is typical for these metaheuristic algorithms when used separately [11]. There had been also an orthogonal PSO algorithm, which included the orthogonal design operator to the simple PSO algorithm in order to avoid falling into local optima. Compared to the traditional PSO algorithm, this approach also gave more effective solutions [12]. When the path space model of the mobile robot is transformed by decomposition of the two-dimensional representation of the route, pairs of particles are capable of exchanging information for the crossover operator. This led to avoiding falling into local optima and giving feasible and reasonable solutions [13]. Very recent approaches took the path planning problem to four dimensions, thus defining the problem as a calculus of variation problem (CVP). Then, the solution to the CVP is effectively provided by applying the PSO algorithm [14].

The Ant Colony Optimization (ACO) algorithm is also a metaheuristic algorithm, which is inspired by the behavior of ants and their process of seeking food [15]. Hybrid approaches and implementations have been very popular, one of them being a combination of ACO and genetic algorithm (GA), referred to as smartPATH [16]. This concrete hybrid algorithm includes improvements of the ACO algorithm and modified crossover operator for the GA part of the algorithm, thus avoiding the inclination towards local minima. Results have shown that this algorithm performs much better that the standard ACO algorithm and also the Bellman-Ford method [16]. A Two-way ACO algorithm has been applied to the global path planning problem in a static environment, such that there are two ant tribes walking in opposite directions from starting and ending point [17]. The heuristic information was then gathered by the initial point, destination point and ants' movements [17]. An endpoint approximation method proved to give efficient results, which consisted of constantly moving the starting and ending point of the grid towards each other, so the convergence obtained an accelerated speed [18]. A type of ant colony system with potential field heuristics provided effective construction of the robot's path planning and avoidance of environmental obstacles [19]. In a late research study, cellular ant colony algorithm consisted of two ant colonies running with different strategies [20]. Then, these paths were evolved by using cellular rules, so the ants could jump to the region which leaded to the solution, thus resulting in a more stable algorithm [20]. A Best-Worst ant colony method also improved the searching processes significantly, and a Max-Min system for limiting the global pheromone intensity provided applicable solutions to the path planning problem [21].

Genetic algorithms (GA) is a technique which makes the best use of the search by implementing the concepts of natural evolution [22]. A combination of the Anytime Planning criteria and multi-resolution search spaces in the genetic algorithm resulted with several parallel evolutions, thus exchanging information between these parallel search threads about the low-cost solutions in the environment and improving the overall convergence [23]. Recent improved GA constituted of several optimizations to this algorithm. A hill-climbing method was firstly used to improve the mutation of the algorithm. Then, particle swarm optimization (PSO) was utilized to speed up convergence, and in the end an emulation takes place with float-point coding involved in the improved GA [24]. A dual population evolution with proportion threshold adaptation when having a fixed length binary path improved the capability of the algorithm to gravitate towards global optimal solutions and accelerated the convergence speed as well [25]. Introducing an adaptive local search operator and applying an orthogonal design method in the process of the initialization of the population and including intergenerational elite mechanism, also gave great accelerations compared to the original GA [26]. There have been research studies that implemented a hybrid of Artificial Potential Fields (APF) and GA, which uses the APF method to determine the collision-free area and avoid the environmental obstacles, and then it applies the fitness function of the GA accelerated by least-square curve fitting [27]. This combination has been applied to soccer robots' path planning and obtained effective results in terms of finding an optimal route [27].

3   Quad-harmony Search Algorithm

Given that the distance between the start and end destination has to be minimized by the algorithm by which a search agent operates, the global path planning in grid-based environments can be defined as a Linear Programming problem [28]. Given a graph (G, A), start node s, goal node t, and cost w_{ij} for every $arc(i,j)$ that exists in A, the problem would be defined as follows:

$$\min \sum_{ij \in A} \omega_{ij} x_{ij}, \tag{1}$$

where $x \geq 0$ and for all i:

$$\sum_i x_{ij} - \sum_j x_{ji} = \begin{cases} -1, & \text{if } i = s \\ 1, & \text{if } i = t \\ 0, & \text{otherwise} \end{cases} . \tag{2}$$

This research paper presents a novel Quad-Harmony Search (QHS) algorithm which is effectively applied to the global path-planning problem. By using the Quad-tree free space decomposition algorithm, this algorithm splits the environmental data of the agent into four equally sized sub-grids in a recursive manner, until it finds a single cell or an obstacle-free sub-grid. In this phase, all of the free regions are labeled with proper numbers and treated as a single node later in the Harmony Search (HS) stage of the algorithm. The HS stage consists of applying the Harmony Search method [29] to find an optimal route, given the data provided from the Quad-tree stage of the algorithm. The search space of the agent is static and the agent is able to move in four different directions, namely labeled as up, right, down and left.

The QT stage of the algorithm used for the QHS is executed in the following manner:

1. Give the necessary discretized environmental data to the mobile robot as input.
2. Check if there is a single cell. If this is the case, determine whether it's a free rectangle or an obstacle. If it's empty, label it with a unique number. Otherwise, label it with -1.
3. Separate the current examined part of the environment into four equally sized regions.
4. Explore these regions and check if they are obstacle-free:

   a. When all of the regions are obstacle-free, label them with the same unique number.
   b. When two adjacent regions are obstacle free, label them with the same unique number and repeat the recursion steps from Step 2.
   c. When 4.a) and 4.b) result with a false outcome, repeat the recursion steps for all of the regions from Step 2.

The result of the application of the QT stage of the algorithm on a sample grid-based environment (Fig. 1) is shown at Fig. 2.

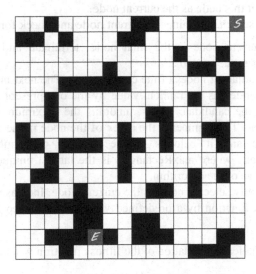

Fig. 1. Sample grid-based environment

2	2	-1	3	5	5	6	6	-1	-1	23	23	-1	-1	28	28
-1	-1	4	-1	5	5	7	-1	-1	22	23	23	27	27	28	28
1	1	1	1	5	5	8	8	24	24	25	-1	29	29	-1	30
1	1	1	1	5	5	-1	9	24	24	25	26	-1	-1	31	-1
10	10	11	-1	-1	-1	-1	-1	32	32	32	32	36	36	-1	38
10	10	-1	12	17	17	18	18	32	32	32	32	-1	37	-1	38
13	13	-1	15	19	19	20	20	33	-1	-1	35	39	39	41	41
-1	14	16	15	19	19	21	-1	33	34	-1	35	40	-1	-1	42
-1	-1	44	-1	-1	-1	51	-1	-1	65	-1	67	73	73	-1	74
43	43	-1	45	50	50	51	-1	66	65	68	67	73	73	75	74
46	-1	-1	48	52	52	54	-1	69	69	71	71	-1	76	-1	77
-1	47	49	48	-1	53	54	55	70	-1	-1	72	-1	76	78	77
-1	-1	-1	-1	-1	-1	60	60	79	79	80	80	84	84	84	84
56	56	57	57	-1	-1	60	60	-1	-1	80	80	84	84	84	84
58	58	-1	-1	-1	61	-1	63	-1	-1	-1	82	85	85	-1	-1
58	58	59	-1	62	61	64	63	81	81	83	82	-1	-1	-1	86

Fig. 2. The labeled grid from Fig. 1 – the output of the QT phase

The results from the QT stage of the algorithm are fed as input to the Harmony Search stage of the QHS algorithm. Here, the input is used to construct an adjacency list which should be used for the graph search of the algorithm. Then, these data are utilized for the purposes of the fitness function of the HS algorithm. The value of the fitness function is then computed as follows:

1. Initialization. Set the start node of the algorithm to the initial point of search of the mobile robot. Set this node as the current node.
2. Adjacent nodes search. Examine the current node and check for its label:

   a. When the current node is the final node, terminate and exit with success. Otherwise, continue to step 2.b.
   b. Examine the adjacent nodes of the current node. The next node to be selected is the node generated as the random value by the one present in the current value (member) of the candidate vector, modulus the maximum number of labelled rectangle areas. This number is the index of the node in the adjacency list of the current node and it belongs to the domain of numbers given by [1; maxRectangles], where maxRectangles is the largest unique number generated by the QT stage of the algorithm.
   c. Set the randomly selected node as the current node. Increase the returning value of the fitness function by a predefined value. Repeat Step 2 for this particular node.

Hence, the mathematical formulation of the fitness function would be interpreted as in Eq. 3.

$$f(.) = \sum_{g \in G} NodeCost(g),$$

(3)

where g is the current examined node, G is the domain set of the g node, and NodeCost(g) stands for the cost of the g node related to its position compared to the destination point, defined with Eq.4.

$$NodeCost(g) = \begin{cases} 1, & g \text{ is before the destination node} \\ 0, & g \text{ is a destination node, or is after.} \\ & \text{the destination node} \end{cases} \tag{4}$$

For the purposes of our research, several types of grids were tested by using the QHS technique. These grids had sizes of 32x32, 64x64 and 128x128, and they had 10%-90% percentage of obstacles. These results from the QHS algorithm were then compared to other two metaheuristic approaches, namely the Ant Colony Optimization (ACO) and Genetic Algorithm (GA). Results from this research are detailed and presented in Fig. 3, Fig. 4 and Fig. 5 for the grid sizes of 32x32, 64x64 and 128x128 respectively. As it can be clearly seen, the QHS algorithm has performed with great acceleration in convergence compared to other metaheuristic approaches. The GA needed a lot more iterations to converge, and always found only a local optima, i.e. it always found little variations of the first obtained feasible path, whereas the ACO algorithm always found a global optima, but needed more iterations to provide an optimal solution.

Also, for all of the test grids used for this research study, a check for optimality of the obtained solutions by HS was performed. This check was made with the help of the well-known Breadth First Search (BFS) algorithm. Lengths of the optimal paths obtained using Breadth First Search and Harmony Search algorithms were the same, and they are presented in Table 1. These results, along with the comparison to the other metaheuristic algorithms, make QHS algorithm the right choice for the global path planning problem of a search agent.

Table 1. Lengths of the optimal paths obtained using Breadth First Search and Harmony Search algorithms

Percentage of obstacles	Grid sizes		
	32x32	64x64	128x128
10%	66	133	74
20%	61	114	71
30%	66	100	89
40%	50	234	66
50%	51	38	53
60%	19	22	30
70%	9	11	47
80%	5	8	28
90%	7	4	59

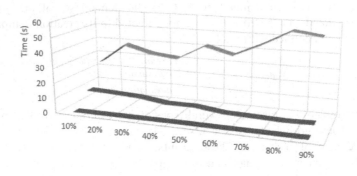

Fig. 3. Comparison of QHS, Ant Colony and Genetic Algorithm for 32x32 grid sizes

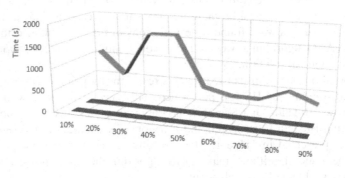

Fig. 4. Comparison of QHS, Ant Colony and Genetic Algorithm for 64x64 grid sizes

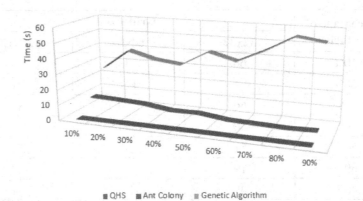

Fig. 5. Comparison of QHS, Ant Colony and Genetic Algorithm for 128x128 grid sizes

4   Conclusion

This paper presented a novel metaheuristic approach to solving the global path planning problem, namely the QHS algorithm. This algorithm was founded on the effective search space division capabilities of the Quad-tree algorithm and the optimization capabilities of the Harmony Search algorithm. Different grid sizes and different percentages of obstacles have been examined and experimental results have been elaborated. This research has shown that this algorithm performs better than other metaheuristic methods, namely the ACO and GA algorithms in terms of time and acceleration of convergence. Hence, this algorithm is a great foundation for further researches in the field of metaheuristic algorithms applied to NP-complete problems, such as the global path planning problem detailed in this paper.

References

1. Lau, B., Sprunk, C., Burgard, W.: Efficient grid-based spatial representations for robot navigation in dynamic environments. Robotics and Autonomous Systems (2012)
2. Aarts, E.H., Korst, J., Van Laarhoven, P.J.: Simulated annealing. Local Search in Combinatorial Optimization, 91–120 (1997)
3. Ho, Y.J., Liu, J.S.: Simulated annealing based algorithm for smooth robot path planning with different kinematic constraints. In: Proceedings of the 2010 ACM Symposium on Applied Computing, pp. 1277–1281. ACM (2010)
4. Liang, Y., Xu, L.: Global path planning for mobile robot based genetic algorithm and modified simulated annealing algorithm. In: Proceedings of the first ACM/SIGEVO Summit on Genetic and Evolutionary Computation, pp. 303–308. ACM (2009)
5. Du, Z.Z., Liu, G.D.: Path Planning of Mobile Robot Based on Genetically Simulated Annealing Algorithm 12, 36 (2009)
6. Wang, H.B., Yang, W.J., Wang, J.H.: Research on Path Planning for Mobile Robot Based on Grid and Hybrid of GA/SA, vol. 479, pp. 1499–1503. Trans. Tech. Publ. (2012)
7. Hussein, A., Mostafa, H., Badrel-din, M., Sultan, O., Khamis, A.: Metaheuristic optimization approach to mobile robot path planning. In: 2012 International Conference on Engineering and Technology (ICET), pp. 1–6. IEEE (2012)
8. Kennedy, J., Eberhart, R.: Particle swarm optimization. In: Proceedings of the IEEE International Conference on Neural Networks, vol. 4, pp. 1942–1948. IEEE (1995)
9. Poli, R., Kennedy, J., Blackwell, T.: Particle swarm optimization. Swarm Intelligence 1(1), 33–57 (2007)
10. Liu, W.K.Z.H.Y., Zhi-lei, C.: Path Planning for Robots Based on Quantum-behaved Particle Swarm Optimization. Microcomputer Information 11, 066 (2010)
11. Huang, H.C., Tsai, C.C.: Global path planning for autonomous robot navigation using hybrid metaheuristic GA-PSO algorithm. In: 2011 Proceedings. SICE Annual Conference (SICE), pp. 1339–1343. IEEE (2011)
12. Qian-Zhi, M., Xiu-Juan, L.: The application of hybrid orthogonal particle swarm optimization in robotic path planning. In: 2010 Sixth International Conference on Natural Computation (ICNC), vol. 7, pp. 3536–3540. IEEE (2010)
13. Li, W., Wang, G.Y.: Application of improved PSO in mobile robotic path planning. In: 2010 International Conference on Intelligent Computing and Integrated Systems (ICISS), pp. 45–48. IEEE (2010)

14. Ma, Y., Zamirian, M., Yang, Y., Xu, Y., Zhang, J.: Path Planning for Mobile Objects in Four-Dimension Based on Particle Swarm Optimization Method with Penalty Function. Mathematical Problems in Engineering (2013)

15. Dorigo, M., Birattari, M., Stutzle, T.: Ant colony optimization. IEEE Computational Intelligence Magazine 1(4), 28–39 (2006)

16. Chaari, I., Koubaa, A., Bennaceur, H., Trigui, S., Al-Shalfan, K.: smartPATH: A hybrid ACO-GA algorithm for robot path planning. In: 2012 IEEE Congress on Evolutionary Computation (CEC), pp. 1–8. IEEE (2012)

17. Xianlun, T.A.N.G., et al.: Ant Colony Optimization Based on Maximum Selection Probability for Path Planning in Unknown Environment. Journal of Computational Information Systems 8(24), 10325–10332 (2012)

18. Wang, P.D., Tang, G.Y., Li, Y., Yang, X.X.: Ant colony algorithm using endpoint approximation for robot path planning, pp. 4960–4965. IEEE (2012)

19. Luo, D.L., Wu, S.X.: Ant colony optimization with potential field heuristic for robot path planning. Systems Engineering and Electronics 32(6), 1277–1280 (2010)

20. Wu, Y.F., Zhang, X.X., Wu, J.Q.: Using Cellular Ant Colony Algorithm for Path-Planning of Robots. Applied Mechanics and Materials 182, 1776–1780 (2012)

21. Qiao, R., Zhang, X.B., Guang-xing, Z.H.A.O.: Global Path Planning of Mobile Robot Based on Improved Ant Colony Algorithm. Journal of Anhui University of Technology (Natural Science) 1 (2009)

22. Goldberg, D.E.: Genetic algorithms in search, optimization, and machine learning (1989)

23. Lucas, D., Crane, C.: Development of a multi-resolution parallel genetic algorithm for autonomous robotic path planning. In: 2012 12th International Conference on Control, Automation and Systems (ICCAS), pp. 1002–1006 (2012)

24. Liu, C., et al.: Dynamic path planning for mobile robot based on improved genetic algorithm. Chinese Journal of Electronics 19(2), 2010–2014 (2010)

25. Hua, J.M.W.H.Z., Xingzhe, X.: Applying improved genetic algorithm to global path planning for mobile robot. Computer Applications and Software 8, 033 (2011)

26. Yan, X.: An Improved Robot Path Planning Algorithm. TELKOMNIKA (Telecommunication, Computing, Electronics and Control) 10(4), 629–636 (2012)

27. Xu, X., Xie, J., Xie, K.: Path planning and obstacle-avoidance for soccer robot based on artificial potential field and genetic algorithm. In: The Sixth World Congress Intelligent Control and Automation, WCICA 2006, vol. 1, pp. 3494–3498 (2006)

28. Schrijver, A.: Theory of linear and integer programming. Wiley (1998)

29. Geem, Z.W., Kim, J.H., Loganathan, G.: A new heuristic optimization algorithm: harmony search. Simulation 76(2), 60–68 (2001)

Evaluation of the Implemented System of Issuing B Integrated Permits Based on Open Government Concept

Milos Roganovic[1], Ivan Bisevac[1], Bratislav Predic[1], Dolores Ananieva[2],
Ilco Trajkovski[2], and Dejan Rancic[1]

[1] Faculty of Electronic Engineering, University of Niš, Serbia
{milos.roganovic,ivan.bisevac,bratislav.predic,
dejan.rancic}@elfak.ni.ac.rs
[2] Asseco SEE Macedonia, Macedonia
{dolores.ananieva,ilco.trajkovski}@asseco-see.mk

Abstract. This paper provides an overview of current initiatives for the development of e-government originating from Europe and United States of America. This study emphasizes the importance of transformation of e-government, from government-oriented services (e-Government 1.0), through the user-oriented (e-Government 2.0), to an open, transparent administration that involves citizens and legal entities in the process decision-making. In this paper we propose configurable framework for fast development of systems that covers defined process of exchanging information and documents among citizens and legal entity on one side and employer in government institution on other side. Proposed framework is build around atomic building block named state with defined requirements, connections between blocks and loops. Based on proposed framework we developed system for issuing B integrated permits. Developed system serves as a confirmation of successful implementation of the open government concept.

Keywords: e-government 2.0, web 2.0, open government.

1 Introduction

In recent years with the advancement of internet technology and expansion of mobile devices, information has become available almost everywhere. Volumes of government data are constantly increasing which makes publishing and managing governmental data a very tricky and demanding task. Using widely available technology [1] [2] we will try to increase efficiency of public government institution. Going from user oriented (e-Government 2.0) [3] one step forward to transparent government that involves citizens and legal entities in process, we achieve more efficient communication in both direction. Also transparency involve some kind of global knowledge into common government processes. Looking from user perspective, before user start some common procedure (e.g. issuing some permit) he can find a lot instructions that government purpose on web, but these approach include live examples of similar

V. Trajkovik and A. Mishev (eds.), *ICT Innovations 2013*,
Advances in Intelligent Systems and Computing 231,
DOI: 10.1007/978-3-319-01466-1_12, © Springer International Publishing Switzerland 2014

procedures that other user has start before. All kind of permits has some kind of restrictions that need to be accomplished by issuer, and also in these aspects transparency can help government to react in real time according to citizens complaints. We try to identify all common things that is usual for specific kind of government procedures that require frequent interaction among employees in government, citizens and legal entities. Our focus is to develop such a framework that is suitable for defining business process on high level of abstraction which will produce in final step a system that could be published on Web and become widely available.

The remainder of this paper is organized as follows. Section 2 presents new initiative that is propagated from EU. Section 3 explains proposed framework for defining e-government business process. Section 4 gives implementation details and provides example of concrete product. Finally, Section 5 draws some conclusions and outlines future work.

2   New Initiative

Europe 2020 [4] is a new strategy in the European Union that will be applied the next decade. Strategy was adopted on 17 Jun 2010 and aims to create a smart, sustainable and inclusive economy of the European Union which will create new employment opportunities, better productivity and social cohesion. The Strategy sets out five key areas that need to reach a high level of development up to the year 2020, namely: employment, innovation, education, social inclusion and climate / energy. Concrete action at national and EU level are necessary to support this strategy. Each Member expected to adopt individual national targets belong to the defined areas. Within the strategy defined seven initiatives (Digital Agenda, Innovation, Youth on the Move, Efficient use of resources, Industrial policy in an era of globalization, Agenda for new skills and jobs, the platform against poverty) that must be implemented to achieve smart, sustainable and inclusive economy.

The first initiative was developed, adopted Digital Agenda [5] as an EU strategy for the development of digital economy by the end of 2020th year. Digital Agenda focuses on the technologies of the 21st century, and Internet services that Europe will allow the creation of new jobs, promote economic prosperity and improve quality of life of citizens and legal entities. Within the Agenda are defined seven priority areas for action: creation of a single digital market, the greater interoperability, enhancing confidence and security in the Internet, much faster internet access, more investment in research and development, improving digital literacy and digital inclusion and use of information and communications technologies to address challenges facing society (climate change, aging population).

Objective 1: Better access to information and active participation of citizens and legal entities in creation of the policy, creation of services oriented to needs, collaborative work of users to create services, reuse of public sector information, increase transparency, involvement of citizens and legal entities in decision-making,

Objective 2: Mobility in the single market, which is achieved by using electronic services for the establishment and starting a business, study, employment, change of residence, cross-border services for businesses;

Objective 3: To reduce administrative barriers, improvement of organizational processes and the promotion of "green government" that supports low CO_2 emissions;

Objective 4: To create the necessary key factors and   prerequisites in order to meet the expectations of the declaration date, the use of open platforms and the introduction of interoperability, creating a unique identification of citizens and legal persons of the EU, an innovative e-Government;

Considering ISA program of EU [6], and according to action plan of Serbia [7], we proposed framework for defining e-government business processes as solution for building web based applications that are mentioned in action plan.

3 Proposed Framework for Defining e-Government Business Process

Our proposed framework includes several steps in defining business process.

3.1 Identifying Roles of the System

In these step we classify users of the system by roles, each role could be public or specific type. We define all possible role types of the system; some roles could be overlapped or in real situation user could belong to more than one role.

3.2 Defining Distinguished States of the System

States are defined by name and requirements that has to be accomplished before state can be transformed to connected states. Each state has assigned roles that can access to defined state. Notifications are also part of the state, it represent an action that is performed each time when process current state become defined state. Typical notification is sending an e-mail message to user that belongs to particular role, and is also connected to state.

3.3 Defining Order of the Steps

After defining all states we specify connectors that represent possible actions that transform particular state to maximum one state defined before.

3.4 Defining Loops in Process

Loops are specific type of connectors that connect state to another state that is chronologically earlier.

Based on proposed framework we define several processes as illustrated below.

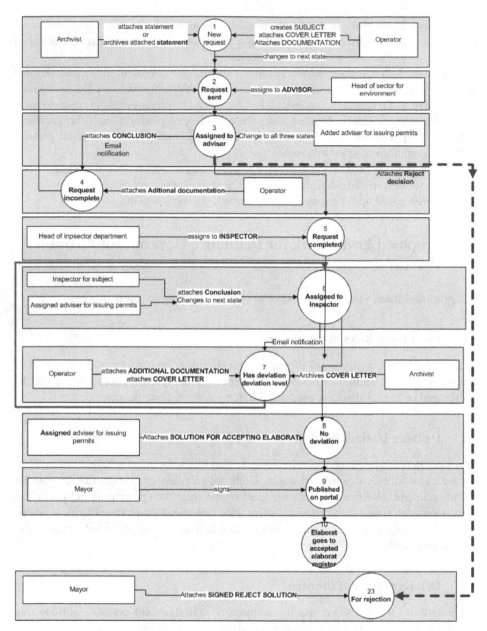

Fig. 1. Defining e-government business process using proposed framework

4   Electronic Permits Issuing System

Based on proposed framework we develop system for electronic issuing of B integrated permits, widely available on address http://ekoloska-dozvola.mk/

Fig. 2. Electronic issuing of B integrated permits

Electronic permits issuing system described in this paper covers whole country Macedonia but can be extended to larger area. It's divided on smaller entities (municipalities) in which employees are assigned to roles. Flexible user and role management enables making areas of responsibilities of different participants involved in the process, including both citizens and government representatives, thus improving communication between interested parties. User management is simplified and easy to use, only two types of roles are able to manipulate with these part of the system: Global administrator, and local administrators.

Operator in charge:	Operator1 : Bratislav Predić
Reference number	36
Municipality	Чаир
Company name	Компанија
Company legal status	јкп
Company and land ownership	Власник компаније
Number of employees	5
Company representative	Представник
Designed capacity	5
Request type	New installation
Information for the authority in charge	Информација за одговорно лице
Main activity code	444

Public hearing
No notes

Reject application

Fig. 3. An example of unfulfilled conditions for transferring process from current state

Ultimate goal (issuing permit) is achieved through iterations. Every iteration has defined constraints that must be satisfied in order to change to next state. For example in every state after adding electronic document archivist should archive it.

Fig. 4. Notification e-mail to specific user according to state transformation

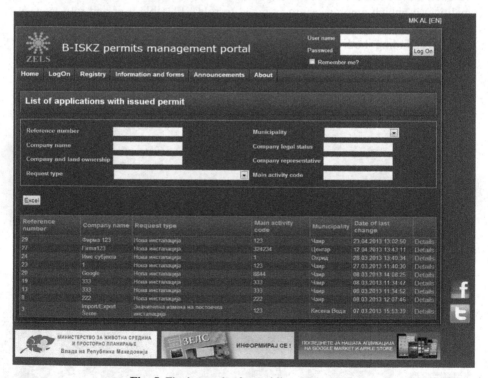

Fig. 5. Final state that has public role attached

User involved in current iteration has access to status of permit through the system. Also there are email notifications about changes. They can be parameterized to speed up whole process and increase efficiency.

The end product of whole process is issued permit which could be accessed by all citizens in permits registry. But citizens are not passive spectators, they could be involved in whole process during public hearing. It's essential for such electronic government system because it gives opportunity to improve quality of process and representatives to get feedback from them.

We find that citizens are generally satisfied with the electronic provision of information (transparency). Electronic government strategies - transparency and interactivity - are important factors that directly affect e-government satisfaction and indirectly affect trust. Individuals who use system for issuing permits are not only critical consumers but also demanding citizens. Citizens are belonging to certain public role, they can access to all processes that reach final state.

Like the evaluation of all other information systems initiatives, the evaluation of e-government in both theory and practice has proven to be important and complex. The complexity of evaluation is mostly due to the multiple perspectives involved, the difficulties of quantifying benefits, and the social and technical context of use. The importance of e-government evaluation [8] is due to the enormous investment put in by governments for delivering e-government services and to the considerable pace of growing in the e-government field. However, despite the importance of the evaluation of e-government services, the literature shows that e-government evaluation is still an immature area in terms of development and management. Our system involves reevaluation over certain time periods in which users permanently connect with their subject.

System for managing the B-integrated licenses is created using ASP.NET MVC [9] technology and Entity framework [10] for data model. In addition were used HTML, CSS, Javascript (jQuery library).

According to proposed list of actions [11], proposed system satisfies security requirements mentioned in Action 1, also framework is reusable component mentioned in Action 2, and system could be also integrated within some other system mentioned in Action 3.

5   Conclusion and Future Work

The transition of e-government, the concept of e-Government 1.0 to Open Government, occurred in the last decade. To short period of a few years, worn appearance new Web-based technologies and increase awareness users about their importance in the decision-making, e-government services grew from 1.0 into an open, transparent and participatory government.

A lot of researches are made in e-government domain in recent years. We try to increase efficiency of public government institution using widely available technology. Our approach has intention to design flexible framework for building business processes of government institution. In future work our focus will be using citizens as sensor nodes to alert government institution about changes that has influence on society in order to improve interaction.

References

1. Murugesan, S.: Understanding Web 2.0. IT Professional 9(4), 34–41 (2007)
2. O'Reilly, T.: What Is Web 2.0: Design Patterns and Business Models for the Next Generation of Software. Communications & Strategies (1), 17 (2007)
3. Eggers, W.: Government 2.0: Using Technology to Improve Education, Cut Red Tape, Reduce Gridlock, and Enhance Democracy. Rowman & Littlefield Publishers, Lanham (2005)
4. Europe (2020) Strategy, http://europa.eu/rapid/pressReleasesAction.do?reference=ip/10/225&format=pdf&aged=1&language=en&guiLanguage=en
5. EU Digital Agenda, http://eur-lex.europa.eu/LexUriServ/LexUriServ.do?uri=com:2010:0245:fin:en:pdf
6. ISA program of the EU, http://ec.europa.eu/isa/
7. Serbian action plan 2012/2013, http://mtt.gov.rs/download/za_clanke/Akcioni%20plan%20za%20realizaciju%20strategije%20razvoja%20informacionog%20drustva%202013-2014.pdf
8. Alshawi, S., Alalwany, H.: E-government evaluation: Citizen's perspective in developing countries. Information Technology for Development 15(3), 193–208 (2009)
9. ASP.NET MVC, http://www.asp.net/mvc
10. Entity framework, http://msdn.microsoft.com/en-us/data/ef.aspx
11. European Interoperability Architecture (EIA), Phase 2 – Final Report: Common Vision for an EIA, http://ec.europa.eu/isa/documents/isa_2.1_eia-finalreport-commonvisionforaneia.pdf

Data Mashups of Debates
in Parliament, Public and Media

Mile Grujovski

Nextsense, Partizanski Odredi 62, Skopje, Macedonia
mile@nextsense.com

Abstract. Debates on draft laws during the lawmaking procedure are primarily conducted in the Parliaments, but the media also gives their view on the subject, and lately discussions are being held online on the social networks. Usually those three viewpoints are identical, however in modern democracies citizens and media can have diametrically opposite positions with parliamentarians. In this paper we describe a prototype system that collects and combines data from three different data sources and presents analytical results to the users. As data sources we use: website of the Parliament, online news media, and Twitter network. Results of the analysis are visually communicated to the users with word cloud and word tree representations. Interoperable application interface is available for integration with external systems.

Keywords: Parliament, Debates, Visualization, Word Cloud, Word Tree.

1   Introduction

The term mashup here refers to web application that combines data from multiple information sources. First source is a website of the Parliament of Republic of Macedonia that provides transcript documents from the debates. Second source is news articles in news media and the third source is posted messages on Twitter network. Transcript documents from the first source represent the view on the subject by the parliamentarians. The second source provides interpretation by the Media and the third source is open online discussions by citizens and experts. The general problem that we are solving in this paper is the design of a system for aggregation of information from those sources on one side, and analysis and visualization of combined results on the other side.

Main objective of the modern parliaments today is to make lawmaking processes as much transparent as possible. Global Center for ICT in Parliaments [3] compiles annual report and defines directions for development in parliaments. One of the main recommendations of the latest report is improving transparency of the legislative processes by enhanced access to information. In this manner the Assembly of Republic of Macedonia publishes online all the documents related to the work of the

V. Trajkovik and A. Mishev (eds.), *ICT Innovations 2013*,
Advances in Intelligent Systems and Computing 231,
DOI: 10.1007/978-3-319-01466-1_13, © Springer International Publishing Switzerland 2014

assembly and its committees [6]. In our solution[1] software agent loads transcripts from Parliament to our database for analysis and processing.

Today the internet and social networks are considered to form a distinct type of opportunity for political participation which considerably diverges from traditional participation channels, they possesses the ability to inform, activate and engage citizens [15]. One unique characteristic of the Twitter micro-blogging network is its real-time notification nature. This characteristic of the network is utilized during the earthquake detection where Twitter messages are used as social sensors [11]. Here we are mainly interested in discussions on the topics of policy and political issues including criticism. For this purpose we use Twitter messages from manually selected set of Twitter users for analysis in our system.

The lack of response from the reader in the traditional written media is overwhelmed in their online counterpart. Besides initiating discussion, online media offers opportunity for readers to comment and discuss on the content. News media provide a resource for political discussion and create opportunities for exposure to conflicting viewpoints, encouraging political talk that might not otherwise occur [12]. Those online resources represent opinion from the journalists in our analysis.

The resulting statistics on what is important for voters on one side and media on other side can be valuable information for the delegates in the parliament. In case when some subject receives greater attention in the online Media and Twitter than in the Parliament discussions, that can indicates inconsistency and potentially diametrically different opinions.

As defined by Liu et al. [9] this combination of different data generates new value from information from external data sources. We use Word Cloud visualization for representation of the results [1][16][2]. This service practically represents a tool for statistical analysis of the discussion documents [5]. This service provides valuable results to our primary users – citizens [7], and: media, journalists and analysts as secondary users.

In Section 2 we will provide a brief overview of similar solutions, Section 3 outlines the main functional modules. Data structures, algorithms and interactivity are covered in Section 4. Section 5 discusses the advantages, disadvantages and future work.

2   Related Work

At 2007 the newspaper New York Times introduced a web based tool called "Transcript Analyzer" that provided visual representation of text records of debates. This tool calculates total number of words for each speaker in the debate. Statistical information like total speaking time for participant and overall total speaking time information is presented. Unfortunately this tool is only available for selected transcript documents of some debates and it does not provide option for uploading document for analysis or any other integration with external sources of transcript documents.

Many Eyes (Viégas, et al., 2007) is another online visualization tool. Data is manually uploaded by the user and then rich visualizations can be generated. This service successfully solves the problem of interpretation of statistical results to the

---

[1] Visual Mashups. http://stenogrami.com

user, but in order to perform analysis from different sources user have to manually compile a single document and upload it to the service where processing is performed.

Tropes[2] is tool for analysis and visualization of textual contents. This tool performs highly efficient semantic analyses and excellent visual representation of the results. We found two disadvantages of this tool: first it is implemented as an application that needs to be installed on the client computer and second semantic analyses require language specific knowledge. TIARA system [8] is using Latent Dirichlet Allocation model for semantic analysis that produces visual text summary of large text corpora. The time-based visual representation of TIARA combined with word cloud visualization could be applied in our solution.

The real-time nature of Twitter messages is successfully applied in earthquake detection (Sakaki, Okazaki and Matsuo, 2010). Here twitter messages are used as social sensors for real-time event. In our work we are primarily interested in political talk in Twitter messages. The category political here is used to distinguish messages on subject of policy or political issues that can also include criticism. Graham [4] is working on a filter for detection of political talk online. Šķilters et al. [13] has analyzed the messages during the Latvian parliamentary election in 2010 and discovered different polarization categories of Twitter messages. By using the polarization factor it is possible to structure and classify messages. In our research we measure the topic of interest by calculating the frequencies of occurrences of words in Twitter messages. The recent research [14] on the use of Twitter by the Canadian politicians has indicated that though many Canadian politicians are using Twitter, it is mostly used to broadcast official party information. Considering this result, we decided to evaluate twitter discussions not from the politicians but from manually selected active users.

Exposure to different views is vital in forming valid opinions and in learning to appreciate the perspectives of others [10]. Both in political theory and empirical work, there is near unanimous agreement that exposure to diverse political views is good for democracy and should be encouraged. Aldo some authors [17] argue that social networking sites does not increase participation in political discussions, we believe that over time this will increase.

All referenced systems operate on a data set that is provided by the user. The biggest advantage of our proposed service is that data is automatically aggregated and combined from different source locations: parliamentary web site, news media and Twitter network. This allows us to create local data repository and execute more complex statistical computations.

3   Service Modules

Functional modules in our service are organized around three main problems: transmission and transformation of data, indexing and storing collected data, and statistical analysis and visualization.

---

[2] http://www.semantic-knowledge.com

3.1   Module for Transferring Data

The main advantage of this solution is the continuous integration with the actual sources of data. That means automatic retrieval and transfer of following data: transcript documents from parliamentary web site, twitter messages from twitter network, and online news articles from selected sources. We define two processing locations as: "source" meaning: web site of the Parliament, Twitter network or Media website, and "service" meaning processing location of the visualization service. The idea for local replication of data is based on following considerations:

1. Visualization service is independently isolated from the original data sources
2. Fast data loading
3. Advanced analytical queries can be executed on a local dataset

A dedicated agent software component is responsible for actual fetching of data from different data sources. We have implemented separate agents for each data source that we use.

3.2   Module for Indexing Data

Index structure in our service is used for executing analytical queries required by the analysis module. The indexed content contains: plain transcript text, article contents and twitter messages. Transcripts documents are published on the parliamentary web site in different file formats: Portable Document Format (PDF), Microsoft Word Binary Format (DOC) or Office Open XML (DOCX). As such document needs first to be converted into readable plain text format. Document content is stored as text in the relational database and also it is indexed by full text indexer.

3.3   Visualizations Generator Module

Our service supports three visualization types: Word Cloud, Word Tree and Word Distribution Chart. Word Cloud representation shows set of words arranged on the viewing area respectively each with a different size. The size and position of each word is proportional to the term frequency. Figure 1 is Word Cloud visualization from transcript document generated from our service.

We base this visualization on the following aesthetic principles:

A_1. Words with higher frequencies are more noticeable
A_2. Words with higher frequencies are positioned toward the center
A_3. Favor symmetry and balance

Word Tree visualization shows a structure of all the words that are followed by a given word or phrase. Tree structure is called suffix tree and is not directly observable without previous computation and analysis. Suffix tree is deduced from the document content and is different for each root word or phrase. Wattenberg and Viégas (2008) define it as a form of keyword-in-context (KWIC) structure. Figure 2 is an example of word tree visualization from real transcript document.

Fig. 1. Word Cloud visualization from our service

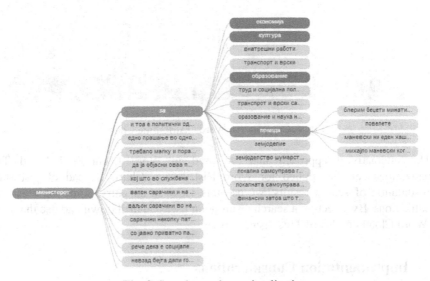

Fig. 2. Sample word tree visualization

The advantage of Word Tree view is that it instantly visually communicates the sentence structure for the given root word or context. On the other side disadvantage is that user needs to enter the root word. In order to surpass this disadvantage we propose that entry in the word is a word from Word Cloud visualization.

In Word Distribution chart an analysis is executed over a set of documents. Let's consider this simple scenario: A user wants to browse the discussions where act on "e-commerce" was discussed, instead of going through each transcript document and evaluate, he can execute a new Word Distribution Chart search and the results show where this subject was most argued. Figure 3 displays this visualization.

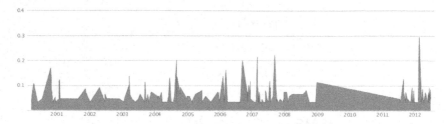

Fig. 3. Word Distribution Chart

In Word Distribution Chart X-axis represents the date of discussion and Y-axis is relevancy of the search results. We propose a further extension of this chart with functionality for comparative analysis. The idea is to analyze more than one word and their distribution in transcript documents. TIARA (Liu et al. 2012) is using similar visual metaphor for time-based visualization. Figure 4 presents results from comparative word distribution chart.

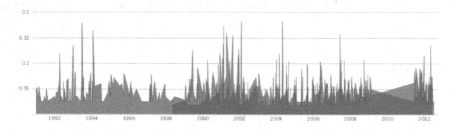

Fig. 4. Comparative Word Distribution Chart

The interactivity approach that we used in Word Cloud and Word Tree visualizations can be applied for Word Distribution Chart. Instead of just static representation of search results, we use this view as an entry point for other two visualizations. By selecting a search result point user can drill down and use that point for Word Cloud and Word Tree visualizations.

4   Implementation Considerations

In a Word Cloud visualization two computational problems are considered: first one is creation of sorted list of top n words from a set of M words; and second is representation of those words on a two dimensional surface according to the aesthetic principles A_1–A_3. Algorithm is presented in Table 1.

Second problem is arrangement of all elements of set N on a two dimensional surface. According to first aesthetic criteria A_1, words with higher frequencies should be more noticeable. For each element n from N, we assign numerical value $s(x)$ where $s(x)$ is size of the word on the surface. Size is proportional to the word frequency and is in a range min $\leq s(x) \leq$ max. Having in mind that all elements in N are sorted by frequency in descending order, for each element n in N ($n \in N$) we apply the algorithm described in Table 2.

Table 1. Algorithm for generating list of most frequent words in text

Step	Description
1	Find all distinct words K from a set of words M ($K \leq M$)
2	Remove all irrelevant words (stop words)
3	For each word k ($k \in K$) calculate and assign number of occurrences – frequency
4	Sort elements in K by frequency, in descending order
5	Create subset N of first n words from K ($n \leq K, N \subset K$)

Table 2. Algorithm for word positioning

Step	Description
1	Assign random x position
2	Place the word on the position x and move it on the y axis as close to the center as possible until center is reached or collision with other element is detected
3	Move the word on the x axis toward the center until element is centrally aligned or collision with other element is detected

In Word Tree visualization we solve problems of creation of suffix tree structure for and calculation of positions of each tree node on the screen. Let's suppose that a transcript document is set S of sentences, and that k is a search keyword entered by the user. Sentences are delimited by set of delimiters D.

Table 3. Algorithm for creating suffix tree structure

Step	Description
1	Separate all sentences S by using delimiter characters from D
2	For each sentence s ($s \in S$) separate part of the sentence that comes after keyword k and create a subset of suffix sentences S_1
3	Create tree node n with keyword k and suffix sentences S_1
4	For each sentence in S_1 use the first word in the sentence to create set of distinct words set K_2
5	Add all elements of K_2 as child nodes to the node n
6	Use each element in K_2 as a keyword and recursively perform steps 2-6

When suffix tree structure is calculated next a simple tree layout algorithm is used to calculate the position of each node in space. When positions are calculated, nodes are drawn on the surface. We draw quadratic Bézier Curve between connected nodes.

Table 4. Algorithm for constructing Word Distribution Chart

Step	Description
1	Execute full-text search on the given keyword k
2	Extend search results R with excluded points
3	Sort points by their x coordinate in ascending order
4	For each point p_i ($1 \leq i \leq n$-1, where n is total number of points) draw a line between points p_i and p_{i+1}

Word Distribution Chart is defined with set of points P each with pair of coordinates x and y. X-axis represents time, and y-axis represents number of word occurrences or relevancy. Table 4 represents the procedure for calculation and drawing of Word Distribution Chart visualization. In case of comparative Word Distribution Chart, set of keywords $K = \{k_1, k_2 \ldots k_n\}$ is used, then for each keyword k ($k \in K$) execute steps 1-4. In this case different line color or style needs to be used for each keyword line.

In the Word Tree visualization we are facing with a problem of limited visible screen area on one side, and large number of tree branches on the other side. Wattenberg and Viégas (2008) considered this problem and they proposed a method of "level in detail" for solving this problem. Here we propose that instead of automatically opening all tree branches, only first level beneath the root node will be expanded and all other branches will be collapsed. By clicking on the tree nodes user can further browse and expand all child nodes and so on.

In a Word Distribution Chart visualization user can select point from the chart and according to our idea use it as an input to other visualization. Selected point will be used for Word Tree visualization.

This project is implemented as an ASP.NET web solution with AJAX enabled web services. An open source library called JavaScript InfoVis Toolkit is used for the tree layout algorithms. Highcharts library provides functionalities for drawing charts. A set of IFilter components are used for converting document formats into plain text. We use an open source full text search and indexing engine library named Lucene.

5   Results and Discussion

At the time of this writing the module for data aggregation has collected 2000 transcript documents, 50000 articles and 50000 twitter messages. Data mashup (Figure 1) shows color coded words allocated in the word cloud by their term frequency, the terms "закон" and "законот" have the same meaning but are shown as separate words. Word tree visualization (Figure 2) gives view of the contextual usage of selected word from word cloud visualization. Algorithm shown in Table 1 when applied on documents from multiple sources can take significant time to finalize. Processing time is less than 5 seconds when analysis period is up to 7 days, and for more than 30 days becomes very slow.

One of the main goals of this system was to aggregate and combine data from three sources. Data is transferred, converted and stored in the processing location. The active integration channel with data sources continuously provides data to the analysis and visualization components. Selected visualizations can be generated by using the existing tools, but it requires manual operations of selecting and combining documents which is an extensive task especially when analysis period is variable. This visualization answers on one place what were the most discussed topics in Parliament, social networks and Medias. Generated information is valuable because it shows what citizens and media like to discussed compared to what parliamentarians are discussing.

The system is executing only statistical analyses on the textual contents, additional semantic analyses should also be performed in order to have more accurate results. The choice of Twitter users is important because from it depends the objectivity of the analysis. In this case users are chosen in such a way to balance the various political options.

6   Conclusions

In this research we tried to use the informational mixture from three data sources to represent different viewpoints in the lawmaking processes. The continuous integration between the data sources and web service plays an important part in this visualization and analysis platform. In addition to that interconnected visualizations provide drill down functionality and analysis from different perspectives. Furthermore, word distribution chart offers a chronological analysis to set of transcript documents. This visualization service can make a contribution in the direction of constructing specialized automated visualizations platforms. On the other side our service deduces and presents statistical results that not directly available by manually reading the documents. Service can be also used for analyzing data for time period from one of more data source. Some disadvantage of this solution are lack of semantic analyses and diverse visualizations. Visualizations can be further extended with functionalities for saving views and debating on the view. Future work will mainly cover the development of semantic analyses, additional visualizations and collaborative functionalities.

References

1. Andersson, A., Larsson, N.J., Swanson, K.: Suffix Trees on Words. In: Combinatorial Pattern Matching, pp. 102–115. Springer (1996)
2. Gambette, P., Véronis, J.: Visualizing a Text with a Tree Cloud. In: International Federation of Classification Societies Conference, pp. 561–569 (2009)
3. Global Centre for ICT in Parliament: World e-Parliament Report 2012. United Nations (2012)
4. Graham, T.: Needles in a haystack: A new approach for identifying and assessing political talk in nonpolitical discussion forums. Javnost - The Public 15(2), 17–36 (2008)

5. Hearst, M.A.: TileBars: visualization of term distribution information in full text information access. In: Proceedings of the SIGCHI Conference on Human Factors in Computing Systems, pp. 59–66. ACM Press/Addison-Wesley Publishing Co. (1995)
6. Inter-Parliamentary Union: Guidelines for Parliamentary Websites (2009)
7. Lathrop, D., Ruma, L., Steele, J.: Open Government. O'Reily Media (2010)
8. Liu, S., Zhou, X.M., Pan, S., Song, Y., Qian, W., Cai, W., Lian, X.: TIARA: Interactive, Topic-Based Visual Text Summarization and Analysis. ACM Transactions on Intelligent Systems and Technology 3(2) (2012)
9. Liu, X., Hui, Y., Sun, W., Liang, H.: Towards Service Composition Based on Mashup. In: 2007 IEEE Congress Services (2007)
10. Mutz, D., Martin, P.: Facilitating Communication across Lines of Political Difference: The Role of Mass Media. American Political Science Review 95(1), 97–114 (2001)
11. Sakaki, T., Okazaki, M., Matsuo, Y.: Earthquake Shakes Twitter Users: Realtime Event Detection by Social Sensors. In: Proceedings of the 19th International WWW Conference (WWW 2010), pp. 851–860. ACM (2010)
12. Shah, D.V., Cho, J., Evelend, W.P., Kwak, N.: Information and Expression in a Digital Age: Modeling Internet Effects on Civic Participation. Communication Research 32(5), 531–565 (2005)
13. Šķilters, J., Kreile, M., Bojārs, U., Brikše, I., Pencis, J., Uzule, L.: The pragmatics of political messages in twitter communication. In: García-Castro, R., Fensel, D., Antoniou, G. (eds.) ESWC 2011. LNCS, vol. 7117, pp. 100–111. Springer, Heidelberg (2012)
14. Small, T.A.: Canadian Politics in 140 Characters: Party Politics in the Twitterverse. Canadian Parliamentary Review, 45–49 (2010)
15. Strandberg, K.: Public deliberation goes on-line?: an analysis of citizens' political discussions on the Internet prior to the Finnish parliamentary elections in 2007. Javnost - the Public 15(1), 71–90 (2008)
16. Wattenberg, M., Viégas, F.B.: The Word Tree, an Interactive Visual Concordance. IEEE Transactions on Visualization and Computer Graphics 14(6), 1221–1228 (2008)
17. Zhang, W., Johnson, T., Seltzer, T., Bichard, S.L.: The Revolution Will be Networked - The Influence of Social Networking Sites on Political Attitudes and Behavior. Social Science Computer Review 28(1), 75–92 (2010)

Service – Oriented Architecture Model for Blood Type Analysis (Smart I (Eye) Advisory Rescue System) in Military Environment

Jurij F. Tasič[1], Ana Madevska Bogdanova[2],
Jugoslav Achkoski,[3] and Aleksandar Glavinov[4]

[1] University of Ljubljana, Faculty of Electrical Engeneering,
Tržaška 25, SI-1000 Ljubljana, Slovenia
jurij.tasic@fe.uni-lj.si
[2] Faculty of Computer Science and Engeneering,
Rugjer Boshkovikj 16, P.O. Box 393 1000 Skopje, Macedonia
ana.madevska.bogdanova@finki.ukim.mk
[3,4] Military Academy "General Mihailo Apostolski",
Vasko Karangelevski bb, 1000 Skopje, Macedonia
jugoslav.ackoski@ugd.edu.mk, aglavinov@yahoo.com

Abstract. This paper proposes a model for telemedical Information System that can be used in a military environment. It is consisted of two modules: off-line Advisory Intelligent Module that obscures timely blood type recognition and an on-line module for distance interpretation of blood tests (determining the blood type) and fast delivering request to transport the injured in the nearest medical facility. The Smart I (eye) Advisory Rescue System (SIARS) will provide timely help to injured military persons or civilians, during or after military operations, and will provide correct blood serum analysis near the battlefield, as well as sending request for safest and fastest route to the nearest medical facility.

Keywords: SOA, Telemedicine, information system, military, modules and blood transfusion.

1 Introduction

Usually, during or right after the field combats, there is a need for blood infusion of the injured persons and it is done by using the neutral, type zero [12]. The idea is to find the way for fast blood serum analysis, and to provide the appropriate blood transfusion even in the local hospitals where no specialists are available, and/or during the transportation to the nearest (safest) medical facility. This can be done with the system of gel-cards [12]. The technician fills the gel-cards with the patient blood sample and obtains its image. The technician is a military medical person in the combat unit that executes military tasks. Dependent on the military doctrine, this person can be a part of a company or a section in a battalion.

V. Trajkovik and A. Mishev (eds.), *ICT Innovations 2013*,
Advances in Intelligent Systems and Computing 231,
DOI: 10.1007/978-3-319-01466-1_14, © Springer International Publishing Switzerland 2014

The interpretation of the gel-cards, according the traditional method for determining the blood type, is done by the specialist in the laboratory in a medical facility. The SIARS will broaden the usage of the Geloscope [12] by introducing additional information to the gel-card digital image of the injured (smart image). The specialist will read the blood type of the wounded person by the information from the smart image and deliver the needed information back to the technician. This scenario is possible only when on-line connection to the specialist exist. In the absence of this kind of connection, the Advisory Intelligent Module - AIM will take place. The SIARS will integrate both scenarios, in order to deliver exact blood type analysis in off-line and on-line mode.

2   Related Work

There are a large number of referred research papers from this area. Analyzing the relevant literature, we have drawn the conclusion that in many technologically developed countries the implementation of SOA in the telemedicine IS in the military domain is driven by the unique reason - to protect and save military person/civilian lives. The society benefits from the technology development by government founded research projects. The "Extending Service Oriented Architectures to the Deployed Land Environment" [1] shows that the interest about service-oriented architecture guides to extended implementation of SOA in information and communication systems, that are part of military and civil domain. Since it is simply implementing Service Oriented Architecture (SOA) [2], [3], [4], [5] base functionality into unchangeable organization infrastructure at the same time there are numerous factors that are more complicated implementing SOA in design of systems in military domain. The implementation of service-oriented architecture in systems for deployed land forces in missions which execute military operations or other kind of missions as a mission for establishing peace or keeping peace lead by NATO, EU or UN, represents a big challenge for its effective implementation in military domain.

In [6], information systems for command and control that are exploited in headquarters in an operational level use service oriented architecture in order to increase capability in exchanging information in military environment. SOA approach allows flexibility in increasing capability through integration and systems interoperability based on using Commercial-Off-The Shelf (COTS) technology and standards. A solution by Thales (UK), based on SOA is presented in the paper. It covers architecture modeling, SOA Governance and summary of a multinational demonstration activity used to implement prototype services. It concludes that SOA solution can be used for increasing the capability in military environment.

Also, as a further work, Thales proposes developing SOA solution of information systems of this kind, in all areas in military environment. The understanding business process domain will be tied to opportunities for the development of services focused on delivering improvements in capability to the customers. In addition, the solution is planned to be used for integration and interoperability with other information systems in military environment.

In [7] are presented advances of Service Oriented Architecture (SOA) for developing information systems and to what extent it will help for effective use of Network Enabled Capability (NEC). Also, advances of SOA are exploited in building and designing military information systems for supporting - NEC. Until now, the Service Oriented Architecture was implemented in integration of specific legacy systems in military domain to provide information to support current way of working. SOA is used only in a field of operational process re-engineering and organizational optimization where adopting of SOA can deliver. Service orientation offers unique possibility in changing the way of exploiting information systems due to Ministry of Defence (MOD) departments, and their operational formations can increase their efficiency. The paper exploits SOA, with the intention that MoD will exploit benefits of NEC on a higher level.

In [8], the accelerated development of the information technology and telecommunications, i.e. the wireless and mobile communications and their convergence in telematics, lead to emerge of a new type of an information infrastructure that holds a potential for supporting a lot of advanced services is health. The paper surveys applications used in the wireless telemedical systems. There are short overviews of the group of successfully developed applications – electronic patient file, emergency telemedicine, teleradiology, home monitoring. The authors express their anticipation that the development of the IT will lead to even more useful applications that will include better civil services or improvement of the existence ones.

In [9] is stated that the immense increase of telemedical implementations, enables the users to obtain better health services with increased attention to each patient and diagnosis on a higher level. In the paper, the authors have presented the implementation of the telemedicine service based on SOA, the cardiology reformulation and it was designed in different stages.

In [10], an effective transformation of management in medicine is presented, achieved by the implementation of the telemedicine system. There are several principles that can't be avoided in order to increase the chances of successfully building of telemedical IS: (1) pragmatical selection of telemedical applications and web sites; (2) the clinician drivers and the users should own the system; (3) telemedicine management and support should follow best-practice business principles; (4) the technology should be as user-friendly as possible; (5) telemedicine users must be well trained and supported, both technically and professionally; (6) telemedicine applications should be evaluated and sustained in a clinically appropriate and user-friendly manner; (7) information about the development of telemedicine must be shared. If telemedicine is to realize its full potential, it must be properly evaluated and the results of any evaluations published, whether the results are positive or negative. The telemedicine is about communication with colleagues and patients across large distances and they should be given the possibility to incorporate their own experience.

3   The Phases of SIARS Model Development

SOA presents the next step of standard for reusing information systems and loosely coupled systems. Moreover, independency of implementation platform means that older hardware and newer software can be replaced and updated without negative implications toward other components of a system as far as communication interface of service is not changed. Following the latest Information and Communication Technology (ICT), the telemedical IS based on SOA for blood transfusion should be designed as a hybrid system (store-and-forward mode and real-time mode).

The model of SIARS development based on SOA will be created in several phases.

In the initial phase of the proposed model, we are focused on a usage of service-oriented architecture for building information systems, with the intention of finding relevance for developing this telemedical model.

The first phase covers the research about the designing of prototype modules. The telemedical system will contain two modules – off-line module – Advisory Intelligent Module (AIM) and on-line module for distance interpretation of blood tests (determining the blood type by the specialist) and fast request delivering to transport the wounded to the nearest medical facilities. The interpretation of the blood type is going to be done with the gel-cards. The conversion of the gel-cards physical appearance with the blood sample and the reagenses into the digital image is enabled with Geloscope that produces the e-image [12].

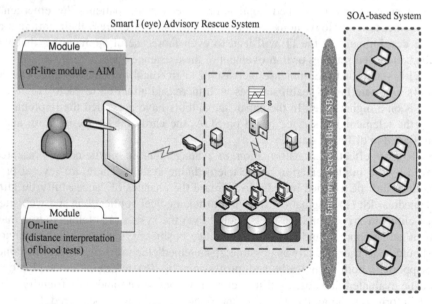

Fig. 1. Model of SOA-based Smart I (eye) Advisory Rescue System [11]

The second phase - developing the off-line module for automatic blood type analysis from gel-cards.

The Advisory Intelligent Module for blood type determination will work independently - analyze the digital image of the gel-cards and produce a recommendation to the technician for the blood transfusion process in the situation when no information about the serological blood type of the injured is available (there is no e-file nor connection to the specialist). The sophisticated machine learning hybrid system will give the accurate information about the patient blood group serological results.

The second phase – developing the on-line module that sends the blood gel-cards digital image to the specialist. The specialist reads and interpret the blood type of the injured. This module will introduce some additional information to the image:

- ID
- Time
- Location
- Coordinates;

To provide accurate estimation of coordinates and location, we are connecting SIARS to GPS and installing digital map from region were military operations are ongoing.

The third phase is dedicating for developing an application that automatically takes the image from the Geloscope and enriches it with the aforementioned data. The application will contain intuitive interface that can be used in this case of limited time. According to the information from the smart-image, the specialist will resolve the blood type of the injured patient and send this information back to the technician. At the same time, the technician will obtain the information about the closest medical facility and the estimated time for the transportation to the nearest hospital for the injured patient, if needed.

The transportation of the injured patient is a critical moment, because there might be enemy zones that disable the safe transportation. We emphasize this possibility, because in this case the obtained information for the shortest path/time for transportation to nearest hospital may be in discrepancy with the real situation. On Figure 2, we propose a solution with inclusion of Tactical Operational Center(TOC).

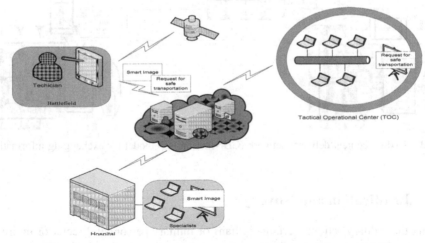

Fig. 2. Scheme for sharing information b/w TOC, technician and specialists

TOC will get the information (ID, time, location and coordinates) and it will send the technician information about the safest route and needed transportation time for the injured patient.

The fourth phase is about developing the protocol for communication between the technician and TOC. If the technician estimates that the patient has serious injuries and if urgent transportation is needed, he sends the information (ID, name (optional), time, location and coordinates) to TOC. The TOC delivers the information in the nearest hospital, so the preparations for the injured will be on time. The return information to the technician will be about the transportation type, place or nearest hospital.

The fifth phase is developing an interface of the application. The system will contain different subsystems, connected to each other needed for creating the information and their delivery to the end users (see Figure 3).

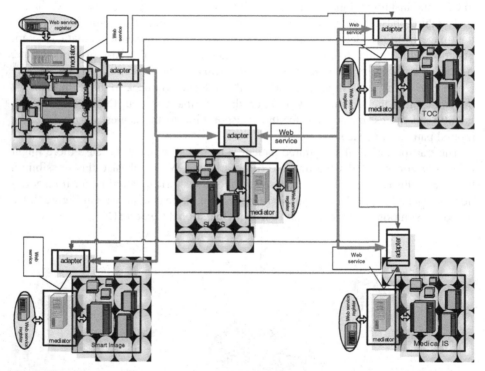

Fig. 3. SIARS's content delivery network (CDN) using P2P model for exchanging information [13]

4   Justification and Novelty

During the military actions, certain civilians or military personnel directly or indirectly involved in the action can be injured too. According the previous statement, the logical conclusion is that in a given moment some of the wounded or injured will need urgent blood transfusion. The transfusion can be done by a blood stocks, and this must be done in a limited period of time. It is very important if the technician has the information about the injured blood type, in order to avoid the transfusion only with the O-type.

5   Expected Results

- Precise and fast determining the patient blood type, to proceed with the procedure of blood transfusion on time with the right blood type
- Developed standardized protocol for information exchange, important for the fast blood type determination
- Defined model and methodology for integration of applications from selected ICT information systems that does not support SOA. This approach will override the technology dependency and it will allow integration of different platform information systems (DCOM, CORBA, RPC, RMI)
- Safety mechanisms integration, selection and description of safety standards in the ICT, to achieve high level of access control and user
- Defined metrics for service evaluation in the telemedicine IS with SOA
- Verification of the telemedicine IS functionality according the defined goals.

6   Conclusion

The Telemedicine information system can be implemented in every unit that performs military operations. Every unit that operates in an open field, no matter the nature of the unit, includes medical technicians, soldiers with the essential medical training. Technician role is to give the first aid to the injured. In the scope of the proposed telemedical system, the technicians will have certain equipment (device or terminal) that will enable to transmit, analyze, or process the signal about the blood type determination and realization of fast and safe transportation to the nearest hospital if needed.

According to the technician capabilities, this system can be used in a civil variant (Crisis Management, Rescue, Recovery and Protection), dependent of the users need.

The successful implementation of the system opens new opportunities for telemedical use in the military environment, focused on decreasing the number of potential victims due to the late or wrong blood transfusion.

The system will contribute in bringing the development of the telemedicine in a higher level. This will confirm the thesis that military increasingly uses the most sophisticated and up-to-date technologies for its needs. As a result, the design and implementation of the system will lead to improvement of the blood transfusion process (the time limitation factor) and more human lives will be saved.

References

1. Medlow, D.: Saab Systems: Extending Service Orientated Architectures to the Deployed Land Environment. In: Military Communications and Information Systems Conference (MilCis), Australia (2009), http://www.milcis.com.au/

2. OASIS: Reference Model for Service Oriented Architecture 1.0 (October 2006), http://docs.oasis-open.org/soa-rm/v1.0/
3. OASIS: Reference Architecture Foundation for Service Oriented Architecture Version 1.0 Committee Draft 02 (2009), http://docs.oasis-open.org/soa-rm/soa-ra/v1.0/soa-ra-cd-02.pdf
4. Pulier, E., Taylor, H.: Understanding Enterprise SOA. Manning (2006)
5. Hurwitz, J., Bloor, R., Baroudi, C.: Marcia Kaufman. Service Oriented Architecture or Dummies. Wiley (2007)
6. Radcliffe, S., Trotman, L., Duncan H.: Supporting Capability Evolution Using a Service Oriented Architecture Approach in a Military Command and Control Information System, http://nectise.com/pdfs/2_Stewart%20Radcliffe.pdf
7. Brehm, N., Gómez, J.M.: Secure Web Service-based resource sharing in ERP networks. International Journal on Information Privacy and Security (JIPS) 1, 29–48 (2005)
8. Pattichis, C.S., Kyriacou, E., Voskarides, S., Pattichis, M.S., Istepanian, R., Schizas, C.N.: Wireless telemedicine systems: an overview. IEEE Antennas and Propagation Magazine 44, 143–153 (2002)
9. Guillén, E., Ubaque, J., Ramirez, L., Cardenas, Y.: Telemedicine Network Implementation with SOA Architecture: A Case Study. In: Proceedings of the World Congress on Engineering and Computer Science WCECS 2012, pp. 6–9 (2012)
10. Yellowlees, P.M.: Successfully developing a telemedicine system. Journal of Telemedicine and Telecare 11, 331–335 (2005)
11. Achkoski, J., Trajkovik, V., Davcev, D.: Service-Oriented Architecture Concept for Intelligence Information System Development. In: The Third International Conferences on Advanced Service Computing SERVICE Computation 2011 (IARIA), pp. 32–37 (2011)
12. Meza, M., Breskvar, M., Kosir, A., Bricl, I., Tasic, J., Rozman, P.: Telemedicine in the blood transfusion laboratory: remote interpretation of pre-transfusion tests. Journal of Telemedicine and Telecare 13(7), 357–362 (2007)
13. Achkoski, J., Trajkovik, V.: Usage of Service-Oriented Architecture for Developing Prototype of Intelligence Information System. Tem Journal 1, 234–245 (2012)

Performance Impact of Ionospheric and Tropospheric Corrections of User Position Estimation Using GPS Raw Measurements

Enik Shytermeja[1], Alban Rakipi[1], Shkëlzen Cakaj[2], Bexhet Kamo[1], and Vladi Koliçi[1]

[1] Faculty of Information Technology (FTI), Polytechnic University of Tirana, Tiranë, Albania
{eshytermeja,arakipi,bkamo,vkolici}@fti.edu.al
[2] Post and Telecommunication of Kosovo (PTK), Prishtinë, Kosovo
shkelzen.cakaj@fulbrightmail.org

Abstract. Global Positioning Systems (GPS) refers to satellite-based radio-positioning systems and time-transfer systems that provide three-dimensional course, position, and time information to suitably equipped users. At present, GPS system is world-wide used for positioning and navigation, attracting great attention from the scientific, professional and social community.GPS satellites are orbiting Earth at altitudes of 20.200 km and the GPS signal is mostly affected by the atmospheric effects. The scope of this paper is to investigate the performance impact of the atmospheric correction models in the overall positioning accuracy. Real GPS measurements were gathered using a single frequency receiver and post–processed by our proposed innovative adaptive LMS algorithm. We integrated Klobuchar and Hopfield correction models enabling a considerable reduction of the vertical error.

Keywords: GNSS systems, GPS raw measurements, Ionosphere, Troposphere, Klobuchar model, TEC, Hopfield model, PVT algorithm, VDOP.

1 Introduction

The Global Navigation Satellite Systems (GNSS) technique has evolved to become the most widely available positioning tool used by both civilians and scientists [1].

GPS is a satellite-based navigation radio system which is used to verify the position and time in space and on the Earth [2].The GPS satellites are orbiting the Earth at altitudes of about 20.200 km and it is generally known that the atmospheric effects on the GPS signals are the most dominant spatially correlated biases. The atmosphere causing the delay in GPS signals consists of two main layers: ionosphere and troposphere [3].

The Ionosphere is the band of the atmosphere from around (50 – 1000 km) above the earth's surface and is highly variable in space and time, with certain solar-related ionospheric disturbances [4]. Ionosphere research attracts significant attention from the GPS community because ionosphere range delay on GPS signals is a major erro-source in GPS positioning and navigation. The ionospheric delay is a function of the

V. Trajkovik and A. Mishev (eds.), *ICT Innovations 2013*,
Advances in Intelligent Systems and Computing 231,
DOI: 10.1007/978-3-319-01466-1_15, © Springer International Publishing Switzerland 2014

total electron content (TEC) along the signal path and the frequency of the propagated signal, mostly affecting the vertical component of user's position (VDOP). There are different statistical model available for the correction of ionospheric range error in single frequency applications. However we can distinguish two of them such as: the Klobuchar model for GPS [5] or the NeQuick model [6] foreseen for use in European GALILEO system. In our algorithm we employed the widely used Klobuchar model because of its simple structure and convenient calculation.

The troposphere is the band of the atmosphere from the earth's surface to about 8 km over the poles and 16 km over the equator [7]. The tropospheric propagation delay is directly related to the refractive index (or refractivity). The signal refraction in the troposphere is separated into two components: the dry and the wet component, where the dry or hydrostatic component is mainly a function of atmospheric pressure and gives rise to about 90% of the tropospheric delay.

There are different mathematical models that can be used to correct the tropospheric error such as Saastamoinen and Hopfield Model [8].In verifying the effects of relative tropospheric delay in user's position estimation, we employed the most common and precise method, called Hopfield correction model.

In the next section, we will investigate in details the impact of Klobuchar and Hopfield model for ionospheric and tropospheric errors respectively, implemented in our algorithm in post-processing mode. Therefore, the recording of the observed satellite-to-receiver pseudoranges is required and this is achieved from the use of a single frequency receiver.

This work is divided in five major sections. In the first section we give a short introduction regarding the atmospheric errors. In the second section we describe in details the process of data collection and the tools used for measurements, then in the third section the implementation of our proposed PVT algorithm is presented. The fourth and fifth sections are dedicated to the obtained results, visualized with different kinds of plots. Finally we conclude our work with comments and conclusions.

2   Experimental Data Set

In this section we explain the GPS data collection process and the implementation of a post-processing adaptive PVT algorithm, where we included mathematical Ionospheric and Tropospheric correction models aiming to an improved accuracy of user's position estimation.

The single frequency measurements of the GPS pseudoranges were recorded using the SAT-SURF receiver, manufactured by ISMB [9]. Later on all the gathered data such as pseudoranges, Time Of Week, Week Number, the satellite's coordinates Xs, Ys and Zs, Ionosphere and Troposphere coefficients, were post-processed by the SAT-SURFER software. SAT-SURFER gets GPS raw data measurements, displays them on the screen and logs these data in different files allowing any further post-processing activity. Our real experiment and GPS data gathering was conducted outside our laboratory in the Lingotto Campus at nearly 1.30 PM. The GPS raw measurements displayed from SAT-SURFER in its GUI format are illustrated in Fig. 1.

The major error contribution in the overall user position accuracy comes from the Ionosphere layer, affecting mostly the vertical component and increasing in such way VDOP (Vertical Dilution of Precision) [10].The ionospheric parameters taken from the SAT-SURFER log files are ilustrated in Fig.2.

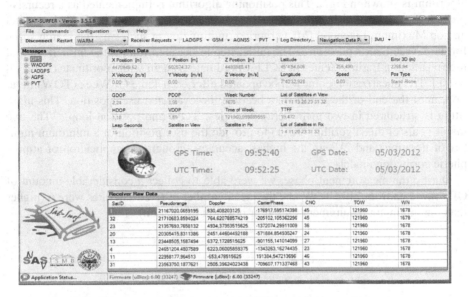

Fig. 1. SAT-SURFER Graphical User Interface (GUI)

PosID	TOW	WN	Alpha0	Alpha1	Alpha2	Alpha3	Beta0	Beta1	Beta2	Beta3
640	319283	1675	1.30E-08	0	-5.96E-08	5.96E-08	110592	-65536	-262144	393216
640	319283	1675	1.30E-08	0	-5.96E-08	5.96E-08	110592	-65536	-262144	393216
640	319283	1675	1.30E-08	0	-5.96E-08	5.96E-08	110592	-65536	-262144	393216
640	319283	1675	1.30E-08	0	-5.96E-08	5.96E-08	110592	-65536	-262144	393216
640	319283	1675	1.30E-08	0	-5.96E-08	5.96E-08	110592	-65536	-262144	393216
640	319283	1675	1.30E-08	0	-5.96E-08	5.96E-08	110592	-65536	-262144	393216
1390	320033	1675	1.30E-08	0	-5.96E-08	5.96E-08	110592	-65536	-262144	393216
1390	320033	1675	1.30E-08	0	-5.96E-08	5.96E-08	110592	-65536	-262144	393216
1390	320033	1675	1.30E-08	0	-5.96E-08	5.96E-08	110592	-65536	-262144	393216
1390	320033	1675	1.30E-08	0	-5.96E-08	5.96E-08	110592	-65536	-262144	393216
2113	320783	1675	1.30E-08	0	-5.96E-08	5.96E-08	110592	-65536	-262144	393216
2113	320783	1675	1.30E-08	0	-5.96E-08	5.96E-08	110592	-65536	-262144	393216
2113	320783	1675	1.30E-08	0	-5.96E-08	5.96E-08	110592	-65536	-262144	393216
2113	320783	1675	1.30E-08	0	-5.96E-08	5.96E-08	110592	-65536	-262144	393216

Fig. 2. Ionospheric Correction parameters taken from SAT-SURFER log file

The α and β are the input data of our adaptive positioning algorithm necessary for the mitigation of ionospheric error in the user's position estimation. It will be later shown that we achieve a considerable improvement of the vertical component and a decreased VDOP, after the application of this correction in our main algorithm.

3   Implementation of Our Proposed Positioning Algorithm

In this section, we propose an innovative adaptive PVT algorithm compiled in MATLAB® environment. The specific computation flow diagram of our positioning algorithm is shown in Fig.3. This positioning algorithm is implemented as a recursive procedure with several iterations based on Least Square Mean (LSM) [11] solution and on Maximum Likehood (ML) criterion, minimizing in such way the search space for the " True User's position ". The initial step in our algorithm is the initialization of user position in Earth's Center in ECEF Coordinate System with coordinates LP = [0 0 0 0]. The linearization point will be updated after each Time Of Week (TOW) iteration, until the end of the iteration to become the evaluated user position. This algorithm is structured in two main iteration cycles: TOW and position loop.   The two major goals of this algorithm are: 1) to provide the user position in a minimum number of iterations and 2) with the highest accuracy through the application of atmospheric correction models.

During the measurement phase, we were able to collect a considerable amount of GPS raw data from 2122 TOW and the mean number of fixed satellites were 6 (higher than the minimum requirement for a user position estimation).

Fig. 3. Computational flow diagram of our Positioning algorithm

During the measurement phase, we were able to collect a considerable amount of GPS raw data from 2122 TOW and the mean number of fixed satellites were 6 (higher than the minimum requirement for a user position estimation).

4   Ionospheric and Tropospheric Correction Models

The focus of this section is to evaluate the ionospheric and tropospheric effect on GPS positioning solution. Prior to the correction models loop in our MATLAB® code, it is important to exclude from our computation the satellites with a C/N_0 lower than a predefined threshold. The threshold for C/N_0 was chosen to be equal to the mean value of the C/N_0 column in the data logs and was set to be $C/N0 > 37$ dBHz. This criterion is implemented in a Weighted Matrix and enables an increased accuracy in the user's position estimation.

```
Weight = Struct(TOW).Data(i).CNO;
W = diag(Weight(i));
```

The pseudoranges are affected by errors, which can be modeled as Gaussian random variables:

- With zero mean,
- Independent and identically distributed,
- with variance σ_{UERE}^2 .

The errors affecting the pseudoranges can be expressed by the following equation:

$$\rho = sqrt\ [(x_i - x_{Lp})^2 + (y_i - y_{Lp})^2 + (z_i - z_{Lp})^2] - c * t_{Lp} + c * T_a \quad (1)$$

where:

- $T_a = T_{Iono} + T_{Trop}$ – is the sum of Ionospheric and Tropospheric error contributions, respectively;
- $[x_i, y_i, z_i]$ – are the coordinates of user's unknown position;
- $[x_{Lp}, y_{Lp}, z_{Lp}]$ – are the coordinates of linearization point, which in our case is the position obtained from the last algorithm iteration;

We will describe in further details these two types of corrections that we take into account in the error affected pseudoranges.

4.1   Ionospheric Corrections

We propose the implementation of Ionospheric corrections based on the Klobuchar model [5].The Klobuchar model implemented in our algorithm uses as input:

- Receiver generated terms:

— λ_u - User Geodetic Latitude WGS 84 (semi – circles)
— ϕ_u - User Geodetic Longitude WGS 84 (semi – circles)

— *E* - Elevation angle between the user and the satellite , measure clockwise positive from the true north (semi- circles)
— *A* - Geodetic azimuth angle of the satellite
— *GPS time* - Receiver's computed system time.

- Satellite transmitted terms:

— α_n - coefficients of a cubic equation representing the amplitude of the delay.
— β_n - coefficients of a cubic equation representing the period PER of the model.

We designed a function *ionogen.m* to calculate the delay caused by Ionosphere layer, which was called in our main PVT algorithm. The main inputs of the ionospheric correction function are:

- *PER* is the period of the cosine function and implicates the interval of the ionospheric activity in daytime. It is expressed by the following formula, whose inputs are taken from the ionosphere log file:

$$PER = \beta_0 + \beta_1 * lat_m + \beta_2 * lat_m^2 + \beta_3 * lat_m^3 \tag{2}$$

where lat_m - is the geomagnetic latitude of the Earth's projection of the ionospheric intersection point (mean ionospheric height assumed to be 350 km).

- *Amplitude of the model*

$$AMP = \alpha_0 + \alpha_1 * lat_m + \alpha_2 * lat_m^2 + \alpha_3 * lat_m^3 . \tag{3}$$

The inputs of the Klobuchar model were taken by loading the Elevation and Azimuth angles for each TOW and number of fixed satellites, using the following lines in MATLAB ®code:

```
azimuth   = Struct(TOW).Data(i).Azimuth;
elevation = Struct(TOW).Data(i).Elevation;
```

We observed that these coefficients are constant even for different TOW (Fig. 3) and this result is due to the fact that ionospheric parameters do not change in a short measurement time.

4.2   Tropospheric Corrections

As described in previous section, the signal refraction in the troposphere is separated into two components: the dry and the wet component, where the dry component contributes about 90 % of the total tropospheric delay. The tropospheric delay is approximated by using the Hopfield model [8], whose inputs in our algorithm are:

- *T temperature* in ^{0}C.
- *P pressure* in hPa.
- H_u *humidity ratio* in % .

- *R Earth radius:* R = 6371 km.
- *E Satellite Elevation angle.*

We designed a function in MATLAB® named *tropogen.m* to calculate the delay caused by the Troposphere layer, represented by the following relations:

- Total Tropospheric error contribution

$$\Delta \rho_{Tropospheric}(E) = \Delta \rho_{dry}(E) + \Delta \rho_{wet}(E) \tag{4}$$

where $\Delta \rho_{wet} = K_w(I(h_w) - b)$ and $\Delta \rho_{dry} = K_d(I(h_d) - b)$.

- Humidity ratio in % in dry and wet conditions is given by these equations:

$$H_w = 11000 \quad \text{and} \quad H_d = 40136 + (148.72 * T) .$$

5   Plots and Results

In this section are presented the results of our work, which are visualized with different kinds of plots. In Figure 4 are plotted in Cartesian coordinates the true position of the user's receiver and the cloud of points which represent the estimated positions, outputs of our PVT algorithm for all GPS epochs.

We can easily observe that after applying the proposed ionospheric and tropospheric correction models, the error in the vertical component (height z) is significantly reduced.

Figure 5 shows the estimated positions and the true position in Geographical coordinates for a better understanding of the atmospheric residual errors. The Klobuchar model reduces the vertical error with a value equal to 33.7 m. The Tropospheric Hopfield model applied in our adaptive PVT algorithm, gives a slight correction to the vertical error in the amount of 1.5724 m. This was an expected outcome because Tropospheric error's impact is lower compared to the Ionospheric one, in the total error contribution. These important results are summarized in Table 1.

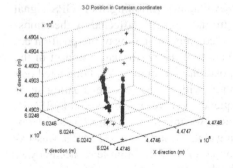

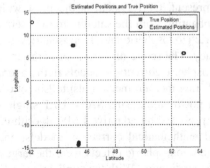

Fig. 4. 3-D estimated positions in Cartesian Coordinates

Fig. 5. Estimated positions and True position in Geographical coordinates

Table 1. Summary Table of the different trials computed for the PVT solution

	Latitude	Longitude	Height (m)
Without correction	45.0351^0	7.6704^0	45.1066
Ionosphere correction	45.0351^0	7.6704^0	11.4066
Troposphere correction	45.0351^0	7.6704^0	43.5342
Iono + Tropo correction	45.0351^0	7.6704^0	9.8342

Finally, the user's position estimated by our adaptive LMS algorithm for all GPS epochs or TOWs is:[45.0351^0 N 7.6704^0 E 9.8342 m], where we implemented a mathematical coordinates conversion function [XYZ to North East Height(m)].This estimated position is very close to the true position and this is illustrated in Fig. 6.

Fig. 6. Estimated user's position from our adaptive PVT algorithm in google map

6   Comments and Conclusions

At present, GPS system is the most widely available positioning tool. GPS signal propagating from the satellite to the users on ground is greatly affected by the atmospheric (Ionospheric and Tropospheric) effects, reducing in such way the position accuracy. The scope of this paper is the investigation of the performance impact of the atmospheric correction models in the overall positioning accuracy. We conducted real GPS raw measurements outside Lingotto campus with the scope of evaluating static user positioning, using SAT-SURF single frequency receiver manufactured by ISMB. We proposed an adaptive LMS algorithm, where we integrated Klobuchar and Hopfield mathematical correction models, enabling data post-processing. As it was expected, ionospheric effects were the largest source of error for high level accuracy of GPS positioning. With the application of Klobuchar model for Ionospheric correction, we achieved a reduction of the vertical error with a value equal to 33.7 m; however this model did not affect significantly the horizontal positioning. On the other hand, the integration of Hopfield Tropospheric model in our positioning algorithm, gave a slight improvement of the vertical error of 1.5724 m and mostly due to its dry component. An important factor enabling the fast and accurate convergence of the estimated positions to the true one, was the implementation of Weighted Matrix with the C/N_0 threshold set at 37 dB/Hz. This criteria enabled the discard of GPS pseudoranges that contributed to an increased positioning error. In this paper, we were concerned about the positioning performance of our algorithm for a static user and not taking into consideration the user's motion. However we are strongly confident that this issue can be overcome by restricting the C/N_0 threshold and increasing the number of iterations of the positioning algorithm.

In the future work, we will focus on the mitigation of other error's contribution such as relativistic, ephemerides and satellite clock errors. We will also investigate the positioning performance achieved after the application of EGNOS and differential corrections, using double frequency GPS receivers for PPP applications.

References

1. McNeff, J.G.: The Global Positioning System. IEEE Transactions on Microwave Theory and Techniques 50, 645–652 (2002)
2. Warnant, R., Kutiev, I., Marinov, P., Bavier, M., Lejeune, S.: Ionospheric and geomagnetic conditions during periods of degraded GPS position accuracy: 2.RTK events during disturbed and quiet geomagnetic conditions. Advances in Space Research 39, 881–888 (2007); Satirapod, C., Chalermwattanachai, P.: Impact of Different Tropospheric Models on GPS Baseline Accuracy: Case Study in Thailand. Journal of Global Positioning Systems 4, 36–40 (2005)
3. Hofmann-Wellenhof, B., Lichtenegger, H., Collins, J.: Global Positioning System: Theory and Practice. Springer, Heidelberg (1997)
4. Filjar, R., Kos, T., Kos, S.: Klobuchar-Like Local Model of Quiet Space Weather GPS Ionospheric Delay for Northern Adriatic. Journal of Navigation 62, 543–554 (2009)
5. Hochegger, G., Nava, B., Radicella, S.M., Leitinger, R.: A family of ionospheric models for different uses. Phys. Chem. Earth 25, 307–310 (2000)
6. Langley, R.B.: Propagation of the GPS Signals. In: Kleusberg, A., Teunissen, P.J.G. (eds.) GPS for Geodesy, 2nd edn. LNES, vol. 60, pp. 111–150. Springer, Heidelberg (1998)
7. Xinlong, W., Jiaxing, J., Yafeng, L.: The applicability analysis of troposphere delay error model in GPS positioning. Aircraft Engineering and Aerospace Technology 80, 445–451 (2009)
8. SAT-SURF The Training Board for GNSS – User Manual, SAT-SURF-1-NAV-08, Issue 1.0 (October 27, 2008)
9. Leva, J.L.: Relationship between navigation vertical error, VDOP, and pseudo-range error in GPS. IEEE Transactions on Aerospace and Electronic Systems 30, 1138–1142 (1994)
10. He, Y., Bilgic, A.: Iterative least squares method for global positioning system. Adv. Radio Sci. 9, 203–208 (2011)

Web Service Performance Analysis in Virtual Environment

Marjan Gusev, Sasko Ristov, and Goran Velkoski

Ss. Cyril and Methodious University,
Faculty of Information Sciences and Computer Engineering,
Rugjer Boshkovikj 16, 1000 Skoipje, Macedonia
{marjan.gushev,sashko.ristov}@finki.ukim.mk, velkoski.goran@gmail.com

Abstract. Virtualization is a technique that allows several guest operating systems (OSs) to run on a single physical server and share its hardware resources (CPU, RAM, Storage, Network, etc). However, the virtualization implements an additional layer in the stack and thus a performance decrease is expected. In this paper, we analyze the performance behavior of two simple web services (WS) Concat and Sort. The former is memory demanding WS which mostly utilizes the main memory, while the latter is memory demanding WS and utilizes both the main memory and the CPU. The WSs are hosted on two different environments: host (bare metal) and guest (virtualized). We realized several experiments varying the load with different number of concurrent messages and their size to determine the regions where the performance decreases due to virtualization. Despite the expectation that virtual environment will reduce the performance, the results show that it even improves the average performance of 4 to 5%.

Keywords: Apache Tomcat, JAVA, Virtualization, VMware.

1 Introduction

Nowadays, each enterprise data center uses virtualization, either to host different operating systems, or to enable a multitenant environment (when more than one applications use the system) or fault tolerance (when one application is hosted on more than one server). Each virtualization software operates on top of a layer of system software, called hypervisor or VMM (virtual machine monitor), inserted between the guest OS and the underlying hardware [2]. Virtualization improves the resource utilization by providing integrated operating platform applications based on heterogeneous and autonomous resources aggregation [11]. It helps the companies with software licenses since most of the licenses are bound with the number of CPUs. By using virtualization, the companies can create a certain virtual machine instances with maximum allowed CPUs to comply with the license. Most of the cloud service providers use virtualization to provide flexible and cost-effective resource sharing [16].

There are different levels of virtualization: Full Virtualization, Paravirtualization, OS-level Virtualization, and Native Virtualization [15]. Several commercial

V. Trajkovik and A. Mishev (eds.), *ICT Innovations 2013*,
Advances in Intelligent Systems and Computing 231,
DOI: 10.1007/978-3-319-01466-1_16, © Springer International Publishing Switzerland 2014

virtualization software solutions exist on the market, and the most frequently used examples are VMware's ESX [14] and Microsoft's Hyper-V. There are also open source virtualization softwares, such as KVM and XEN.

Despite all the benefits of the virtualization, it implements an additional layer in the stack which usually reduces the performance. The cloud virtual platform reduces the WS performance more than 30% compared to on-premise platform using the same runtime environment and hardware infrastructure [10]. The performance decrease is emphasized for input / output bound applications (more than 110% overhead) and a little (about 10% overhead) for CPU bound applications [7].

In this paper, we analyze the performance behavior generated by VMware's ESX using the same hardware and the same runtime environment. The expectations are based on the hypothesis that the virtual environment will decrease the performance. We perform series of experiments for computation intensive and memory demanding WSs. These WSs are hosted either on the host environment or the guest (virtual) environment deployed on separated physical servers where the virtual environment uses all of the physical server resources. We are also aiming to find out which server load (a certain number of concurrent messages with a particular size) provides minimal and maximal performance decrease if WSs are migrated on a virtual environment.

The rest of the paper is organized as follows. Related work in the area of performance in virtualized environment is presented in Section 2. Section 3 describes the realized experiments, infrastructure and platform environments. In Section 4, we present the results of the experiments and analyze how the message size and the number of concurrent messages impact the WS performance in the virtual (guest) and host platform environments. The conclusion and future work are specified in Section 5.

2   Related Work

Research results about WS performance can be found in many papers covering different domains. WS performance is analyzed and several tools for performance prediction are proposed. For example, Li et al. [5] present a CloudProphet that can accurately predict the response time of an on-premise web application if migrated to a cloud virtual environment. CloudGuide explores which cloud configurations meet the performance requirements and cost constraints and also can find new configuration when workload changes [6].

WSs can be simulated and tested for various performance metrics before they are deployed on Internet servers, which give results close to the real environment [13]. Ristov and Tentov [9] analyzed the web server performance parameters response time and throughput via WSs with two main input factors message size and number of messages. Here we extend this research to compare the WS performance with the same input factors in the host and guest environments.

The virtual environments are also worse than the host environment for cache intensive algorithms when the data exceeds the cache size [8]. However, there is such a phenomenon that some algorithms run faster in a virtual rather than in a host environment. For example, cache intensive algorithms, such as dense matrix matrix multiplication, can generate smaller number of cache misses in a virtual environment if cache memories are private per core [3].

3   The Methodology

This section describes the testing methodology including identification of environment, infrastructure and platform, test plan and design implementation details. Several steps were performed to create efficient and effective tests and results.

3.1   Test Environment Identification

We conduct the experiments on a traditional client-server architecture using the same hardware infrastructure and runtime environment, but different platforms. Two same web servers are used as hardware infrastructure with Intel(R) Xeon(R) CPU X5647 @ 2.93GHz with 4 cores and 8GB RAM. The other server with the same hardware infrastructure is used as a client and SOAPUI [12] is used to create various server load tests. Client and servers are in the same LAN segment to exclude the network impact [4].

Two different platforms are deployed. *Host* environment consists of traditional Ubuntu OS as depicted in Figure 1 a). *Guest* environment is developed using VMware ESX 4.1 and Ubuntu OS above it as depicted in Figure 1 b). Both environments use all physical resources.

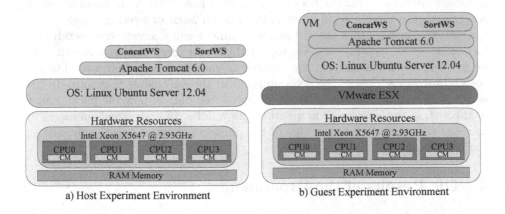

a) Host Experiment Environment

b) Guest Experiment Environment

Fig. 1. Testing Environments

3.2   Test Data

Performance parameter response time is measured for various experiments with different number and sizes of concurrent requests for both platforms. Client is on the same VLAN as the web server, with network response time smaller than 1 ms, and none of the packets are lost during the test. Thus, we can assume that the response time measured with SOAPUI is the same as the server's response time.

The basic goal is to measure the performance peaks caused by virtualization in the guest environment. Test data consists of Concat and Sort WSs. The *Concat WS* accepts two string parameters and returns a string which is a concatenation of the input. This is a memory demanding WS that depends on the input parameter size M with complexity $O(M)$. The *Sort WS* also accepts two string parameters and returns a string that is a concatenation of the two input strings which is then alphabetically sorted using sort function in [1]. This is also a memory demanding service that depends on the input parameter size M. In addition, it is a computation intensive WS with complexity $O(M \cdot log_2 M)$.

Experiments are repeated for parameter sizes M that change values from $0, 1, 2, \cdots, 9KB$ for Concat WS and $0, 1, \cdots, 7KB$ for Sort WS. The server is loaded with a various number of messages (requests) N in order to retain server normal workload mode, that is, 12, 100, 500, 750, 1000, 1250, 1500, 1750 and 2000 requests per second for each value of M.

3.3   Test Plan

The first part of the experiment consists of series of test cases that examine the impact of increasing the message size on the server response time. The second part of the experiment consists of series of test cases that examine the impact of increasing the number of concurrent messages on the server response time. All test cases are performed on: 1) WSs hosted on a host environment; and 2) WSs hosted on a virtual (guest) environment. Each test case runs for 60 seconds; N messages are sent with M bytes each, with variance of 0.5. The accent is set on server response time in regular mode, but not on burst or overload mode.

Our hypothesis is that the virtual environment will decrease the overall performance for all experiments. On top of this, we expect the response time to be increased while increasing the number of messages and their size. The research problem addresses the determination of the parameter that impacts the server performance the most. We are eager to find out: 1) how the performance is affected, is the parameter in question the number of concurrent messages or message sizes and 2) does the type of the platform (host or guest) influence on the performance.

4   The Results of the Experiments

This section describes the results of testing the performance impact on virtualization layer. We also analyze the results to understand the performance impact of different message sizes and the number of concurrent messages on both WSs.

4.1   Performance of the Host OS

The WS performance is measured while hosted on the host OS with different payload: 1) different message size for constant number of concurrent messages and 2) different number of concurrent messages for a constant message size.

Figure 2 depicts the response time of the Concat WS delivered by the host environment. We observe that the response time is similar regardless of both input parameters and it is in the range of 2 to 4 ms. The response time increases over 4 ms when both parameters are increased to their maximum size in the web server regular mode.

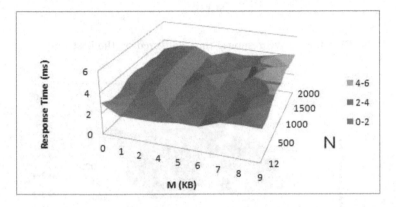

Fig. 2. Concat WS response time delivered by the host OS

Response time of Sort WS in host OS is depicted in Figure 3. We observe different distribution for response time compared to Concat WS. That is, only for small N the response time does not depend on the other parameter M. For $N \geq 100$, the response time mostly depends on the message parameter size M. The response time varies from 2ms to 6.25s.

4.2   Guest OS

We measure the performance of WSs hosted in the guest OS using the same payload as in the previous case, i.e.: 1) different message size for constant number of concurrent messages and 2) different number of concurrent messages for constant message size.

The results for response time of Concat WS hosted on the guest OS are shown in Figure 4. As in the previous case (host OS), we observe very similar distribution of response time as a function of both input parameters M and N.

Figure 5 presents the response time of Sort WS hosted on guest OS. Similar to host environment, we observe that the response time does not depend on the parameter M only for small N. For $N \geq 100$, the response time mostly depends on the message parameter size M. The response time varies from 2ms to 6.86s.

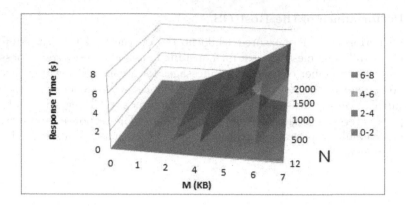

Fig. 3. Sort WS response time delivered by the host OS

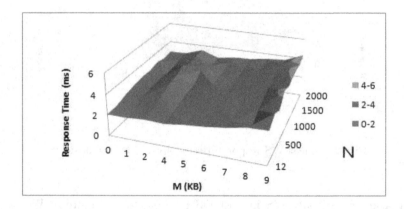

Fig. 4. Concat WS response time while hosted on the guest OS

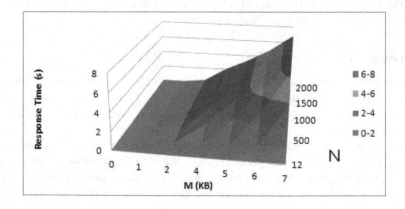

Fig. 5. Sort WS response time while hosted on the guest OS

4.3 Host vs. Guest Performance Comparison

The performance of both WSs (hosted in host and guest OSs) is compared with different payload depending on different message sizes and different number of concurrent messages.

Concat WS Platform Performance Comparison. Figure 6 depicts the guest vs host relative response time comparison of the Concat WS. We observe a phenomenon. That is, the relative response time is smaller than 1 for all values of both input parameters, except when N is small. Two local extremes appear in points $(M, N) \in \{(4, 1250), (4, 1500)\}$. We believe that they appear due to communication time impact for small response time and the effect of the virtualization which is a part of our further research.

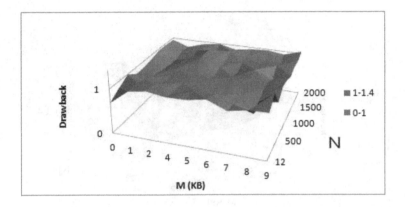

Fig. 6. Guest vs host performance comparison for Concat WS

Table 1 presents the relative performance for each test case and average for Concat WS. Its values are in the range of $[0.69, 1.24]$ and average value is 0.95 which means that the virtual environment provides average 5% better performance than host.

Sort WS Platform Performance Comparison. The relative response time comparison for Sort WS for guest vs host environment is depicted in Figure 7. The results also show partially the phenomenon, i.e., when one of the parameters is small ($M \leq 2$ or $N \leq 100$), the Sort WS provides better performance while hosted in guest OS. Opposite to this, the host environment provides better performance if both parameters are huge.

The relative performance of guest vs. host OS for Sort WS is shown numerically in Table 2. The relative time varies from the worst value of 0.59 to its maximum value of 1.23. The average value is 0.96, i.e., the virtual environment provides average 4% better performance than host.

Table 1. Relative performance of guest compared to host environment for Concat WS

N / M	0	1	2	4	5	6	7	8	9	AVG
12	0.69	1.01	1.07	1.01	1.02	1.09	0.92	1.01	1.03	**1.02**
100	0.86	1.02	1.06	0.96	1.03	1.01	0.88	0.74	1.24	**0.99**
500	0.99	0.97	0.94	0.98	0.95	0.94	0.90	0.80	0.87	**0.91**
750	0.84	0.86	0.91	0.95	0.94	0.93	0.92	0.92	1.04	**0.94**
1000	0.86	0.82	0.94	0.92	0.83	0.89	0.92	0.90	0.99	**0.91**
1250	0.90	0.78	0.99	1.11	0.92	0.82	0.96	0.78	0.91	**0.93**
1500	0.96	0.78	0.87	1.12	0.91	0.91	0.93	0.96	0.98	**0.95**
1750	0.96	0.86	0.93	0.87	1.00	0.91	0.86	0.85	0.96	**0.91**
2000	0.95	0.82	0.94	0.99	0.89	0.95	0.98	0.92	1.05	**0.96**
AVG	**0.89**	**0.88**	**0.96**	**0.99**	**0.94**	**0.94**	**0.92**	**0.88**	**1.01**	**0.95**

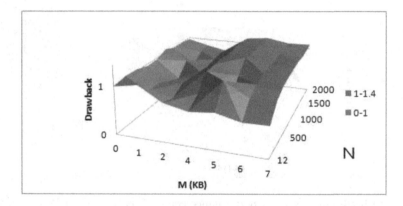

Fig. 7. Guest vs host drawback for Sort WS

Table 2. Relative performance of guest compared to host OS for Sort WS

N / M	0	1	2	4	5	6	7	AVG
12	0.99	1.07	0.79	0.69	0.74	0.63	0.64	**0.79**
100	1.04	0.80	0.72	0.86	1.18	0.91	0.93	**0.92**
500	1.03	0.81	0.87	1.17	0.71	1.02	1.03	**0.95**
750	0.96	0.69	0.83	1.16	0.98	1.09	1.07	**0.97**
1000	0.97	0.59	0.81	1.17	1.23	1.09	1.08	**0.99**
1250	1.08	0.86	0.78	1.12	1.16	1.06	1.09	**1.02**
1500	0.99	0.66	0.81	1.15	1.20	1.10	1.07	**1.00**
1750	0.99	0.84	0.85	1.16	1.16	1.10	1.08	**1.02**
2000	0.96	0.85	0.75	1.12	1.17	1.08	1.10	**1.00**
AVG	**1.00**	**0.80**	**0.80**	**1.07**	**1.06**	**1.01**	**1.01**	**0.96**

5 Conclusion and Future Work

In this paper we realized the performance analysis and comparison of web server performance by analyzing the response time. Two WSs were tested for different loads varying the main input factors: the message size and the number of concurrent messages. The experiments are realized on the same WSs hosted on bare-metal and virtualized servers on the same hardware and runtime environment.

The performance of both WSs is different. Memory demanding WS (Concat WS) shows similar performance for each value of the input parameters. Its performance decreases only when both input parameters are huge. The performance of both computation intensive and memory demanding WS (Sort WS) directly depends on the input parameter message size. The response time is stable when one of the input parameters is small. The conclusion is the same for both platforms.

We also defined quantitative performance indicators to determine the risk of migrating the services in virtual environment for various message size and number of concurrent messages. Opposite to the hypothesis, the conclusion is that the performance increases for about 5% for memory demanding and about 4% for both memory demanding and computation intensive WS if it is migrated in a virtual environment. However, the latter provides smaller performance in a virtual environment if both input parameters are huge.

Our plan for further research is to continue with performance analysis of other hypervisors on different hardware platforms, web servers and runtime environment. We will analyze how these WSs perform in dynamic, elastic and scalable and multitenant cloud environment.

References

1. Bentley, J.L., McIlroy, M.D.: Engineering a sort function. Softw. Pract. Exper. 23(11), 1249–1265 (1993)
2. Crosby, S., Brown, D.: The virtualization reality. Queue 4(10), 34–41 (2006)
3. Gusev, M., Ristov, S.: The optimal resource allocation among virtual machines in cloud computing. In: Proceedings of The 3rd International Conference on Cloud Computing, GRIDs, and Virtualization (Cloud Computing 2012), pp. 36–42 (2012)
4. Juric, M.B., Rozman, I., Brumen, B., Colnaric, M., Hericko, M.: Comparison of performance of web services, WS-Security, RMI, and RMI-SSL. J. Syst. Softw. 79(5), 689–700 (2006)
5. Li, A., Zong, X., Kandula, S., Yang, X., Zhang, M.: CloudProphet: towards application performance prediction in cloud. SIGCOMM Comput. Commun. Rev. 41(4), 426–427 (2011)
6. Liew, S.H., Su, Y.Y.: Cloudguide: Helping users estimate cloud deployment cost and performance for legacy web applications. In: 2012 IEEE 4th International Conference on Cloud Computing Technology and Science (CloudCom), pp. 90–98 (December 2012)

7. Lloyd, W., Pallickara, S., David, O., Lyon, J., Arabi, M., Rojas, K.: Migration of multi-tier applications to infrastructure-as-a-service clouds: An investigation using kernel-based virtual machines. In: 2011 12th IEEE/ACM International Conference on Grid Computing (GRID), pp. 137–144 (2011)
8. Ristov, S., Kostoska, M., Gusev, M., Kiroski, K.: Virtualized environments in cloud can have superlinear speedup. In: Proceedings of the Fifth Balkan Conference in Informatics, BCI 2012, pp. 8–13. ACM (2012)
9. Ristov, S., Tentov, A.: Performance impact correlation of message size vs. Concurrent users implementing web service security on linux platform. In: Kocarev, L. (ed.) ICT Innovations 2011. AISC, vol. 150, pp. 367–377. Springer, Heidelberg (2012)
10. Ristov, S., Velkoski, G., Gusev, M., Kjiroski, K.: Compute and memory intensive web service performance in the cloud. In: Markovski, S., Gusev, M. (eds.) ICT Innovations 2012. AISC, vol. 257, pp. 215–224. Springer, Heidelberg (2013)
11. Sahoo, J., Mohapatra, S., Lath, R.: Virtualization: A survey on concepts, taxonomy and associated security issues. In: Proceedings of the 2010 Second International Conference on Computer and Network Technology, ICCNT 2010, pp. 222–226. IEEE Computer Society (2010)
12. SoapUI: Functional testing tool for web service testing (June 2013), http://www.soapui.org/
13. Tripathi, S., Abbas, S.Q.: Performance comparison of web services under simulated and actual hosted environments. Int. J. of Computer Applications 11(5), 20–23 (2010); published By Foundation of Computer Science
14. VMware: VMware ESX (June 2013), http://www.vmware.com/
15. Walters, J.P., Chaudhary, V., Cha, M., Gallo Jr., S.: A comparison of virtualization technologies for HPC. In: Proceedings of the 22nd International Conference on Advanced Information Networking and Applications, AINA 2008, pp. 861–868. IEEE Computer Society (2008)
16. Wang, G., Ng, T.S.E.: The impact of virtualization on network performance of amazon EC2 data center. In: Proceedings of the 29th Conference on Information Communications, INFOCOM 2010, pp. 1163–1171. IEEE Press (2010)

Bayesian Multiclass Classification of Gene Expression Colorectal Cancer Stages

Monika Simjanoska, Ana Madevska Bogdanova, and Zaneta Popeska

Ss. Cyril and Methodious University, Faculty of Computer Sciences and Engineering,
Rugjer Boshkovikj 16, 1000 Skopje, Macedonia
m.simjanoska@gmail.com,
{ana.madevska.bogdanova,zaneta.popeska}@finki.ukim.mk

Abstract. Recent researches of Colorectal Cancer (CRC) aim to look for the answers for its occurrence in the disrupted gene expressions by examining colorectal carcinogenic and healthy tissues with different microarray technologies. In this paper, we propose a novel generative modelling of the Bayes' classification for the CRC problem in order to differentiate between colorectal cancer stages. The main contribution of this paper is the solution of the distinguishing problem between the critical CRC stages that remained unsolved in the published materials - distinguishing the stage I with stage IV, and stage II with stage III. The Bayesian classifier enabled application of the 'smoothing procedure' over the data from the third stage, which succeeded to distinguish the probabilities of the mentioned stages. This results are obtained as a continuation of our previous work, where we proposed methodologies for statistical analysis of colorectal gene expression data obtained from the two widely used platforms, Affymetrix and Illumina. Furthermore, the unveiled biomarkers from the two platforms were used in our generative approach for modelling the gene expression probability distribution and were used in the Bayes' classification system, where we performed binary classifications. This novel approach will help in producing an accurate diagnostics system and precising the actual stage of the cancer. It is of great advantage for early prognosis of the disease and appropriate treatment.

Keywords: Microarray Analysis, Machine Learning, Bayes' Theorem, Colorectal Cancer, Classification.

1 Introduction

Colorectal cancer is the fourth most common cause of death from cancer worldwide. The incidence, mortality and prevalence research showed that it mostly occurs in the developed regions with a total incidence of 1,234,000 cases in 2008 [1]. Recently, the colorectal cancer (CRC) occurrence is considered to be tightly connected to the gene expression phenomena. The whole genome gene expression has been observed with different types of microarray technologies in order to detect increased or decreased gene expression levels of particular genes. Gene

V. Trajkovik and A. Mishev (eds.), *ICT Innovations 2013*,   177
Advances in Intelligent Systems and Computing 231,
DOI: 10.1007/978-3-319-01466-1_17, © Springer International Publishing Switzerland 2014

expression profiling by microarrays is expected to advance the progress of personalized cancer treatment based on the molecular classification of subtypes [2].

In our previous research, we analysed colorectal gene expression from the two commonly used microarray chips, Affymetrix and Illumina. In our previous work we concluded that even though some scientists claim the two platforms produce equal outputs when examining same tissue; when considering a particular cancer, the analysis showed that both of them require different statistical approach. Therefore, we proposed methodologies for distinguishing significant genes, i.e., the biomarkers, for tissues probed with both microarrays, respectively [3]. The two biomarker sets were appropriately preprocessed for the prior distribution modelling and therefore applied in the Bayes' theorem to compute the posterior probabilities for each of the carcinogenic and the healthy class. The procedure reliability has been confirmed with the classification of new and unknown patients for the classifier, who are already diagnosed with CRC.

However, once we obtained very accurate Bayesian binary classificator, we confronted the challenge of producing Bayesian multiclass classificator capable of predicting the patient's current CRC stage. Current staging tests as: CAT scan, Magnetic Resonance Imaging, PET scan, Surgery, Complete Blood Count, etc. [4], are based either on imaging, or, blood tests and the analysis may last longer and may evoke additional stress to the patients. We believe that this type of classification is essential since the results are obtained immediately and it does not require additional microarray analysis.

The rest of the paper is organized as follows. In Section 2 we present the latest work related to the multiclass classification of the CRC stages. In Section 3 we present the methodology used to extract information from the biomarkers for the different CRC stages and the classification process itself. The experiments and the results are given in Section 4, and eventually, we present our conclusions and plans for future work in the final Section 5.

2   Related Work

In this section we briefly present some work related to the problem of gene signature revealing and the use of appropriate classifier to diagnose CRC.

Eschrich et al. [5] state that even though the Dukes' staging system, A to D, is the gold standard for predicting CRC prognosis; however, accurate classification of intermediate-stage cases, C and B, is problematic. Therefore, they propose molecular staging neural network classifier based on a core set of 43 genes that seem to have biologic significance for human CRC progression in order to discriminate good from poor prognosis patients. Another prove that stage II and III, according to the American Joint Committee on Cancer TNM staging system, are problematic for prognosis prediction is presented by Salazar et al. [6]. They present the development and validation of a gene expression signature of 18 genes that is associated with the risk of relapse in patients with stage II or III CRC. Their classifier identifies two thirds of patients with stage II colon cancer who are at sufficiently low risk of recurrence who may be safely managed without adjuvant chemotherapy. Similarly, Donada et al. [7] examined 120 stage II colon cancer patients in order to investigate the combined role of clinical, pathological and molecular parameters to identify those stage II patients who better benefit from adjuvant therapy. Farid in his research [8], compared the unsupervised artificial neural networks (ANNs) to the histopathological TNM staging system and proved that ANNs were significantly more accurate for diagnosis and survival prediction than the TNM staging system. Frederiksen at al. [9] used a nearest neighbour classifier to classify normal, and Dukes' B and C samples with less than 20% error, whereas Dukes' A and D could not be classified correctly.

The microarray experiments from patients diagnosed with different cancer stages that are used in this paper, are also applied in different researches. Here we present part of the literature related to those sets.

Laibe et al. [10] profiled both stage II and stage III carcinomas. They realized that expression profile of stage II colon carcinomas distinguishes two patterns, one pattern very similar to that of stage III tumors, based on a 7-gene signature. The function of the discriminating genes suggests that tumors have been classified according to their putative response to adjuvant targeted or classic therapies. Tsukamoto et al. [11] performed gene expression profiling and found that the overexpression of OPG gene may be a predictive biomarker of CRC recurrence and a target for treatment of this disease. Hong et al. [12] aimed to find a metastasis-prone signature for early stage mismatch-repair proficient sporadic CRC patients for better prognosis. Their best classification model yielded a 54 gene-set with an estimated prediction accuracy of 71%. Another problem of limited discrimination for Dukes stage B and C disease is presented by Jorissen et al. [13]. They conclude that metastasis-associated gene expression changes can be used to refine traditional outcome prediction, providing a rational approach for tailoring treatments to subsets of patients. Finally, three of the five microarray data sets used in this paper, have also been used by Schlicker et al. [14]. They model the heterogeneity of CRC by defining subtypes of patients with homogeneous biological and clinical characteristics and match these subtypes to cell lines for which extensive pharmacological data is available, thus linking targeted therapies to patients most likely to respond to treatment.

3 The Methodology

In this section we present the methodology used for finding significant gene signature and its application in the Bayesian multiclass classification.

3.1 Microarray Experiments

Colorectal stages systems are designed to enable physicians to stratify patients in terms of expected predicted survival, to help select the most effective treatments, to determine prognoses, and to evaluate cancer control measures [15]. The microarray experiments we used in this paper are retrieved from Gene Expression Omnibus database [16] using the following GEO accession IDs: GSE37892,

GSE21510, GSE9348, GSE14333 and GSE35896. The experiments have been performed using the Affymetrix Human Genome U133 Plus 2.0 Array which contains 54675 probes, but the unique genes observed are 21050. All data is organized into four CRC stages [17]:

- *Stage I* - In this stage cancer has grown through the superficial lining, i.e., mucosa of the colon or rectum, but has not spread beyond the colon wall or rectum. This set contains gene expression from 137 patients.
- *Stage II* - In this stage cancer has grown into or through the wall of the colon or rectum, but has not spread to nearby lymph nodes. The set contains gene expression from 257 patients.
- *Stage III* - In this stage cancer has invaded nearby lymph nodes, but is not affecting other parts of the body yet. The set contains gene expression from 182 patients.
- *Stage IV* - In this stage cancer has spread to distant organs. This set contains gene expression from 81 patients.

In order to unveil the biomarker genes in Section 3.2, we used the microarray experiment with GEO accession ID GSE8671, where 32 carcinogenic and 32 adjacent normal tissues were probed with the same Affymetrix platform.

3.2   Biomarkers Selection

The biomarkers selection methodology consists of few steps necessary for producing reliable results. Once we have retrieved both CRC and healthy tissues data, we use the following procedure which reduces the number of genes in every step:

- **Normalization.** As a suitable normalization method we use Quantile normalization, since it makes the distribution of the gene expressions as similar as possible across all samples [18] and we are interested in the genes that show significant changes in their expression.
- **Filtering methods.** In order to remove the genes with almost ordered expression levels, we used an entropy filter which measures the amount of information, i.e., disorder about the variable.
- **Paired-sample t-test.** Considering both the carcinogenic and healthy tissues are taken from the same patients and that the whole-genome gene expression follows normal distribution [19], we used a paired-sample t-test.
- **False Discovery Rate.** This method solves the problem of false positives, i.e., the genes which are considered statistically significant when in reality there is not any difference in their expression levels.
- **Volcano Plot.** Previous methods identify different expressions in accordance with statistical significance values and do not consider biological significance. In order to display both statistically and biologically significant genes we used the volcano plot visual tool.

3.3   Modelling the Prior Distributions

The biomarkers revealed in Section 3.2 showed very high precision while diagnosing both carcinogenic and healthy patients [3]. This intrigued us to test their ability to correctly classify patients into the different cancer stages we defined in Section 3.1. In order to apply the biomarkers in the Bayes' theorem, at first we must model the prior distributions for each CRC stage distinctively. Considering the little variation in the biomarkers probability distributions among the CRC stages, we used the following preprocessing procedure:

- **Round-up threshold method.** Some genes, due to noise, are negatively expressed. One way to remove the genes with negative expression is to transform all gene expression values below some threshold cut-off value to that threshold value [20]. This method is known as Round-up threshold method. In order to avoid eventual gene accumulation at one point, and thus, sustain the prior distribution shape, we chose a whole interval instead of particular value. Therefore, we map any expression value below the threshold value of 2 into the interval [0,2].
- **Normalization.** Even though the noisy gene expression values from the experiments have been previously normalized using the Quantile Normalization, we additionally used the normalization in (1) so that the overrepresented genes will be leading factor in the histogram distribution shape. Let $S_i(j,k)$ be the current stage i, for a particular gene j and a given patient k, where $i \in \{1,2,3,4\}$, $j \in \{1,...,m\}$ and $k \in \{1,...,n\}$. The number of biomarkers is m and n is the number of patients. Then the normalized gene expression is calculated as:

$$N_i(j,k) = |\frac{S_i(j,k) - \mu}{\sigma}|, \tag{1}$$

where μ and σ are the mathematical expectation and the standard deviation of $S_i(j, 1 : k)$, respectively.
- **Smoothing method.** As discussed in Section 2, stage II and III are problematic and difficult to be correctly classified because od their similarity. In this paper we propose additional smoothing method applied only on stage III gene expression data. Hereupon, we used Moving Average smoothing method, a lowpass filter, to remove the short term fluctuations.
- **Hypothesis testing.** Once we used the previous methods our data is ready for the generative modelling of the stages' prior distributions. In order to eliminate the possibility of randomly picking up the patients whose distributions does not represent the real stage's distribution, we choose the training set according to the skewness factor, i.e., the training set consists of the patients whose floored skewness factor is most common at the particular stage. The number of patients involved in the training set is nearly $\frac{3}{4}$ from the total number of patients in each stage. Our generative model fits to four types of distributions: Normal, Lognormal, Gamma, and Extreme Value. The distribution parameters are estimated using the Maximum Likelihood Estimation

(MLE) method, with a confidence level of $\alpha = 0.01$. Then we perform the Chi-square goodness-of-fit test of the default null hypothesis that the data in the tissue (vector) comes from the particular distribution with mean and variance estimated from the MLE method, using the same significance level of $\alpha = 0.01$. Once we have obtained the probabilities from the testing for each gene distinctively, we choose the distribution whose probability is highest and we assign it to the particular gene in each of the four stages.

3.4   Multiclass Bayesian Classification

As we modelled the prior distributions of all four CRC stages, we are now able to use them in the Bayes' theorem and to calculate the posterior probability for each patient to belong to each of the four classes. Given the prior distributions we can calculate the class conditional densities, $p(\boldsymbol{x}|C_i)$, as the product of the continuous probability distributions of each gene from $\boldsymbol{x}$ distinctively:

$$p(\boldsymbol{x}|C_i) = \prod f_1 f_2 \cdots f_n \, . \tag{2}$$

Since we have unequal number of patients in all four classes, considering the total number of 657 tissues, we defined the prior probabilities $P(C_i)$, to be $P(C_1) = 0.2085$, $P(C_2) = 0.3912$, $P(C_3) = 0.2770$ and $P(C_4) = 0.1233$. We consider these prior probabilities to be *test case I*. In order to assume equality in the probability of patient to be diagnosed with any of the four stages, we define *test case II*, where the prior probabilities are $P(C_1) = P(C_2) = P(C_3) = P(C_4) = 0.25$. Therefore, we calculate the posterior probability $P(C_i|\boldsymbol{x})$, as:

$$p(C_i|\boldsymbol{x}) = \frac{p(\boldsymbol{x}|C_i) * P(C_i)}{\sum_1^4 p(\boldsymbol{x}|C_i) * P(C_i)} \, . \tag{3}$$

The tissue $\boldsymbol{x}$ is classified according to the rule of maximizing the a posteriori probability (MAP):

$$C_i = \max p(C_i|\boldsymbol{x}) \, . \tag{4}$$

4   Experiments and Results

In this section we present the experiments and the obtained results.

In Section 3.2 we presented the methodology for biomarkers revealing from 32 carcinogenic and 32 healthy tissues whose gene expression was measured using the Affymetrix microarray technology. Comparing the two types of tissues, 138 genes showed significant changes in their gene expressions. Since, they showed great ability in distinguishing CRC from healthy patients, we used them in this paper to test whether the same precision will be obtained when classifying different CRC stages.

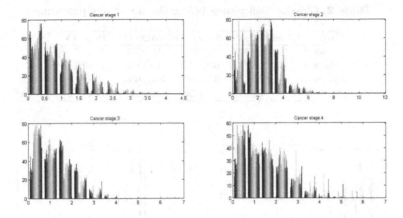

Fig. 1. The four stages after normalization

Once we retrieved gene expression data from patients diagnosed with different CRC stages we excluded all genes except the 138 biomarkers. Following the Round-up threshold method explained in Section 3.3, we handled the negative gene expression values. The results in Table 1 are from the classification of the CRC stages using the Bayesian classifier we developed in [3]. *Test Case I* and *II* refer to the prior probabilities we defined in our research [3] for both carcinogenic and healthy class. The results show that all CRC stages are classified as carcinogenic with high percent of correctness. Hereafter, our aim is to design a highly accurate Bayesian classifier with the ability to classify between CRC stages.

In order to emphasize the stages prior distributions we used the normalization method presented in (1). The results presented in Figure 1 show that stage I and stage IV have many similarities in common, as well as stage II and stage III. This is not an unexpected phenomena, since we presented some problematic classifications in Section 2. At the beginning of this research, the classification results, presented in Table 2, didn't show any problems in discriminating between stage I and stage IV; however, stage II and stage III could not be properly recognized. As a solution to this problem, we propose additional smoothing method, applied only on gene expression data from stage III. Figure 2 presents the visual changes in the distribution of stage III data.

Table 1. Bayesian Binary Classification Sensitivity

Input	Test Case I	Test Case II
Stage I	0.971	0.846
Stage II	0.969	0.876
Stage III	0.967	0.83
Stage IV	0.988	0.84

Table 2. Classification results before the smoothing procedure

Input/Class	Stage I	Stage II	Stage III	Stage IV
Stage I	**71.53%**	0.73%	18.25%	9.49%
Stage II	11.28%	**69.65%**	15.18%	3.89%
Stage III	26.37%	**37.36%**	35.16%	1.09%
Stage IV	20.99%	7.41%	8.64%	**62.96%**

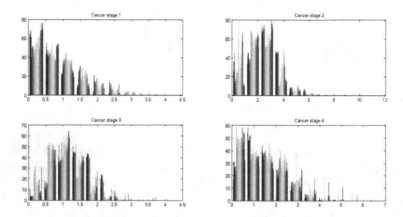

Fig. 2. The four stages after smoothing

Table 3. Bayesian Multiclass Classification Sensitivity

Input	Test Case I	Test Case II
Stage I	0.74	0.76
Stage II	0.54	0.51
Stage III	0.73	0.71
Stage IV	0.64	0.69

Table 4. Classification results after the smoothing procedure

Input/Class	Stage I	Stage II	Stage III	Stage IV
Stage I	**73.72%**	2.18%	13.86%	10.21%
Stage II	9.33%	**53.69%**	34.24%	2.72%
Stage III	4.94%	14.83%	**72.52%**	7.69%
Stage IV	20.98%	7.40%	7.40%	**64.19%**

Eventually, using carefully chosen training set of patients, we applied our generative approach for modeling the prior distributions of each class as described
in Section 3.3. Applied the class conditional densities in the Bayes' theorem
as defined in Section 3.4, produced the results from the Bayesian multiclass
classification presented in Table 3. *Test Case I* and *II* refer to the different

probabilities we defined in Section 3.4. The comparison of the percentage of correctly classified patients against the other classes is presented in Table 4.

5   Conclusion

In this paper we confronted the challenge of designing a multiclass Bayesian classifier, capable of discrimination between different colorectal cancer stages. The medical analysis shows that it is very important to discriminate between the cancer stages, in order to give the right treatment to the patient. For that purpose we used the revealed CRC biomarkers, and performed series of preprocessing procedures to produce applicable data for Bayesian classification. The results showed that Bayes' theorem can be used for problems where even details determine the class.

The main contribution of this paper is the solution of the distinguishing problem between the critical CRC stages, that remained unsolved in the published materials of [5, 6, 9] - stage I with stage IV, and stage II with stage III. We applied a 'smoothing procedure' over the data from the third stage, which succeeded to distinguish the probabilities between the aforementioned stages. The developed Bayesian classification methodology is a result of a sound mathematical and statistical theory implementation and the produced results are reliable.

In our future work we aim to test the methodology presented in this paper on gene expression data obtained from other microarray technologies, and therefore, derive general conclusion over the multiclass classification.

References

1. GLOBOCAN (2008),
   http://globocan.iarc.fr/factsheets/cancers/colorectal.asp
2. Jain, K.: Applications of biochips: From diagnostics to personalized medicine. Current Opinion in Drug Disc. & Develop. 7(3), 285–289 (2004)
3. Simjanoska, M., Bogdanova, A.M., Popeska, Z.: Bayesian posterior probability classification of colorectal cancer probed with affymetrix microarray technology. In: Proceedings of the 36th International Convention, MIPRO, CIS Intelligent Systems (2013)
4. NCI: Colon cancer treatment (2013), http://www.cancer.gov/cancertopics/pdq/treatment/colon/Patient/page2
5. Eschrich, S., Yang, I., Bloom, G., Kwong, K.Y., Boulware, D., Cantor, A., Coppola, D., Kruhøffer, M., Aaltonen, L., Orntoft, T.F., et al.: Molecular staging for survival prediction of colorectal cancer patients. Journal of Clinical Oncology 23(15), 3526–3535 (2005)
6. Salazar, R., Roepman, P., Capella, G., Moreno, V., Simon, I., Dreezen, C., Lopez-Doriga, A., Santos, C., Marijnen, C., Westerga, J., et al.: Gene expression signature to improve prognosis prediction of stage ii and iii colorectal cancer. Journal of clinical oncology 29(1), 17–24 (2011)
7. Donada, M., Bonin, S., Barbazza, R., Pettirosso, D., Stanta, G.: Management of stage ii colon cancer-the use of molecular biomarkers for adjuvant therapy decision. BMC Gastroenterology 13(1), 1–13 (2013)

8. Ahmed, F.E.: Artificial neural networks for diagnosis and survival prediction in colon cancer. Molecular Cancer 4(1), 29 (2005)

9. Frederiksen, C.M., Knudsen, S., Laurberg, S., Ørntoft, T.F.: Classification of dukes' b and c colorectal cancers using expression arrays. Journal of Cancer Research and Clinical Oncology 129(5), 263–271 (2003)

10. Laibe, S., Lagarde, A., Ferrari, A., Monges, G., Birnbaum, D., Olschwang, the COL2 Project, S.: A seven-gene signature aggregates a subgroup of stage ii colon cancers with stage iii. OMICS: A Journal of Integrative Biology 16(10), 560–565 (2012)

11. Tsukamoto, S., Ishikawa, T., Iida, S., Ishiguro, M., Mogushi, K., Mizushima, H., Uetake, H., Tanaka, H., Sugihara, K.: Clinical significance of osteoprotegerin expression in human colorectal cancer. Clinical Cancer Research 17(8), 2444–2450 (2011)

12. Hong, Y., Downey, T., Eu, K.W., Koh, P.K., Cheah, P.Y.: A metastasis-pronesignature for early-stage mismatch-repair proficient sporadic colorectal cancer patients and its implications for possible therapeutics. Clinical & Experimental Metastasis 27(2), 83–90 (2010)

13. Jorissen, R.N., Gibbs, P., Christie, M., Prakash, S., Lipton, L., Desai, J., Kerr, D., Aaltonen, L.A., Arango, D., Kruhøffer, M., et al.: Metastasis-associated gene expression changes predict poor outcomes in patients with dukes stage b and c colorectal cancer. Clinical Cancer Research 15(24), 7642–7651 (2009)

14. Schlicker, A., Beran, G., Chresta, C.M., McWalter, G., Pritchard, A., Weston, S., Runswick, S., Davenport, S., Heathcote, K., Castro, D.A., et al.: Subtypes of primary colorectal tumors correlate with response to targeted treatment in colorectal cell lines. BMC Medical Genomics 5(1), 66 (2012)

15. OConnell, J.B., Maggard, M.A., Ko, C.Y.: Colon cancer survival rates with the new american joint committee on cancer sixth edition staging. Journal of the National Cancer Institute 96(19), 1420–1425 (2004)

16. Gene expression omnibus (2013), http://www.ncbi.nlm.nih.gov/geo/

17. MayoClinic: Colon cancer (2013), http://www.mayoclinic.com/health/colon-cancer/DS00035/DSECTION=tests-and-diagnosis

18. Wu, Z., Aryee, M.: Subset quantile normalization using negative control features. Journal of Computational Biology 17(10), 1385–1395 (2010)

19. Hui, Y., Kang, T., Xie, L., Yuan-Yuan, L.: Digout: Viewing differential expression genes as outliers. Journal of Bioinformatics and Computational Biology 8(supp. 01), 161–175 (2010)

20. Tamayo, P., Slonim, D., Mesirov, J., Zhu, Q., Kitareewan, S., Dmitrovsky, E., Lander, E.S., Golub, T.R.: Interpreting patterns of gene expression with self-organizing maps: Methods and application to hematopoietic differentiation. Proceedings of the National Academy of Sciences 96(6), 2907–2912 (1999)

On the Choice of Method for Solution of the Vibrational Schrödinger Equation in Hybrid Statistical Physics - Quantum Mechanical Modeling of Molecular Solvation[*]

Bojana Koteska[1], Dragan Sahpaski[1], Anastas Mishev[1], and Ljupčo Pejov[2]

[1] University Sts Cyril and Methodius, Faculty of Computer Science and Engineering, Skopje, Macedonia
{bojana.koteska,dragan.sahpaski,anastas.mishev}@finki.ukim.mk
[2] University Sts Cyril and Methodius, Faculty of Natural Sciences & Mathematics Institute of Chemistry, Skopje, Macedonia
ljupcop@pmf.ukim.mk

Abstract. Several numerical methods for solution of vibrational Schrödinger equation in the course of hybrid statistical physics – quantum mechanical modeling of molecular solvation phenomena were applied, tested, and compared. The mentioned numerical methods were applied to compute the anharmonic OH stretching vibrational frequencies of the free and solvated hydroxide anion in diluted water solutions on the basis of one-dimensional vibrational potential energies computed at various levels of theory, including density functional theory based methodologies, as well as methods based on many-body perturbation theory. The tested methods included: i) simple Hamiltonian matrix diagonalization technique, based on representation of the vibrational potential in Simons-Parr-Finlan (SPF) coordinates, ii) Numerov algorithm and iii) Fourier grid Hamiltonian method (FGH). Considering the Numerov algorithm as a reference method, the diagonalization technique performs remarkably well in a very wide range of frequencies and frequency shifts. FGH method, on the other hand, though showing a very good performance as well, exhibits more significant (and non-uniform) discrepancies with the Numerov algorithm, even for rather modest frequency shifts. Particular aspects related to HPC-implementation of the numerical algorithms for the applied methodologies were addressed.

Keywords: Fourier grid Hamiltonian method, Numerov algorithm, diagonalization of Hamiltonian matrix, solvation, intermolecular interactions, anharmonic O-H vibrational frequency shifts, Monte-Carlo simulation.

1 Introduction

Most of the processes relevant to chemistry, biochemistry, but also for industrial processes and device functioning occur in condensed phases. Due to the variety of

---

[*] This paper is based on the work done in the framework of the HP-SEE FP7 EC funded project.

V. Trajkovik and A. Mishev (eds.), *ICT Innovations 2013*,
Advances in Intelligent Systems and Computing 231,
DOI: 10.1007/978-3-319-01466-1_18, © Springer International Publishing Switzerland 2014

intermolecular interactions occurring in condensed phases, varying in both type and strength, the overall force field that the molecules experience when embedded in a condensed phase is rather complex. At the current state of development of experimental techniques, there are practically only indirect ways to judge on the type and strength of these interactions. In particular, intramolecular vibrational frequencies are a rather significant indicator of both the type and the strength of noncovalent intermolecular interactions. From experimental side, these can be measured rather precisely with the available spectroscopic techniques (IR, Raman, etc.). From theoretical side, the available quantum chemical modeling codes enable straightforward computation of *harmonic* intramolecular vibrational frequencies (and not of the "real", *i.e.* inherently anharmonic values).

Therefore, often the experimental studies devoted to the subject have been accompanied by quantum chemical computations of the "predicted" vibrational frequency shifts which are in turn compared to the experimental spectroscopic ones. The computed "frequency shifts" (e.g. with respect to the free molecule values) are in most cases the harmonic values. As discussed before [1-3], very often it comes out that the "predicted (harmonic) frequency shifts" are in fortuitous agreement with the experimental data. Aside from that fact, numerous theoretical results have often been used to "confirm" the experimental evidence for various aspects of the intermolecular interactions. The most straightforward remedy to the situation would be to compute the more realistic anharmonic frequencies and frequency shifts. It has been shown that anharmonic contributions to the overall observed vibrational frequency shifts may be as high as 30 – 40 % [1-4]. In cases when it is possible to approximate the relevant intramolecular stretching mode as a localized one, the easiest way to compute its anharmonic vibrational frequency is to carry out a cut through the complete vibrational PES. This can be achieved by moving in an appropriate way only the atoms that are relevant to the mode in question. In such way, the one-dimensional (1D) vibrational potential is obtained, which may be further used to solve the vibrational Schrödinger equation. Solving the vibrational Schrödinger equation for an arbitrary anharmonic potential is, however, not a unique procedure. Numerous algorithms have been proposed to accomplish this task. In the present study, we carefully test the performances of three simple approaches for solution of the vibrational Schrödinger equation: the method of diagonalization of Hamiltonian matrix, the Fourier grid Hamiltonian approach [5, 6], and the standard Numerov algorithm, as implemented in the LEVEL 8.0 code [7].

2   Computational Details and Algorithms

We have generated the simulated liquid structure by a series of Monte-Carlo (MC) simulations, carried out in the *NPT* ensemble, implementing the Metropolis sampling algorithm, at $T = 298$ K, $P = 1$ atm. Actually, we have carried out several particular simulations. In the first one, one hydroxide anion plus 83 water molecules were placed in a cubic box with side length of 13.35 Å, imposing periodic boundary conditions - this simulation mimics exactly the conditions correspond to the experimental data in [8]. As this system is small from the viewpoint of long-range corrections (LRC) to the interaction energy, we have proceeded with a subsequent series of MC simulations using larger unit cells (e.g. 2, 3, 4 and 5 OH⁻ ions with 162, 243, 324 and 405 water molecules, and also in a separate series of simulations we have added 1, 2, 3, 4 and 5 Na^+ counterions in the corresponding simulation boxes). Details about these additional simulations are given in our previous paper devoted to the subject [9].

In the present study, we focus our attention to the simplest periodic system that we have considered (1 hydroxide ion plus 83 water molecules). In all MC simulations carried out in the present study, intermolecular interactions were described by a sum of Lennard-Jones 12-6 site-site interaction energies plus Coulomb terms, as explained in details in [9]. Water molecules were modeled using the SPC model potential parameters [10], while the charge distribution in the case of hydroxide anion was described with a slightly modified charged-ring (CR) description, proposed in [11]. All MC simulations consisted of a thermalization phase of at least $2.52 \cdot 10^7$ MC steps, subsequently followed by averaging (simulation) phase of at least $1.26 \cdot 10^8$ MC steps. Series of statistically-uncorrelated configurations from the MC simulation runs were used to calculate the "in-liquid" O-H stretching potential of the hydroxide ion by a quantum-mechanical (QM) approach. The degree of mutual correlation between MC-generated configurations was judged by analysis of the energy autocorrelation function, as explained in [12]. Subsequent to MC simulations, QM calculations were carried out for 100 point-charge embedded supermolecular clusters containing the central OH⁻ ion and all of the water molecules residing in the first hydration shell, as determined from the analysis of the corresponding $O_{ion}...O_{water}$ radial distribution function. This fully – QM region modeling the solution was embedded in a set of point charges placed at the positions of water hydrogen and oxygen atoms as generated by MC simulations, up to 9 Å from the hydroxide anion. Further, a series of single-point energy calculations were carried out to obtain the one-dimensional (1D) *anharmonic* O-H stretching potential energy function, varying the O-H distance from 0.850 to 1.325 Å (in steps of 0.025 Å), moving simultaneously both the oxygen and hydrogen atom, while keeping the center of mass of the OH⁻ ion fixed (*i.e.* mimicking as closely as possible the actual OH stretching mode). O-H stretching potential energy functions were computed at Hartree-Fock (HF), second-order Möller-Plesset perturbation theoretic (MP2) and Density Functional Theory (B3-LYP) levels of theory for all selected 100 oscillators, using the standard Pople-style 6-31++G(d,p) basis set was employed for the orbital expansion, solving the HF and the Kohn-Sham (KS) SCF equations iteratively. All quantum-chemical calculations were performed with the Gaussian03 series of codes [13].

As mentioned before, we have applied and tested three methods for solution of the vibrational Schrödinger equation: diagonalization technique, Numerov algorithm and the Fourier grid Hamiltonian methodology. The diagonaization technique was carried out by least-squares fitting of the energies computed by the HF, DFT and MP2 methods to a fifth-order polynomial in ΔrOH ($\Delta r = r - re$), cutting the resulting potential energy functions after fourth order and transforming into Simons-Parr-Finlan (SPF) type coordinates [14]:

$$\rho = 1 - r_{OH,e}/r_{OH} \tag{1}$$

where rOH,e is the equilibrium, i.e. the lowest-energy, value. The one-dimensional vibrational Schrödinger equation was solved variationally, using 15 harmonic oscillator eigenfunctions as a basis set. The fundamental anharmonic O-H stretching frequencies (corresponding to the $|0> \rightarrow |1>$ transitions) were computed from the energy difference between the ground ($|0>$) and first excited ($|1>$) vibrational states.

The second approach was based on Numerov method, also known as Cowell's method. This approach is actually an implicit second-order method for approximate solution of second-order differential equations of the form:

$$y''(x) = f(x, y) \tag{2}$$

with initial conditions: $y(x_0) = y_0; y'(x_0) = y_0'$ with the integrand f(x,y) being independent on y'. If one sums the Taylor series approximations for y(x + h) and y(x - h), substituting f(x,y) for y''(x) and the expression:

$$\frac{[f(x+h, y(x+h)) - 2f(x, y) + f(x-h, y(x-h))]}{h^2} \tag{3}$$

for the second derivative with respect to x, the following final result is obtained:

$$y(x+h) + y(x-h) = 2y(x) + h^2 f(x, y) + \frac{h^2}{12}[f(x+h, y(x+h)) - 2f(x, y) + f(x-h, y(x-h))] \tag{4}$$

Adopting the notation:

$$x_n = x_0 + nh ,$$

$$f_n = f(x_n, y_n) \tag{5}$$

and denoting as y_n the approximation to $y(x_n)$, the following recursive formula is derived:

$$y_{n+1} = 2y_n - y_{n-1} + h^2(f_n + (f_{n+1} - 2f_n + f_{n-1})/12) \tag{6}$$

To begin the recursion, however, two successive starting values of y are required, one of which is y_0 while the other one is approximated by a suitable method. In the present study, we have used the implementation of Numerov method in the LEVEL 8.0 code by Le Roy and coworkers [7].

The third technique used in the present study is Fourier grid Hamiltonian method [5, 6]. It is actually a discrete variable representation-based technique (DVR), in which the continuous range of coordinate values ρ is represented by a grid of discrete values ρ_i. A uniform discrete grid of ρ values was actually used throughout the present study:

$$\rho_i = i\Delta\rho \tag{7}$$

where $\Delta\rho$ is the uniform spacing between the grid points. The state function $|\psi\rangle$ is represented as a vector on a discretized grid of points in coordinate or momentum space, alternatively:

$$|\psi\rangle = \psi^\rho = \sum_i |\rho_i\rangle \cdot \Delta\rho \cdot \psi(\rho_i) = \sum_i |\rho_i\rangle \cdot \Delta\rho \cdot \psi_i^\rho$$

$$|\psi\rangle = \psi^k = \sum_i |k_i\rangle \cdot \Delta k \cdot \psi(k_i) = \sum_i |k_i\rangle \cdot \Delta k \cdot \psi_i^k \qquad (8)$$

$|\varphi\rangle$ being basis functions. Denoting a unitary matrix which performs a fast Fourier transformation (FFT) between the two representations by F, we can write:

$$\psi^k = F\psi^\rho \qquad (9)$$

The kinetic energy operator is diagonal in momentum representation, while the potential energy operator is diagonal in coordinate representation. Therefore, one can use the FFT approach to solve the stationary Schrödinger equation by the FGH method as developed by Marston et al. [5, 6]. Defining a column vector ϕ_n of the form:

$$\phi_n = \begin{bmatrix} 0 \\ 0 \\ \vdots \\ 1 \\ \vdots \\ 0 \\ 0 \end{bmatrix} - n\text{ - th row} \qquad (10)$$

the n-th column of the Hamiltonian matrix can be written in the following way implementing a forward and reverse FFT:

$$H_{in} = \left[\left(F^{-1}TF + V\right)\phi_n\right]_i \qquad (11)$$

All FGH calculations in the present study were carried out by the FGH1 code [6].

3   Results and Discussion

As described in the Computational methodology section, to obtain the one-dimensional vibrational potential energy function ($V = f(r_{OH})$) for each OH oscillator in a (different) particular aqueous environment, a series of pointwise QM energy calculations were performed. These were carried out varying the O-H distances from 0.850 to 1.325 Å with a step of 0.025 Å. Nuclear displacements were generated keeping the center-of-mass of the vibrating hydroxide ion fixed, to achieve as close as possible mimic of the real O-H stretching vibration. Subsequent time-independent Schrödinger equations with the computed $V = f(r_{OH})$ potentials were solved by the described methodologies.

The first algorithm (involving direct diagonalization of the Hamiltonian matrix) allows rather quick computation of vibrational energy levels (and their differences, corresponding to the anharmonic vibrational frequencies). Using the SPF-type coordinates is particularly convenient, as it allows only a small number of the harmonic oscillator basis functions to be used for the diagonalization and to still achieve a very good convergence. This property makes the current approach computationally cheap. We have already used this algorithm in a number of our previous works for various purposes [1-3 and references therein]. However, as it has not been used widely in the literature, it would be highly desirable to compare its performances with those of other algorithms. The "golden standard", that has been used widely in the literature for the purpose of solving the radial Schrödinger equation, is the Numerov algorithm. It has been implemented in the publicly available LEVEL [7] code written by Le Roy and collaborators. The Fourier grid Hamiltonian method [5, 6], on the other hand, has been claimed to be the simplest method for solution of both time–independent and time-dependent Schrödinger equation. It has become more and more popular in recent years.

Anharmonic frequencies corresponding to the fundamental |0> →|1> vibrational transition for the free OH⁻ ion computed with the three different methods for solution of the vibrational Schrödinger equation are given in Table I, together with the experimental gas-phase data [15].

Table 1. Calculated anharmonic frequencies for the isolated hydroxide ion with the three different algorithms for solution of the vibrational Schrödinger equation at the three different levels of theory (see text for details). Experimental fundamental frequency is listed as well.

OH⁻(g)	HF	B3LYP	MP2	Exp. [1]
	ν / cm^{-1}			
Diagonalization	3868	3552	3652	
Numerov	3913	3566	3667	3556
FGH	4112	3702	3792	

In Table 2, selected anharmonic frequencies of fundamental |0> →|1> vibrational transitions for the solvated OH⁻ ion in different in-liquid environments are presented, together with the frequency shifts. Complete list of values computed at the three levels of theory are available from the authors upon request. As mentioned before, Numerov algorithm seems to be the most widely used general-purpose approach to solution of the vibrational Schrödinger equation. Therefore, we regard it as a sort of benchmark towards which the performances of other methods are compared. In Figs. 1 and 2, the anharmonic frequency shifts of the aqueous hydroxide ion with respect to the gas-phase value computed by diagonalization and FGH approach *vs.* the shifts obtained by the LEVEL code are plotted. It can be seen that the diagonalization

---

[1] The gas-phase value is taken from [15].

method in SPF-type coordinates works remarkably well for frequency shifts up to about 300 cm^{-1}. Remarkable discrepancies between this method and the Numerov algorithm appear only for a very large upshift of about 600 cm^{-1} - when the OH stretching potential is very steep. The performance of this relatively simple and efficient method is, however, excellent for most of the cases relevant to the present study. The general performance of the FGH method may be characterized as quite good (Fig. 2). However, the disagreement between this method and the Numerov approach is more nonuniform and discrepancies are seen for even moderate frequency shifts.

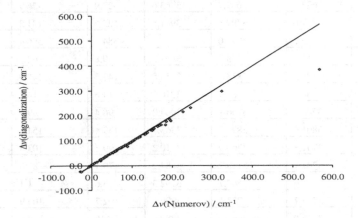

Fig. 1. Anharmonic frequency shifts computed by the diagonalization technique plotted *vs.* the corresponding values computed by Numerov algorithm

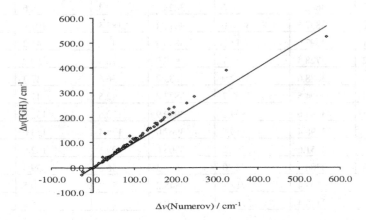

Fig. 2. A plot of the anharmonic frequency shifts computed by the FGH technique *vs.* the corresponding values computed by Numerov algorithm

Table 2. The computed frequencies and frequency shifts with the three algorithms at MP2 level of theory (see text for details)

Method	Diag.	Numerov	FGH	Diag.	Numerov	FGH
Oscil- lator			$v \, / \, cm^{-1}$			$\Delta v \, / \, cm^{-1}$
1	3697.7	3711.6	3838.5	45.7	45.1	46.9
2	3699.6	3713.3	3839.6	47.6	46.8	48.0
3	3720.9	3735.2	3865.4	68.9	68.6	73.8
4	3628.1	3639.7	3763.6	-23.9	-26.9	-28.0
5	3777.6	3797.8	3941.8	125.6	131.2	150.1
6	4036.7	4233.0	4316.5	384.7	566.4	524.9
7	3711.8	3728.9	3867.0	59.8	62.3	75.4
8	3794.5	3813.9	3955.8	142.5	147.3	164.2
9	3666.6	3680.7	3810.2	14.6	14.1	18.6
10	3748.4	3766.1	3904.2	96.4	99.5	112.6
11	3801.4	3821.2	3963.7	149.4	154.6	172.1
12	3751.3	3769.9	3910.2	99.3	103.3	118.6
13	3731.1	3748.2	3884.2	79.1	81.6	92.6
14	3655.6	3667.7	3792.3	3.6	1.1	0.7
15	3754.7	3772.5	3909.4	102.7	105.9	117.8
16	3646.0	3660.0	3789.9	-6.0	-6.6	-1.7
17	3733.6	3750.2	3884.4	81.6	83.6	92.8
18	3758.1	3774.6	3908.3	106.1	108.0	116.6
19	3682.3	3696.2	3823.7	30.3	29.6	32.1
20	3837.5	3860.0	4011.4	185.5	193.4	219.8
21	3697.6	3711.7	3840.3	45.6	45.2	48.7
22	3728.4	3755.7	3918.9	76.4	89.2	127.3
23	3688.6	3703.9	3835.2	36.6	37.3	43.6
24	3694.5	3707.7	3837.0	42.5	41.1	45.4
25	3784.6	3804.8	3949.9	132.6	138.3	158.2
26	3798.4	3819.5	3967.0	146.4	153.0	175.4
27	3751.6	3769.2	3906.6	99.6	102.6	115.0
28	3708.6	3726.0	3863.3	56.6	59.4	71.7
29	3703.4	3717.1	3845.2	51.4	50.5	53.6
30	3725.4	3742.4	3880.6	73.4	75.8	89.0

4   Conclusion and Directions for Future work

In the present work, the anharmonic OH stretching frequencies and the corresponding frequency shifts were computed using three methods for solution of the vibrational Schrödinger equation: i) method of diagonalization of the Hamiltonian matrix in

SPF-type coordinates, ii) Numerov method, iii) Fourier grid Hamiltonian method. This was done for a series of 100 in-liquid OH oscillators in the case of aqueous hydroxide ion, which have been obtained from a Monte Carlo simulation. Calculations were carried out on the basis of one-dimensional OH stretching vibrational potentials computed at three levels of theory: B3LYP, HF and MP2/6-31++G(d,p). The main aim was to make careful comparison between the methods. For most of the oscillators that have been considered, for frequency shifts up to about 300 cm^{-1}, the simple diagonalization technique performs very well. The computed frequency shifts with the diagonalization method are in excellent agreement with those obtained with the Numerov algorithm. The Fourier grid Hamiltonian approach, on the other hand, leads to frequency shifts which are in larger discrepancy with those computed by Numerov algorithm, even for moderate frequency shifts, and the discrepancies are also quite nonuniform.

References

1. Kocevski, V., Pejov, L.: On the assessment of some new meta-hybrid and generalized gradient approximation functionals for calculations of anharmonic vibrational fre-quency shifts in hydrogen-bonded dimers. J. Phys. Chem. A. 114, 4354–4363 (2010)
2. Pejov, L., Hermansson, K.: On the nature of blueshifting hydrogen bonds: ab initio and den-sity functional studies of several fluoroform complexes. J. Chem. Phys. 119, 313–324 (2003)
3. Pejov, L.A.: gradient-corrected density functional and MP2 study of phenol-ammonia and phenol-ammonia(+) hydrogen bonded complexes. Chem. Phys. 285, 177–193 (2002)
4. Silvi, B., Wieczorek, R., Latajka, Z., Alikhani, M.E., Dkhissi, A., Bouteiller, Y.: Critical analysis of the calculated frequency shifts of hydrogen-bonded complexes. J. Chem. Phys. 111, 6671–6678 (1999)
5. Marston, C.C., Balint-Kurti, G.G.: The Fourier grid Hamiltonian method for bound state eigenvalues and eigenfunctions. J. Chem. Phys. 91, 3571–3576 (1989)
6. Johnson, R.D.: FGHD1 – program for one-dimensional solution of the Schrödinger equation, Version 1.01
7. Le Roy, R. J.: LEVEL 8.0: A Computer Program for Solving the Radial Schröodinger Equa-tion for Bound and Quasibound Levels, University of Waterloo Chemical Physics Research Report CP-663 (2007), http://leroy.uwaterloo.ca/programs/
8. Corridoni, T., Sodo, A., Bruni, F., Ricci, M.A., Nardone, M.: Probing water dynamics with OH-. Chem. Phys. 336, 183–187 (2007)
9. Mitev, P., Bopp, P., Petreska, J., Coutinho, K., Ågren, H., Pejov, L., Hermansson, K.: Different structures give similar vibrational spectra: The case of OH– in aqueous solution. J. Chem. Phys. 138, 64503 (2013)
10. Berendsen, H.J.C., Postma, J.P.M., van Gunsteren, W.F., Hermans, J.: Interaction models for water in relation to protein hydration. In: Pullman, B. (ed.) Intermolecular Force, pp. 331–342. Reidel, Dordrecht (1981)
11. Ufimtsev, I.S., Kalinichev, A.G., Martinez, T.J., Kirkpatrick, R.J.: Chem. Phys. Lett. 442, 128–133 (2007)

12. Coutinho, K., de Oliveira, M.J., Canuto, S.: Sampling configurations in Monte Carlo simu-la-tions for quantum mechanical studies of solvent effects. Int. J. Quantum Chem. 66, 249–253 (1998)
13. Frisch, M.J., et al.: Gaussian 03 (Revision C.01). Gaussian, Inc., Pittsburgh (2003)
14. Simons, G., Parr, R.G., Finlan, J.M.: New alternative to dunham potential for diatomic-molecules. J. Chem. Phys. 59, 3229–3234 (1973)
15. Owrutsky, J.C., Rosenbaum, N.H., Tack, L.M., Saykally, R.J.: The vibration-rotation spec-trum of the hydroxide anion (OH-). J. Chem. Phys. 83, 5338–5339 (1985)

Taxonomy of User Intention and Benefits of Visualization for Financial and Accounting Data Analysis

Snezana Savoska[1] and Suzana Loshkovska[2]

[1] Faculty of Administration and Information System Management,
University St. Kliment Ohridski - Bitola, Partizanska bb, 7000 Bitola, Republic of Macedonia
snezana.savovska@uklo.edu.mk
[2] Faculty of Computer Science and Engeneering, Ss. Cyril and Methodius University in
Skopje, Rugjer Boshkovikj 16, 1000 Skopje, Republic of Macedonia
suze@feit.ukim.edu.mk

Abstract. We propose a new multidimensional and multivariate taxonomy for information visualization. This taxonomy takes into consideration users' intentions and the benefits of visualization for analysts, businessmen and managers who use financial and accounting data. To explain the proposed taxonomy, a taxonomic framework has been defined. The framework contains three groups of attributes classified according to visual techniques and their capabilities. We have analyzed and coded several multidimensional and multivariate visualization techniques. Creating this kind of a taxonomic model for visualization of multidimensional and multivariate financial or accounting data implies a possibility for introducing an automatic selection of a visualization technique and the best visual representation.

Keywords: data visualization, taxonomy, financial and accounting data, user's design model, multi-dimensionality.

1 Introduction

Many classifications and taxonomy methods used in data and information visualization usually start from data itself and used techniques. There are many different approaches for creating taxonomies of the three visualization types (information, data and scientific), according to both users and developers' preferences. Taking into consideration the financial and accounting data, the conventional data representation is made dependent on the end users' preferences.

It is very difficult to handle everyday information overflow, which is the subject of analysis by managers and other users. Therefore a proactive policy for preparing visual reports and effective presentations is required. So, we propose a coherent review and conceptual framework that can provide design of desired techniques classification and selection, depending on users' intentions, their capabilities and the benefits granted to the end users.

This study focuses on the use of the most popular visualization techniques for multidimensional and multivariate (mdmv) data visualization, as data is usually multidimensional and its attributes can be of different nature. The creation of the taxonomic model can help in defining the usefulness of each technique for a particular user group according to both specific information and analytical knowledge.

The implementation of the systems for visual representation of mdmv data [27] will lead to more efficient, faster and better prepared visual information in the future. Additionally, it can reduce the time needed for bridging the gap between the necessary and the expected results for a particular users group. In this paper, we also propose some strategies for creating mdmv data visualization and representation automation, which is used by financial and accounting end users.

The paper is organizes as follows. The second section is the survey of the mdmv visualization area. The next section is dedicated to the taxonomy dimensions and the following one explains in detail the proposed taxonomy. The fifth section discusses results and effects of taxonomy usage. The conclusion depicts remarks of future work.

2   Related Works

Several classifications have been proposed for the visualization so far. They assume taxonomy of the data, including its characteristics, the number of independent variables, variables and data types, etc. Keim at all, [15,16,17,18] presented taxonomy by the manner of display techniques: pixel-oriented, geometric projections, icon-based, hierarchical and graph-based. Later, this taxonomy was extended with two orthogonal criteria: the distortion technique and the interaction technique.

Buja [3] proposed taxonomy for high-level multidimensional data visualization, which distinguishes static and interactive views. The root of this taxonomy is the division of data visualization into data rendering and data manipulation. The data rendering is categorized according to the basic plot type, which can serve as a start for further subdivision. Data manipulation is operating with individual plots and organization of multiple plots on the display. The visual representation is classified in three styles: scatter-plot matrix, functional transformation and glyphs.

Chuah and Roth [9] classify the visualization depending on implementation tasks. The meaning in this case is: decomposition of user interface and data interaction system. Zhou and Feiner [28] set visual tasks taxonomy, trying to automate the design process in the visual presentation. They tried to link the high-level presentation intent with the low-level visual techniques. Taking into account the visual aspect, they proposed three types of visual perceptions and knowledge principles: visual organization, visual signaling, and visual transformation. Tweedie [24] describes visualization taxonomy that explains how fairly valid assumptions can support problem solving. According to this taxonomy, the three aspects of externalization are considered. They include underlining data used for representation, the form of interactivity and the input and output information explicitly represented by the visualization. The further division depends on the purpose of the externalization.

Shneiderman [20] incorporated the visualization tasks into the taxonomy, but the diversity of these tasks for different application areas are not taken into consideration. He introduced a visual mantra, which means the visualization process has three phases: overview, zooming and filtering, and detail on user's demands. Shneiderman proposed seven tasks in the information visualization and seven types of visualized data. OLIVE, the taxonomy proposed by Shneiderman's students defines the eight data types: time, 1D, 2D, 3D, multidimensional, trees, networks and workspace.

Sometimes, scientific data has to be visualized with the techniques used for physical data visualization. For this reason, it is necessary to have a different kind of understanding about how users make their visualization. This would lead to better data understanding, focus and evaluation of visual representations, no matter if it is scientific or information visualization [23]. The purpose of the proposed models is creating a spreadsheet with automatic selection of visual techniques that will lead to the desired visual display. Bertin [2] gives the overall matching analysis between data characteristics, graphic variables and human perception. He also defines raw/derived data, as well as constructed and converted data.

The future classifications are made in the direction of defining the statistical data, converted data, such as the box plot, histograms, scatter plots and summary statistics [26], multidimensional trees as well as 3D structures [13]. Card and Mackinlay [4] in their taxonomy described the seven data types: physical data, 1D, 2D, 3D, multidimensional data, trees and networks. Card, Mackinlay and Shneiderman [5] proposed an automated approach based on the similar ideas where the system performs optimization. Card and Mackinlay [4] recommended three phases in the process of data visualization: marking, graphical property determination and elements that demand human control. Other taxonomies involve the graphic rules for plot defining [27], user objectivities in the process of design and optimal representation selection [6]. There is a taxonomy that involves design methodologies, depending on the user preferences and needs [10]. Another taxonomy leads to a catalog development, which means creation of rules that define how the users can link the visualization techniques with specific tasks and data types. Wehrend and Lewis [25] proposed the use of cross-matrix, linking the two classifications: the ones of the objects and the ones of the operations.

In all previously mentioned classifications, the visualized data type plays the major role. For this reason, the visual design that demands taxonomy leads to automated algorithms for the programmers and users. This is called model-based design [22] and is less dependent on the data type. The taxonomy that avoids the data-centric aspect is the Data State Reference Model [7]. According to Chi's taxonomy, each visualization technique is broken down into four data stages, three types of data transformation and four types of within-stage operators [8]. With this taxonomy, it is possible to determine the dependence between visualization models and visualization techniques.

Other taxonomy takes into consideration two aspects of analysis from the user and developer side. It is especially useful for GIS systems. This taxonomy is the Operator Interaction Framework taxonomy. It offers two classifications – developers' and users' [6]. Liu [14] mentioned the mental models for visual reasoning and interaction, which links the interactive representation with the mental models and external visualization. There are proposed many other taxonomies with its own classification method, and they all aim at visualization in specific areas of application, such as medicine, history [11], finance [1], etc. There is even a "periodic system of visualization methods" which pretended to be the basic concentrated concept of successfully used methods for data visualization until now. However, none of this is focused on user's selection of desired techniques and his' intention, gained effects and interaction with data. For these reasons, we propose a newel taxonomic framework for mdmv visualization, which is focused on the specific group of users of financial and accounting data.

3   Taxonomy Dimensions

In this paper, we propose taxonomy for mdmv data visualization for financial and accounting data. To understand the proposed taxonomic framework, it is necessary to define the dimensions according to which the visualization techniques are analyzed and presented. We define the dimension called user intention, which is a prior goal of the visualization [21] from the users' perspective. Users may want to see only the mdmv data on one or more screens, to accept or reject the hypothesis, or to explore data to bring new conclusions. The second dimension defines the effects of visualization techniques that consist of the following properties: visibility (or overview), interpretability of the visualization and the possibility of insight in data.

The third dimension is connected with the interaction possibility. The interactivity means a possibility of data filtering attribute selection, and interactivity with data itself. The next dimension refers to the user groups. Each combination of these four dimensions is a single vector in the four-dimensional space. If one needs to visualize these four dimensions in the 3D space, we need to use some additional signs for the fourth dimension. The available opportunities are the colors or some specific graphics primitives. The set of axes for the dimensions in the 3D Cartesian coordinate system and the fourth dimension can be the object of consideration and discussion. It can also be an object of optimization, aiming to gain the best view of the proposed taxonomy.

4   The Proposed Taxonomic Model

If we present the first three dimensions on the axes in the 3D Cartesian system, the selected technique for mdmv data visualization maybe presented as an independent variable. For this purpose, it is useful to make a coding of all values, which the independent variables can take for all three axes.

The first dimension, user intention, can take discrete data values: data overview; hypothesis confirmation and insight, delve into the data and making new decisions. Their coding is shown on the Table 1. The second dimension is the visualization technique effects. This dimension may take the values: visibility; interpretability and delve into data (insight in data). Within their own division, the attributes may take quantitative, ordered, or nominal data values. The values can be discrete or continuous. The proposed coding for this dimension is shown on Table 2.

The visibility is coded on the number of used visual displays. Another attribute is the selection of the navigation object through dimensions and measures, the granularity of the presented data and the opportunities to display the relations between

Table 1. The first dimension coding – User intention

1	**Data Overview**	UI/DO
2	**Hypothesis Confirmation**	UI/HC
3	**Delve into (the data) and making new decisions**	UI/DID&MND

Table 2. Coding of dimension VTE -Visualization technique's effects

Visualization techniques effect	Type of variable	Variable rang	Used code
Visibility	One screen or n-screen	data are shown on one, two or n-screens	$S_1, S_2, ... S_n$
	Object selection (slider, tab or radio button, combo box, command button)	Possibility to select data with object	SL, TB, RB, CB, COM, NO
	Analytical or aggregated data	Analytical data are shown or data aggregations are shown	AN, AG
	a) Relations are visible b) Relations aren't visible	The relations between data are or are not shown	RV, RNV RELV, RELUNV
Interpretability	The level of data understanding	The data understanding is at the: Low, Middle or High level	LL,ML,HL
	Relationship understanding level	Strong capability, middle, weak or no possibility for correlation discovery	SC, MLC, WC, NoC
	The aggregation understanding	There is visible: clustering, classification, association, rule detection, there is not visible rule	NoVIS, VCLU, VCLA, VASS, VRD
Insight in data	Possibility for ordinary statistical, mathematical data analysis, no possibility	Statistical or mathematical data analysis possibility or no possibility	IDAS, IDAM, IDAN
	Possibility to discover correlation	Correlation discovering level (1-5)	ICORR 0-5
	Possibility for cluster analysis	The level of clusters (1-5)	ICLU 0-5
	Possibility for classification	Possible classification level (1-5)	ICLS 0-5
	Possibility for pattern recognition	The possibility level of pattern recognition (1-5)	IPR 0-5
	Possibility for discovering associations	Association discovering possibility	IAD 0-5

data in the displays. The interpretability may be broken down into three main components. Opportunities of insight or delve into the data is reviewed on the basis of the six parameters. Because the third dimension is connected with interaction possibility, we can define three tasks: selecting data with filtering, selecting the attributes for analysis, and the interaction with data. The coding of the third dimension is shown on Table 3. The interaction possibilities are classified according to data selection at the beginning.

The second interaction possibility can be ranked according to the possibility of selection of time slabs or selecting data with embedded objects that select dimensions, measures or time periods. The third attribute is the possibility of interaction with data. The attribute selection is defined as pre-prepared with queries or in the visualization itself using nested drill-down possibilities, with selection of attributes for analysis using some selection objects or selection with slider or pointer. The interaction possibility with the data ranks from lack of interaction possibility to possibility of selection of analytical or aggregated data. According to the interaction techniques, the possibilities are ranked from simple navigation to zooming possibilities, filtering, distortion as well as linking and brushing.

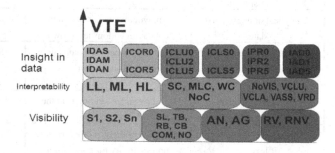

Fig. 1a. The second dimension visual representation

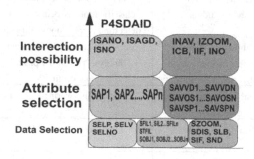

Fig. 1b. The third dimension visual representation

The fourth dimension represents the users groups. The defined user groups are given in Table 4, along with their specific tasks and prior IT knowledge. It is important to point out that the analytical tasks are highlighted for analytical workers and they have the highest level of data decomposition needs, while top managers need data extracted from summary statistics. Managers from the middle level require aggregate data and perform data analysis of selected data. Managers from operating level require simple screens with actual values and opportunities to compare attributes. This includes alerts and displays with opportunities to select attributes. All user groups require specific and different time periods, so opportunities for interaction with data is important.

Analysis under the proposed taxonomy and possibility for selection and interaction

Technique	Possibility for data selection			Attribute selection		Interaction with data	
	I	II	III	I	II	I	II
Dashboard (THWS&Z)	SELV	SFIL2	SND	SAP3	SAVOS3	ISAGD	INAV/IZOOM

Analysis according to the proposed taxonomy with the properties visibility, interpretability and insight in data

Technique	Visibility				Interpretability			Insight in data					
	I	II	III	IV	I	II	III	I	II	III	IV	V	VI
Dashboard (THWS&Z)	S2	SL	AG	RELUNV	HL	NoC	NoVIS	IDAN	ICORR0	ICLU2	ICLS0	IPR2	IAD1

Fig. 2. Analysis based on the proposed taxonomy of time histograms with more displays and possibility select and zoom

5   The Results

The proposed taxonomy classifies the user's purpose and benefits of visual analysis of the financial and accounting data based on several criteria: the user intention, visual effects, and the possibility of selection and interaction with data. Presented strategy, among other things, can serve as the basis for automation of the choice of visualization techniques for specific purposes. Indeed, the implication of such a division would provide a high degree of specification of individual visualization techniques for specific purposes and specific opportunities. Each attribute can be encoded by a weight factor to make optimization mechanisms. The mechanisms would determine the optimal contribution of the techniques for solving specific problems. Surely this is far from easy task and requires complex mathematical calculations, but still worth exploring because of the growing flood of data collected and the need for their efficient and rapid analysis.

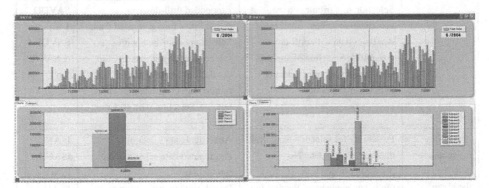

Fig. 3 Visualization with dashboards – selection time histograms on more displays and possibility select and zoom

The choice of visualization techniques, according to the criteria of the proposed taxonomy, as well as the possibilities for automatically selecting the most appropriate visualization techniques are presented. These tasks will be incorporated in the software tools to increase the speed and performance of the analysts, provide faster detection of exceptions, subdivisions etc. The incorporation of such algorithms will automatically lead to the creation of easy to use interactive data visualization. This visualization can have multiple purposes: creation of control systems of current operations (as dashboard), creation of visual data cards with the results in the time (as scorecard), delve into the relation between data, etc. Evaluation of some techniques with this taxonomy is presented in the related authors' work [19]. Example of coding

Table 3. Possibility for selection data, attributes and interaction with data

Interaction possibility	Type of variable (Nominal, Ordered, Quantitative)	Variable rang	Used code
Selection of subsets from the visualization dataset	With previous data preparation, In the visualization screen, No selection possibility	The visualization is prepared with already selected data set, on a whole data set and selection is enabled, Selection is not enabled	SELP, SELV, SELNO
	Enabled selection – filter for data/ There is only time filter/ Enabled with some object (slider, tab, combo or radio button)	There is data filter for all dimensions, Only for time period, Select data set with given object for data selection	SFIL 1-n, STFIL, SOBJ 1-n
	Zoom, Selection, Distortions, Linking and brushing	Data can be selected with zooming, distortion, linking and brushing or interactive filtering	SZOOM, SDIS, SLB, SIF, SNO
Selection of desired data attributes	With previous selected attribute (query or alias)	Previous prepared data set with selected attribute – number of selected attributes	SAP 1-n
	Selection of attributes on the visualization screen: Aggregated Analytic- data with drill-down possibility, Selection of attribute with object selection, Selection with slider or pointer	Embedded drill-down possibilities for aggregated data, Selection of attribute with selection of object dedicated to the desired attribute (radio button, tab, combo box or check box…), Selection with slider or pointer	SAVDD 1-n, SAVOS 1-n, SAVSP 1-n
Possibility of interaction with data in the given visualization	Possibility for selection analytical data/aggregated data/ No possibility interaction	Analytical data selection/ Aggregated data selection/ No interaction	ISAND, ISAGD, ISNO
	Only data navigation, Possibility for zooming, for linking and brushing, for interactive filtering	Possibility for: navigation only, zooming, linking& brushing, interactive filtering	INAV, IZOOM, ILB, IIF

the visualization with dashboards as mdmv technique is shown on the Fig. 2 and the visual representation on Fig. 3.

For all mdmv techniques this kind of tables can be created. They are basis for creating of precise software tools, which can be developed with prediction analysis or optimization algorithms. Maybe the most important organizational issue in this case is staff's training for the visualization possibilities, which sometime means effective and rapid gaining of information or transforming tabular data representation in visual ones, which means much information in shorter time.

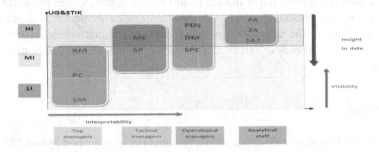

Fig. 4. Taxonomy space for defined variables as the base for classification

Table 4. User groups with specified tasks and level of information and analytical knowledge

Users' group	Specified tasks	Information knowledge	Used code
Top managers	Strategic management and planning activities	High level	SMHI, SMMI, SMLI
	Planning and control activities	Middle level	PCHI, PCMI, PCLI
	Region management activities	Low level (informational and analytical knowledge)	RMHI, RMMI, RMLI
Tactical (middle level) managers	Sector's planning and control	High level of	SPHI, SPMI, SPLI
	Management by exception	Middle level of	MEHI, MEMI MELI
		Low level of (information and knowledge)	LEHI, LEMI, LELI
Operative managers	Standard procedures control	High level of informational and analytical knowledge	SPCHI, SPCMI, SPCLI
	Operation management	Middle level of informational knowledge	OPHI, OMMI, OMLI
	Problem detection and solving	Low level of information and knowledge	PDSHI, PDSMI, PDSLI
Analytical staff	Specific analytic task	High level	SATHI, SATMI, SATLI
	Exception analysis	Middle level	EAHI, EAMI, EALI
	Perception analysis	Low level (information and knowledge)	PAHI, PAMI, PALI

6   Conclusions

We proposed taxonomy of the visualization techniques for mdmv data analysis to support the creation of a coherent and comprehensive conceptual framework that can allow the user classification by user intention and benefits of visualization for financial and accounting data.

The first variable is classified according to user intention. The next variable is the visual effects of visualization and the third one is the interaction possibility. Then, these variables are broken down in details and their possible values are defined. The possible values are presented in the tables that can be easily understood and used to create algorithms for the automation of visual data representation and visual analysis through interactive displays. Visual representation of the techniques is presented in the Fig. 4.

References

1. Ahokas, T.: Information Visualization in a Business Decision Support System (2008), http://www.doria.fi/bitstream/handle/10024/38926/informat.pdf?sequence=1 (January 12, 2013)
2. Bertin, J.: Semiology of Graphics. University of Wisconsin Press, Madison (1983)
3. Buja, A., Cook, D., Swayne, D.F.: Interactive high-dimensional data visualization. Journal of Computational and Graphical Statistics 5, 78–99 (1996)
4. Card, S.K., Mackinlay, J.: The structure of the Information Visualization Design Space. IEEE Xplore (2009), http://dl.acm.org/citation.cfm?id=857632 (January 11, 2013)
5. Card, S.K., Mackinlay, J.D., Shneiderman, B.: Readings in Information Visualization: Using Vision to Think. Morgan Kaufmann Publishers, San Francisco (1999)
6. Chengzhi, Q., Chenghu, Z., Tao, P.: Taxonomy of Visualization Techniques and Systems – Concerns between Users and Developers are Different. Asia GIS CP, Wuhan, China (2003)
7. Chi, E.H., Riedl, J.T.: An Operator Interaction Framework for Visualization Systems. In: Proc. IEEE S. Inf. Vis., pp. 63–70. IEEE (1998)
8. Chi, E.H.: A Taxonomy of Visualization Techniques using the Data State Reference Model. In: Proc. IEEE S. Inf. Vis., pp. 69–75. IEEE (2000)
9. Chuah, M., Roth, S.: On the semantics of interactive visualization. In: Proceedings of IEEE Visualization (Vis 1996), pp. 29–36. IEEE (1996)
10. Espinosa, O.J., Hendrickson, C., Garrett, J.H.: Domain Analysis: A Technique to Design A User-Centered Visualization Framework. In: Proc. IEEE S. Inf. Vis, pp. 44–52. IEEE (1999)
11. Foni, A.E., Papagiannakis, G., Magnetat-Thalmann, N.: A Taxonomy of Visualization Strategies for Cultural Heritage Application. ACM Journal on Computing and Cultural Heritage 3, 1–21 (2010)
12. Inselberg, A.: Parallel Coordinates: Visual Multidimensional Geometry and Its Application. Springer Science+Business Media (2009)
13. A Periodic Table of Visualisation Methods (May 12, 2012), http://www.visual-literacy.org/periodic_table/periodic_table.html

14. Liu, Z., Stasko, J.T.: Mental Models, Visual Reasoning and Interaction in Information Visualization: A Top-down Perspective. IEEE Trans. Vis. Comp. Graph., 999–1008 (2010)

15. Keim, D.A., Mansmann, F., Thomas, J.: Visual Analytics: How Much Visualization and How Much Analytics? SigKDD Explorations Journal 11, 5–8 (2009)

16. Keim, D.A., Mansmann, F., Schneidewind, J., Ziegler, H.: Challenges in Visual Data Analysis. In: International Conference on Information Visualization (IV), pp. 9–16. IEEE Press (2006)

17. Keim, D.A.: Visual Exploration of Large Data Sets. Communications of the ACM 44, 38–44 (2001)

18. Keim, D.A.: Information visualization and visual data mining. IEEE Tran. Vis. & Com. Graph., 1–8 (2002)

19. Savoska, S., Loskovska, S.: Evaluation of Taxonomy of User Intention and Benefits of Visualization for Financial and Accounting Data Analysis. In: The 7th International Conference of IS & Grid Technology, Sofia (2013) (in print)

20. Shneiderman, B.: The eyes have it: A task by data type taxonomy for information visualization. In: IEEE Symposium on Visual Language, pp. 336–343 (1996)

21. Tagerden, D.P.: Business Information Visualization. CAIS. 1, Paper 4 (1999)

22. Tory, M., Möller, T.: Rethinking Visualization: A High-Level Taxonomy. In: Proceedings of the INFOVIS, pp. 151–158. IEEE (2004)

23. Tory M. Möller T.: A Model-based visualization Taxonomy. In: INFOVIS 2002 (2002), ftp://142.58.111.2/ftp/pub/cs/techreports/2002/ CMPT2002-06.pdf (December 1, 2012)

24. Tweedie, L.: Characterizing interactive externalizations. In: Proc. of the ACM Human Factors in Computing Systems Conference (CHI 1997), pp. 375–382. ACM Press (1997)

25. Wehrend, S., Lewis, C.: A problem-oriented classification of visualization technique. In: Proc. of IEEE Visualization (Vis 1990), pp. 139–143. IEEE (1990)

26. Wilkinson, L.: The Grammar of Graphics, 2nd edn. Springer-Verlag Science+Business Medi Inc., New York (1999)

27. Wong, P.C., Bergeron, R.D.: 30 Years of Multidimensional Multivariate Visualization. In: Scientific Visualization, Overviews, Methodologies, and Techniques, pp. 3–33. National Science Foundation, IEEE (1997)

28. Zhou, M., Feiner, S.: Visual task characterization for automated visual discourse synthesis. In: Proc. of the ACM Human Factors in Computing Systems Conference (CHI 1998), pp. 392–399. ACM Press (1998)

Supervisory Coordination
of Timed Communicating Processes

Jasen Markovski*

Eindhoven University of Technology,
PB 513, 5600MB, Eindhoven, The Netherlands
j.markovski@tue.nl

Abstract. We propose a synthesis-centric approach to coordination of
timed discrete-event systems with data and unrestricted nondetermin-
ism. We employ supervisory controllers to exercise the desired coordination, which are automatically synthesized based on the models of the
system components and the coordination rules. We develop a timed process theory with data that supports the modeling process and we provide
for time abstractions that allow us to employ standard synthesis tools.
Following the synthesis of the discrete-event controller that preserves
safe behavior of the supervised system, we analyze the timed behavior
by employing timed model checking. To interface the synthesis tool and
the model checker, we develop a compositional model transformation.

Keywords: supervisory control theory, timed communicating processes,
partial bisimulation, model-based systems engineering.

1 Introduction

Supervisory control theory [16] studies automated synthesis of models of supervisory controllers that ensure safe functioning of the system at hand. The theory
alleviates the difficulties experienced by applying traditional software engineering techniques for control software development [9]. The control requirements
change frequently during the design process, promoting control software development as an important bottleneck in production of high-tech systems.

We proposed to employ supervisory controllers to coordinate discrete-event
system behavior in [4,13]. Based upon the observed signals, the supervisory
controllers make a decision on which activities are allowed to be carried out
safely, and send back control signals to the actuators. By assuming that the
controller reacts sufficiently fast, one can model this *supervisory control feedback
loop* as a pair of synchronizing processes [16]. We refer to the model of the
machine as *plant*, which is coordinated by the model of the controller, referred
to as *supervisor*. The synchronization of the supervisor and the plant, or the
supervised plant, specifies the coordinated system behavior.

The activities of the machine are modeled as discrete events, which are disabled or enabled by the supervisor. Traditionally, the supervisor disables events

* Supported by Dutch NWO project ProThOS, no. 600.065.120.11N124.

V. Trajkovik and A. Mishev (eds.), *ICT Innovations 2013*,
Advances in Intelligent Systems and Computing 231,
DOI: 10.1007/978-3-319-01466-1_20, © Springer International Publishing Switzerland 2014

by not synchronizing with them [16] and, consequently, it comprises the complete history of the supervised system. The events are split into *controllable events*, which can be disabled by the supervisor in order to prevent potentially dangerous or otherwise undesired behavior, and *uncontrollable events*, which must never be disabled by the supervisor. Controllable events model activities over which control can be exhibited, like interaction with the actuators of the machine, whereas uncontrollable events model activities beyond the control of the supervisor, like observation of sensors or interaction with the user or the environment. Moreover, the supervised plant must also satisfy the *coordination requirements*, which model the safe or allowed behavior of the machine.

Initially, we developed a process theory for supervisory coordination to distinguish between the different flows of information. Namely, the supervisor observes the plant by means of history of observable events [3], by observing state signals [4], or shared data variables [13]. The feedback sent to the plant is a set of allowed controllable events. Modeling both information flows simultaneously by synchronous parallel composition is an oversimplification, especially when the information flows comprise different types of information, e.g., state- or data-based observation versus controllable events. The former still prevails in modern state-of-the-art approaches, like [7,18].

We extend the process theory with timing information in the form of timed delays, that respect the established principles of *time determinism* and *time interpolation* [15]. The former states that passage of time should not decide a choice by itself, whereas the latter allows splitting of a delay to several subsequent delays with the same accumulative duration. Naturally, there exist timed extensions of supervisory control theory, like [5,17,6], but our approach differs in that we abstract from timing information, allowing the usage of standard synthesis tools. Following the synthesis, the timed behavior of the supervised system is analyzed by employing timed model checking.

To model the supervisory control loop, we propose an appropriate timed extension of the behavioral preorder *partial bisimulation*. The preorder is employed to state a refinement relation between the supervised and the original plant, allowing controllable events to be simulated and requiring uncontrollable event to be bisimulated. The proposed relation properly captures that uncontrollable events cannot be disabled, while preserving the timed behavior and the branching structure of the supervised system modulo (bi)simulation, cf. [3]. We opt for data-based coordination requirements, which are given in terms of global invariants that depend on the allowed data assignments. They suitably capture the informal requirements written in specification documents used by the industry [12], while supporting compact operational semantics [13].

2   Timed Communicating Processes

To model timed systems, we extend the process theories $BSP_{\parallel}$ of [3], TCP* of [4], and the communicating processes with data of [13], thus obtaining *timed communicating processes (with data)*. The resulting process theory encompasses *successful termination options*, which model that the plant can successfully terminate its execution; *generic communication action prefixes with data assignments*, which model activities of the plant and update data variables; *timed delays*, which model the timed behavior, *guarded commands*, which condition labeled transitions based on data assignments and support supervision; *sequential composition*, an auxiliary operator required for unfolding of iteration; *iteration*, which specifies recurring behavior; and ACP-style *parallel composition with synchronization* [16] and *encapsulation*, which model a flexible coupling in the feedback control loop. We present only a set of core operators and additional process operations can be easily added in the vein of [2,4].

Data elements are given by the set D, data variables are given by V, and data expressions involving standard arithmetical operations are denoted by F. By $\alpha \colon V \to F$ we denote the assignment of the variables needed for the valuations. The arithmetical operations are evaluated with respect to $e_\alpha \colon F \to D$. The guarded commands are given as Boolean formulas, whereas the atomic propositions are formed by the predicates from the set $\{<, \leq, =, \neq, \geq, >\}$ and the logical operators are given by $\{\neg, \wedge, \vee, \Rightarrow\}$, denoting negation, conjunction, disjunction, and implication, respectively. We use B to denote the obtained Boolean expressions, which are evaluated with respect to a given valuation $v_\alpha \colon B \to \{\text{false}, \text{true}\}$, where false denotes the logical value false, and true the logical value true. We update variables by a (partial) update function $f \colon V \rightharpoonup D$. The set of actions is given by A, formed over a set of channels K, i.e, $A \triangleq \{c!_m?_n \mid m, n \in \mathbb{N}, c \in K\}$. By $c!_m?_n$ we denote a generic communication action between m sender and n receiver parties. We write $c!_n$ for $c!_n?_0$ and $c?_n$ for $c!_0?_n$ for $n \in \mathbb{N}$ and $c \in K$, and we write $c!$ for $c!_1$ and $c?$ for $c?_1$. The durations of the timed delays are taken from the set of positive reals $\mathbb{R}_{>0}$.

The set of process terms T is given by the grammar:

$$T ::= 0 \mid 1 \mid a[f].T \mid t.T \mid \phi :\to T \mid \partial_H(T) \mid T + T \mid T \cdot T \mid T^* \mid T \parallel T$$

where $a \in A$, $f \colon V \rightharpoonup F$, $t \in \mathbb{R}_{>0}$, $\phi \in B$, and $H \subseteq A$. Each process $p \in T$ is coupled with a global variable assignment environment that is used to evaluate the guards and keeps track of updated variables, notation $\langle p, (\alpha, \rho) \rangle \in T \times \Sigma$ for $\Sigma = (V \to F) \times V$. Here, α holds the variable assignments, whereas the predicate $\rho \subseteq V$, which is employed for synchronization, keeps track of the updated variables. We write $\sigma = (\alpha, \rho)$ for $\sigma \in \Sigma$, when the components of the environment are not explicitly required. The initial environment $\sigma_0 = (\alpha_0, \text{dom}(\alpha_0))$, where $\text{dom}(g)$ denotes the domain of the function g, provides the initial values of all variables that the process comprises.

The theory has two constants: 0 denotes delayable deadlock that cannot execute any action, but allows passage of time, whereas 1 denotes the option to successfully terminate, also allowing passage of time. The action-prefixed process with variable update, corresponding to $a[f].p$, executes the action a, while updating the data values according to f, and continues behaving as p. The time delay prefix $t.p$ allows passage of time specified by the duration of the delay t and continues behaving like p. The guarded command, notation $\phi :\to p$, specifies a logical guard $\phi \in B$ that guards a process $p \in T$. If the guard is successfully evaluated, the process continues behaving as $p \in T$ or, else, it deadlocks. The encapsulation operator $\partial_H(p)$ blocks all communication actions in H that are considered as incomplete, e.g., if we were to enforce broadcast communication between k processes over channel $c \in K$, then $H = \{c!_m?_n \mid m \neq 1,\ m+n \neq k\}$. The sequential composition $p \cdot q$ executes an action of the first process, or if the first process successfully terminates, it continues to behave as the second. The unary operator p^* represents iteration that unfolds with respect to the sequential composition. The alternative composition $p + q$ makes a nondeterministic choice by executing an action of p or q, and continues to behave as the remainder of the chosen process, or allows passage of time. The binary operator $p \parallel q$ denotes parallel composition, where the actions can always be interleaved and communication can take place over common channels.

We give semantics in terms of timed labeled transition systems, where the states are induced by the process terms and the corresponding data assignments. The dynamics is given by a successful termination option predicate $\downarrow \subseteq T \times \Sigma$, an action transition relation $\longrightarrow \subseteq (T \times \Sigma) \times A \times (T \times \Sigma)$, and a time delay transition relation $\Longrightarrow \subseteq (T \times \Sigma) \times \mathbb{R}_{>0} \times (T \times \Sigma)$. We employ infix notation and we write $\langle p, \sigma \rangle \downarrow$ for $\langle p, \sigma \rangle \in \downarrow$, $\langle p, \sigma \rangle \xrightarrow{a} \langle p', \sigma' \rangle$ for $(\langle p, \sigma \rangle, a, \langle p', \sigma' \rangle) \in \longrightarrow$, and $\langle p, \sigma \rangle \xRightarrow{t} \langle p', \sigma' \rangle$ for $(\langle p, \sigma \rangle, t, \langle p', \sigma' \rangle) \in \Longrightarrow$. Also we write $\langle p, \sigma \rangle \nRightarrow$ if there does not exist $t \in \mathbb{R}_{>0}$ and $\langle p', \sigma' \rangle \in T \times \Sigma$ such that $\langle p, \sigma \rangle \xRightarrow{t} \langle p', \sigma' \rangle$.

We introduce several auxiliary operations needed for concise presentation of the operational rules. We write $f|_C$ for the restriction of the function f to the domain $C \subseteq \mathrm{dom}(f)$, i.e., $f|_C \triangleq \{x \mapsto f(x) \mid x \in C\}$. We write $f\{g\}$ for the function $f|_{\mathrm{dom}(f) \setminus \mathrm{dom}(g)} \cup g$. We define $\downarrow$, $\longrightarrow$, and $\Longrightarrow$ using structural operational semantics [2], depicted by the operational rules in Fig. 1, where symmetrical rules are not depicted and their number is only stated in brackets.

We comment on some of the rules that reflect our design choices. Rule 2 states that the action prefix enables action transitions, whereas the target assignment updates the variables in the domain of the partial variable assignment function with the evaluation of the corresponding data expression. Rule 3 states time interpolation, meaning that every delay can be split as two positive delays. Rules 6 – 10 state that action transitions resolve nondeterministic choices, whereas passage of time does not. Synchronizing of action transitions is possible for actions that stem from the same channel as depicted by rule 25. The resulting communication action must account for the accumulative number of sender and receiver communication parties, all of which are separate components in the parallel composition due to the interleaving semantics of Fig. 1 [2].

The behavioral relation that we employ is a timed extension of partial bisimulation for processes with data [3]. Here, we adapt the approach of [2] to handle the variable assignments appropriately. Every partial bisimulation relation is parameterized with a bisimulation action set that states which actions should be bisimulated, whereas the rest are only simulated.

A relation $R \subseteq T \times T$ is said to be a timed partial bisimulation with respect to a bisimulation action set $B \subseteq A$, if for all $(p, q) \in R$ and $\sigma \in \Sigma$, it holds

$$1 \ \frac{}{\langle 1, \sigma\rangle \downarrow} \qquad 2 \ \frac{}{\langle a[f].p, (\alpha, \rho)\rangle \xrightarrow{a} \langle p, (\alpha\{\{X \mapsto e(f(X)) \mid X \in \mathrm{dom}(f)\}\}, \mathrm{dom}(f))\rangle}$$

$$3 \ \frac{t > 0, \ s > 0}{\langle (t+s).p, \sigma\rangle \xRightarrow{t} \langle s.p, \sigma\rangle} \qquad 4\,(5) \ \frac{\langle p, \sigma\rangle \downarrow}{\langle p+q, \sigma\rangle \downarrow} \qquad 6\,(7) \ \frac{\langle p, \sigma\rangle \xrightarrow{a} \langle p', \sigma'\rangle}{\langle p+q, \sigma\rangle \xrightarrow{a} \langle p', \sigma'\rangle}$$

$$8 \ \frac{\langle p, \sigma\rangle \xRightarrow{t} \langle p', \sigma\rangle, \ \langle q, \sigma\rangle \xRightarrow{t} \langle q', \sigma\rangle}{\langle p+q, \sigma\rangle \xRightarrow{t} \langle p'+q', \sigma\rangle} \qquad 9\,(10) \ \frac{\langle p, \sigma\rangle \xRightarrow{t} \langle p', \sigma\rangle, \ \langle q, \sigma\rangle \not\xRightarrow{}}{\langle p+q, \sigma\rangle \xRightarrow{t} \langle p'+q, \sigma\rangle}$$

$$11 \ \frac{\langle p, \sigma\rangle \downarrow, \ \langle q, \sigma\rangle \downarrow}{\langle p \cdot q, \sigma\rangle \downarrow} \qquad 12 \ \frac{\langle p, \sigma\rangle \downarrow, \ \langle q, \sigma\rangle \xrightarrow{a} \langle q', \sigma'\rangle}{\langle p \cdot q, \sigma\rangle \xrightarrow{a} \langle q', \sigma'\rangle} \qquad 13 \ \frac{\langle p, \sigma\rangle \xrightarrow{a} \langle p', \sigma'\rangle}{\langle p \cdot q, \sigma\rangle \xrightarrow{a} \langle p' \cdot q, \sigma'\rangle}$$

$$14 \ \frac{\langle p, \sigma\rangle \downarrow, \ \langle q, \sigma\rangle \xRightarrow{t} \langle q', \sigma\rangle}{\langle p \cdot q, \sigma\rangle \xRightarrow{t} \langle q', \sigma\rangle} \qquad 15 \ \frac{\langle p, \sigma\rangle \xRightarrow{t} \langle p', \sigma\rangle}{\langle p \cdot q, \sigma\rangle \xRightarrow{t} \langle p' \cdot q, \sigma\rangle}$$

$$16 \ \frac{}{\langle p^*, \sigma\rangle \downarrow} \qquad 17 \ \frac{\langle p, \sigma\rangle \xrightarrow{a} \langle p', \sigma'\rangle}{\langle p^*, \sigma\rangle \xrightarrow{a} \langle p' \cdot p^*, \sigma'\rangle} \qquad 18 \ \frac{\langle p, \sigma\rangle \xRightarrow{t} \langle p', \sigma\rangle}{\langle p^*, \sigma\rangle \xRightarrow{t} \langle p' \cdot p^*, \sigma\rangle}$$

$$19 \ \frac{\langle p, \sigma\rangle \downarrow, \ \langle q, \sigma\rangle \downarrow}{\langle p \parallel q, \sigma\rangle \downarrow} \qquad 20\,(21) \ \frac{\langle p, \sigma\rangle \xrightarrow{a} \langle p', \sigma'\rangle}{\langle p \parallel q, \sigma\rangle \xrightarrow{a} \langle p' \parallel q, \sigma'\rangle}$$

$$22\,(23) \ \frac{\langle p, \sigma\rangle \xRightarrow{t} \langle p', \sigma'\rangle, \ \langle q, \sigma\rangle \not\xRightarrow{}}{\langle p \parallel q, \sigma\rangle \xRightarrow{t} \langle p' \parallel q, \sigma\rangle} \qquad 24 \ \frac{\langle p, \sigma\rangle \xRightarrow{t} \langle p', \sigma\rangle, \ \langle q, \sigma\rangle \xRightarrow{t} \langle q', \sigma\rangle}{\langle p \parallel q, \sigma\rangle \xRightarrow{t} \langle p' \parallel q', \sigma\rangle}$$

$$25 \ \frac{\langle p, \sigma\rangle \xrightarrow{c!_k?_\ell} \langle p', (\alpha', \rho')\rangle, \ \langle q, \sigma\rangle \xrightarrow{c!_m?_n} \langle q', (\alpha'', \rho'')\rangle, \ \alpha'|_{\rho' \cap \rho''} = \alpha''|_{\rho' \cap \rho''}}{\langle p \parallel q, \sigma\rangle \xrightarrow{c!_{k+m}?_{\ell+n}} \langle p' \parallel q', (\alpha'\{\alpha''|_{\rho'' \setminus \rho'}\}, \rho' \cup \rho'')\rangle}$$

$$26 \ \frac{\langle p, \sigma\rangle \downarrow, \ v(\phi) = \text{true}}{\langle \phi :\to p, \sigma\rangle \downarrow} \qquad 27 \ \frac{\langle p, \sigma\rangle \xrightarrow{a} \langle p', \sigma'\rangle, \ v(\phi) = \text{true}}{\langle \phi :\to p, \sigma\rangle \xrightarrow{a} \langle p', \sigma'\rangle}$$

$$28 \ \frac{\langle p, \sigma\rangle \xRightarrow{t} \langle p', \sigma\rangle, \ v(\phi) = \text{true}}{\langle \phi :\to p, \sigma\rangle \xRightarrow{t} \langle p', \sigma\rangle} \qquad 29 \ \frac{\langle p, \sigma\rangle \downarrow}{\langle \partial_H(p), \sigma\rangle \downarrow}$$

$$30 \ \frac{\langle p, \sigma\rangle \xrightarrow{a} \langle p', \sigma'\rangle, \ a \notin H}{\langle \partial_H(p), \sigma\rangle \xrightarrow{a} \langle \partial_H(p'), \sigma'\rangle} \qquad 31 \ \frac{\langle p, \sigma\rangle \xRightarrow{t} \langle p', \sigma\rangle}{\langle \partial_H(p), \sigma\rangle \xRightarrow{t} \langle \partial_H(p'), \sigma\rangle}$$

Fig. 1. Operational rules

that: (1) $\langle p, \sigma\rangle \downarrow$ if and only if $\langle q, \sigma\rangle \downarrow$; (2) if $\langle p, \sigma\rangle \xrightarrow{a} \langle p', \sigma'\rangle$ for $a \in \mathsf{A}$, then there exist $q' \in \mathsf{T}$ and $\sigma' \in \Sigma$ such that $\langle q, \sigma\rangle \xrightarrow{a} \langle q', \sigma'\rangle$ and $(p', q') \in R$; (3) if $\langle q, \sigma\rangle \xrightarrow{b} \langle q', \sigma'\rangle$ for $b \in B$, then there exist $p' \in \mathsf{T}$ and $\sigma' \in \Sigma$ such that $\langle p, \sigma\rangle \xrightarrow{b} \langle p', \sigma'\rangle$ and $(p', q') \in R$; (4) if $\langle p, \sigma\rangle \xRightarrow{t} \langle p', \sigma'\rangle$ for $t \in \mathbb{R}_{>0}$, then there exist $q' \in \mathsf{T}$ and $\sigma' \in \Sigma$ such that $\langle q, \sigma\rangle \xRightarrow{t} \langle q', \sigma'\rangle$ and $(p', q') \in R$; and (5) if $\langle q, \sigma\rangle \xRightarrow{t} \langle q', \sigma'\rangle$ for $t \in \mathbb{R}_{>0}$, then there exist $p' \in \mathsf{T}$ and $\sigma' \in \Sigma$ such that $\langle p, \sigma\rangle \xRightarrow{t} \langle p', \sigma'\rangle$ and $(p', q') \in R$.

If R is a timed partial bisimulation relation such that $(p, q) \in R$, then p is timed partially bisimilar to q with respect to B and we write $p \preceq_B q$. It can be shown that $\preceq_B$ is a preorder and pre-congruence for every $B \subseteq \mathsf{A}$ in the vein of [13,3,2]. Finally, we note that $p \preceq_A q$ amounts to timed bisimulation, whereas $p \preceq_\emptyset q$ reduces to timed simulation preorder for processes with data [2].

3 Controllability

To distinguish between controllable and uncontrollable activities of the system, we split the set of channels to controllable $\mathsf{K_C}$ and uncontrollable $\mathsf{K_U}$ channels, respectively, where $\mathsf{K_C} \cap \mathsf{K_U} = \emptyset$ and $\mathsf{K_C} \cup \mathsf{K_U} = \mathsf{K}$. We put $\mathsf{C} \triangleq \{c!_m?_n \mid m, n \in \mathbb{N}, \ c \in \mathsf{K_C}\}$ and $\mathsf{U} \triangleq \{u!_m?_n \mid m, n \in \mathbb{N}, \ u \in \mathsf{K_U}\}$ for the sets of controllable and uncontrollable events, respectively.

For the specification of the plant, we can take every process $p \in \mathsf{T}$ since the underlying system is unrestricted. The supervisor $s \in \mathsf{T}$, however, must satisfy several structural restrictions. It should be a deterministic process that provides for an unambiguous feedback in terms of synchronizing controllable events [3]. Moreover, it should not be allowed to alter the state of the plant in any other way. It does not comprise any variable assignments, but it is able to observe them. The supervisor relies on data observations to exercise supervision [14].

The form of the proposed supervisor can be summarized in (1):

$$s = \left(\sum_{c \in \mathsf{K_C}} \phi_c :\to c![\emptyset].1 + \psi :\to 1 \right)^*, \tag{1}$$

where $\phi_c, \psi \in \mathsf{B}$ for $c \in \mathsf{K_C}$. The supervisor observes the state of the plant, identified by the corresponding data assignments, checks for which activities the guarded commands that implement the control functions are satisfied, and enables the corresponding controllable events by synchronizing with the counterpart (receiver) action transitions in the plant. It does not keep a full history of events as in the original setting of [16], where the guards ϕ_c for $c \in \mathsf{K_C}$ and ψ depict the supervision actions [14].

Given plant $p \in \mathsf{T}$ and supervisor $s \in \mathsf{T}$ of form (1), we specify the supervised plant as $\partial_H (p \parallel s)$, where the encapsulation enforces desired communication and the set $H \subset \mathsf{A}$ comprises unfinished communication events, differing per case. Different from other approaches, e.g., [16,7,18], that employ synchronizing actions over all system components, we employ the supervisor to render the final communication action complete. In the former situation, the relation between the original and the supervised plant can be provided directly, e.g., as in [3], because the labels of the transitions in the supervised and the original plant coincide. We have to rename controllable actions in the original plant as a supervisor is necessitated to make the plant operational, cf.(1) for the form of the supervisor. We employ a specific partial renaming operation $\gamma \colon \mathsf{T} \mapsto \mathsf{T}$ that completes the controllable communication actions with a send action from the supervisor of (1). The operational rules that define this renaming operation γ are given in Fig. 2.

The relation between the supervised and the original plant is specified as:

$$\partial_H (p \parallel s) \preceq_\mathsf{U} \gamma(p). \tag{2}$$

It states that the supervisor restricts only controllable events form the plant, indicated by the simulation relation between the supervised and the original

$$32 \; \frac{\langle p,\sigma\rangle \downarrow}{\langle \gamma(p),\sigma\rangle \downarrow} \qquad 33 \; \frac{\langle p,\sigma\rangle \xrightarrow{c!m?_n[f]} \langle p',\sigma'\rangle,\; c\in \mathsf{K_C}}{\langle \gamma(p),\sigma\rangle \xrightarrow{c!m+1?_n[f]} \langle \gamma(p'),\sigma'\rangle} \qquad 34 \; \frac{\langle p,\sigma\rangle \xrightarrow{u!m?_n[f]} \langle p',\sigma'\rangle,\; u\in \mathsf{K_U}}{\langle \gamma(p),\sigma\rangle \xrightarrow{u!m?_n[f]} \langle \gamma(p'),\sigma'\rangle}$$

Fig. 2. Renaming operation that completes plant controllable communication events

$$35 \; \frac{v_\alpha(\phi)=\text{false}}{\langle p,(\alpha,\rho)\rangle \models \phi \Rightarrow \xrightarrow{q} \!\!\!\!/} \qquad 36 \; \frac{v_\alpha = \text{true},\; \langle p,(\alpha,\rho)\rangle \xrightarrow{q} \!\!\!\!/}{\langle p,(\alpha,\rho)\rangle \models \phi \Rightarrow \xrightarrow{q} \!\!\!\!/} \qquad 37 \; \frac{v(\phi)=\text{true}}{\langle p,\sigma\rangle \models \phi}$$

Fig. 3. Satisfiability of data-based control requirements

plant, whereas no uncontrollable events can be disabled, as the same uncontrollable events are required to be bisimulated and, thus, present in all states of both processes. It can be shown, again in the vein of [3,4], that the traditional notions of language-based controllability of [5,16] for deterministic system and state controllability [14,18] for nondeterministic systems are implied by (2).

We opt for data-based coordination requirements stated in terms of boolean expressions ranging over the data variables. The data-based control requirements, denoted by the set R, have the following syntax induced by R:

$$R ::= \; \xrightarrow{a} \; \Rightarrow \phi \; \mid \; \phi \Rightarrow \xrightarrow{q} \!\!\!\!/ \; \mid \; \phi,$$

for $a \in \mathsf{A}$ and $\phi \in \mathsf{B}$. Now, we say that a given control requirement $r \in \mathsf{R}$ is satisfied with respect to the process $p \in \mathsf{T}$ in the assignment environment $\sigma \in \Sigma$, notation $\langle p,\sigma\rangle \models r$, according to the operational rules depicted in Fig. 3. By $\langle p,\sigma\rangle \xrightarrow{q} \!\!\!\!/$ we denote that $\{\langle p',\sigma'\rangle \mid \langle p,\sigma\rangle \xrightarrow{a} \langle p',\sigma'\rangle\} = \emptyset$. We note that the first and second requirement are logically equivalent as $(\xrightarrow{a} \Rightarrow \phi) \Leftrightarrow (\neg\phi \Rightarrow \xrightarrow{q} \!\!\!\!/)$.

We structure the modeling process in Fig. 4, which extends previous proposals [11] with timing aspects. In Fig. 4 we assume that the formalization of the coordination requirements and the modeling and abstraction of the timed plant has been successfully completed along the lines of [2]. We employ the synthesis tool Supremica [1] to synthesize a supervisory controller based on the data-based coordination requirements and the time-abstracted plant model. Then, we couple the

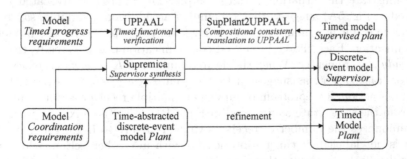

Fig. 4. Core of the proposed framework for supervisory coordination

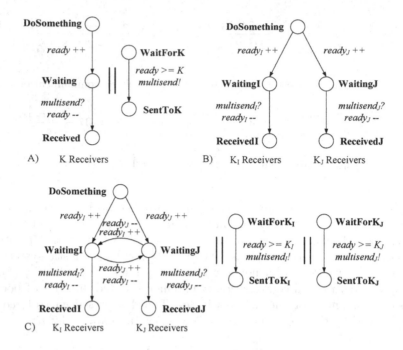

Fig. 5. Translating broadcast multi-communication in UPPAAL: A) Suggested approach in [8]; B) Occurrence of deadlock for multiple receiver actions stemming from the same state; C) Adapted approach with no deadlock for multiple actions

synthesized supervisor with the timed plant to obtain the timed supervised plant. To be able to analyze the timing behavior of the supervised system, we developed the model transformation tool SupPlant2UPPAAL [10]. The tool automatically transforms and refines a Supremica supervised plant to input suitable for the timed model checker UPPAAL [8]. The communication actions in UPPAAL are restricted to broadcast multi-communication with only one sender and multiple receivers that are not forced to participate in the communication, which requires additional effort to conform to rule 25 of Fig. 1.

We illustrate the situation by an example. The approach advocated in [8], depicted in Fig. 5A) suggests that we employ an additional variable, say *ready*, in order to count how many receiver parties are ready to receive the broadcast communication. Each time a broadcast communication becomes ready, the variable *ready* is incremented. When the desired number of K receivers is ready, then the broadcast is sent, as suggested by the guard *ready* $>= K$ associated with the sender action. The problem occurs when multiple receiver actions can originate from the same state as depicted in Fig. 5B). Since the auxiliary transition that announces the number of receivers that are ready to begin the communication is not labeled, it can be interleaved with any other action, which may lead to deadlock. To resolve this situation, we propose to also connect the auxiliary "waiting" states, such that the process can jump between them and always enable the communication as suggested in Fig. 5C). In order to prevent livelock behavior that can occur if the jumps form an indefinite loop, we make the waiting states committed [8], meaning that they do not allow for passage of time. We note that by making the waiting states committed in the situation of Fig. 5B) does not resolve the issue, but it leads directly to undelayable deadlock situation.

We revisit a case study of [12,13] dealing with coordination of maintenance procedures of a high-tech printer and extend it with timing information. Each maintenance procedure has a soft and a hard deadline with respect to number of printed pages. Soft deadline means that the procedure can be executed, whereas hard deadline means that the procedure must be executed to preserve print quality. The coordination problem is to schedule maintenance procedures, while disturbing the printing process as little as possible. We refer to [12,13] for the untimed case study and to [10] for the extended timed models and tools.

The addition of timing information enables us to also study the timing behavior, not possible in previous work [12,13]. For example, we can study whether it is possible that the maintenance procedure is executed within a certain amount of time. To this end, in our model translation to UPPAAL we carry over the component names and state labels from Supremica, cf. [13]. In addition, we introduce a global clock that we name global, to enable queries concerning the elapsed time. For the query above, we can employ the temporal logic formula:

$$E<> \text{ MaintenanceOperation.OperInProg and global} <= 40. \qquad (3)$$

The standard notation E<> stands for: there is some path and eventually there exists a state on that path such that the component MaintenanceOperation, that models the maintenance operation procedure, is in state OperInProg, which denotes that the maintenance operation is executing, and the global clock of the system has counted less then 40 time units. For our demo model, this query is not satisfied, even though we let the system reach soft deadline after 40 time units. The reason that the query is not satisfied is that we also schedule a print job in the time interval $[36, 44]$. Thus, the earliest scheduling of the maintenance procedure is at 44, which can be validated by adapting (3).

4   Concluding Remarks

We presented a synthesis-centric approach to supervisory coordination of timed discrete-event systems based on a process theory for timed communicating processes with data. To define the notion of a supervisor, we provided for a timed extension of the behavioral preorder partial bisimulation. For the implementation of the proposed framework, we employed state-of-the-art tools for supervisory controller synthesis and timed model checking. In order to interface the tools, we developed a model transformation tool. To illustrate our approach, we extended an industrial study dealing with coordination of maintenance procedures in a printing process with timing information and analyzed the timing behavior of the supervised system. As future work, we schedule an extension of the theory with time-based constraints and novel synthesis algorithms.

References

1. Akesson, K., Fabian, M., Flordal, H., Malik, R.: Supremica - an integrated environment for verification, synthesis and simulation of discrete event systems. In: Proceedings of WODES 2006, pp. 384–385. IEEE (2006)
2. Baeten, J.C.M., Basten, T., Reniers, M.A.: Process Algebra: Equational Theories of Communicating Processes, Cambridge Tracts in Theoretical Computer Science, vol. 50. Cambridge University Press (2010)
3. Baeten, J.C.M., van Beek, D.A., Luttik, B., Markovski, J., Rooda, J.E.: A process-theoretic approach to supervisory control theory. In: Proceedings of ACC 2011, pp. 4496–4501. IEEE (2011)
4. Baeten, J., van Beek, D., van Hulst, A., Markovski, J.: A process algebra for supervisory coordination. In: Proceedings of PACO 2011. EPTCS, vol. 60, pp. 36–55. Open Publishing Association (2011)
5. Brandin, B., Wonham, W.: Supervisory control of timed discrete-event systems. IEEE Transactions on Automatic Control 39(2), 329–342 (1994)
6. Cassez, F., David, A., Fleury, E., Larsen, K.G., Lime, D.: Efficient on-the-fly algorithms for the analysis of timed games. In: Abadi, M., de Alfaro, L. (eds.) CONCUR 2005. LNCS, vol. 3653, pp. 66–80. Springer, Heidelberg (2005)
7. Heymann, M., Lin, F.: Discrete-event control of nondeterministic systems. IEEE Transactions on Automatic Control 43(1), 3–17 (1998)
8. Larsen, K.G., Pettersson, P., Yi, W.: UPPAAL in a Nutshell. International Journal on Software Tools for Technology Transfer. 1(1-2), 134–152 (1997)
9. Leveson, N.: The challenge of building process-control software. IEEE Software 7(6), 55–62 (1990)
10. Markovski, J.: Demo models and model transformation tool SupPlant2UPPAAL (2013), http://sites.google.com/site/jasenmarkovski
11. Markovski, J., van Beek, D.A., Theunissen, R.J.M., Jacobs, K.G.M., Rooda, J.E.: A state-based framework for supervisory control synthesis and verification. In: Proceedings of CDC 2010, pp. 3481–3486. IEEE (2010)
12. Markovski, J., Jacobs, K.G.M., van Beek, D.A., Somers, L.J.A.M., Rooda, J.E.: Coordination of resources using generalized state-based requirements. In: Proceedings of WODES 2010, pp. 300–305. IFAC (2010)
13. Markovski, J.: Communicating processes with data for supervisory coordination. In: Proceedings of FOCLASA 2012. EPTCS, vol. 91, pp. 97–111. Open Publishing Association (2012)
14. Miremadi, S., Akesson, K., Lennartson, B.: Extraction and representation of a supervisor using guards in extended finite automata. In: Proceedings of WODES 2008, pp. 193–199. IEEE (2008)
15. Nicollin, X., Sifakis, J.: An overview and synthesis of timed process algebras. In: Huizing, C., de Bakker, J.W., Rozenberg, G., de Roever, W.-P. (eds.) REX 1991. LNCS, vol. 600, pp. 526–548. Springer, Heidelberg (1992)
16. Ramadge, P.J., Wonham, W.M.: Supervisory control of a class of discrete-event processes. SIAM Journal on Control and Optimization 25(1), 206–230 (1987)
17. Saadatpoor, A., Ma, C., Wonham, W.M.: Supervisory control of timed state tree structures. In: Proceedings of ACC 2008, pp. 477–482. IEEE (2008)
18. Zhou, C., Kumar, R., Jiang, S.: Control of nondeterministic discrete-event systems for bisimulation equivalence. IEEE Transactions on Automatic Control 51(5), 754–765 (2006)

Cross-Platform Mobile Development:
Challenges and Opportunities

Suyesh Amatya and Arianit Kurti

Faculty of Technology, Linnaeus University, Växjö, Sweden
sa222dk@student.lnu.se, arianit.kurti@lnu.se

Abstract. Mobile devices and mobile computing have made tremendous advances and become ubiquitous in the last few years. As a result, the landscape has become seriously *fragmented* which brings lots of challenges for the mobile development process. Whilst *native* approach of mobile development still is the predominant way to develop for a particular mobile platform, recently there is shifting towards *cross-platform* mobile development as well. In this paper, we have performed a survey of the literature to see the trends in cross-platform mobile development over the last few years. With the result of the survey, we argue that the *web-based approach* and in particular, *hybrid approach*, of mobile development serves the best for cross-platform development. The results of this work indicate that even though cross platform tools are not fully matured they show great potential. Thus we consider that cross-platform development offers great opportunities for rapid development of high-fidelity prototypes of the mobile application.

Keywords: mobile development, cross-platform, web-based approach, hybrid approach, literature survey, HTML5.

1   Introduction

With the rapid technological advancements in both hardware and software fronts, coupled with broadband internet and World Wide Web, mobile computing has become ubiquitous. People use different kinds of mobile devices (tablets, smartphones, PDAs, etc) for all sorts of different purposes. Just the total *"smartphone"* shipment volumes alone reached 712.6 million units in 2012, up a strong 44.1% than in the year 2011 [1]. This prodigious growth in mobile devices is equally complimented by the growth in mobile content or information that these devices consume. According to the research group Gartner Inc., worldwide mobile app store downloads surpassed 45.6 billion in 2012, nearly double the 25 billion downloads in 2011 which by 2016 will reach 310 billion downloads and $74 billion in revenue [2].

Amidst so much of seeming opportunities, there also lie huge challenges in engineering and developing mobile services and applications. Wide variety of mobile standards and operating systems on different devices mean often unfortunately, one application can work on one mobile device very well, while it does not work on the other [3]. A more prominent challenge than any other is the fragmentation which runs

V. Trajkovik and A. Mishev (eds.), *ICT Innovations 2013*,
Advances in Intelligent Systems and Computing 231,
DOI: 10.1007/978-3-319-01466-1_21, © Springer International Publishing Switzerland 2014

both length (*device fragmentation*) and breadth (*operating system fragmentation*) across the mobile landscape. Devices with different processing, memory, communication, displaying capabilities are examples of device fragmentation. And there are different companies with their own platforms running different operating systems. Apple's iOS, Google's Android, Microsoft's Windows Phone, RIM's BlackBerry OS, Symbian, etc to name a few are the different operating systems that we can find in the mobile devices resulting in operating system fragmentation.

As such, a traditional *native approach* is not always an ideal solution. Hence we investigate the *cross-platform* approaches by performing literature analysis of the papers in the field and argue that these approaches alleviate the aforementioned problems to a great deal.

2   Native Mobile Development and Looking Beyond

In *native approach* of mobile development, developers use a set of development environment and tools in the form of Software Development Kit (SDK) targeted and optimized for specific platform provided by the platform inventors and companies. Choice of a platform relies on how deeply developers want to link the application with the underlying operating system, as capabilities in one operating system may not be available in another. Using an SDK the developer may target a particular operating system and take advantage of its specific capabilities to create an application with those features [4]. Such *native applications* or *native apps* guarantee the best usability, the best features, and the best overall mobile experience but comes with severe restriction of portability and are tied to a specific platform against which they are developed. Also different platforms require different programming languages like Objective-C for iOS, Java for Android, C# for Windows Phone, Java for BlackBerry OS, C++ for Symbian, etc [5]. So targeting multiple platforms means requirement of different skill sets and familiarity with those platforms and writing separate applications for each of them. As a result developing and maintaining applications for multiple platforms become very expensive.

The *web-based approach* has come to fore in recent times to alleviate these problems regarding native development. With the advent of *HTML5*, these *web apps* are built using open standards web technology stack of HTML5/CSS3/JS that run on a standalone mobile web browser and are often now referred to as *HTML5 apps*. HTML5 is a standard and is also used as a blanket term for a family of other related web standards and technologies like CSS3 and JavaScript together with which it represents the complete package or idea that is HTML5 [6]. The browser vendor community has strongly embraced HTML5. In 2011, estimated 336 million units of mobile phones with HTML5 browser support were sold which is expected to surge to 1 billion units in 2013 [7]. Another report estimates more than 2.1 billion mobile devices will have HTML5 browsers by 2016 [8]. These browsers are increasingly supporting many different class of HTML5 features. The features include richer set of tags (date, time, email, etc) for better semantics, offline and local storage capabilities, Geolocation, more efficient connectivity with Web Sockets, first class multimedia (audio and video) support, stunning visuals with SVG, Canvas, WebGL, and CSS3, improved performance and integration with Web Workers [9].

Even though strides are being made on the browsers to bridge the gap, there still is a long way to go as these pure web apps have little or no support for many of the native and device features like camera, microphone, address book, calendar, compound gestures and interaction, etc. And there is also lack of effective packaging and app store distribution. However most of these shortcomings of web-based approach can be overcome by using the *hybrid approach* which combines the best of both the native and web-based approaches. *Hybrid apps* are developed using web technologies (HTML5/CSS3/JavaScript) but built into native apps by wrapping them inside a thin native container that provides access to native platform and device features [10]. Development of such hybrid applications essentially relies on frameworks, such as *PhoneGap* or *Titanium*, that act more or less as a *middleware* or *bridge* and provide the platform-specific implementation of API in the native programming language for the language of the framework to communicate with the native code of different platforms [11] [12]. This means the abstraction layer exposes the device capabilities (native APIs) to the hybrid app as a JavaScript API. The hybrid apps run inside a native container and leverage the device's browser engine and are displayed in a full-screen web view control, and not in a browser [13]. Like native counterparts they can be downloaded from app store and installed on the device. Certainly hybrid apps are much closer to the native apps than the pure web apps as they boast of the power, performance and availability on par with the native applications.

3   Research Question

Inspired from these trends we found that there is a need for a detailed analysis of the existing cross-platform tools and approaches. All this combined with the possibilities that web technologies offers brings a number of opportunities for development and as well some remaining challenges. Therefore the primary aim of this paper is to address the current trends in the cross-platform mobile development. For reaching this aim our approach was built on performing a systematic literature survey and analysis of the outcomes. The remainder of the paper is organized as follows. The following section summarizes the current state of the art when it comes to cross-platform mobile development. A survey of the relevant research papers is conducted to identify the problems and suitable best practices and technologies pertaining to cross-platform mobile applications over the years. Providing analysis, discussion and conclusion about our work and possible future work directions follows this.

4   State of the Art

According to Heitkötter et al. [14], cross-platform development approaches allow developers to implement their apps in one step for a range of platforms, avoiding repetition and increasing productivity. On addition to providing suitable generality of provisioning the apps for several platforms, these approaches also enable developers to capitalize on the specific advantages and possibilities of smartphones/platforms. Palmieri et al. [15] provides a pragmatic comparison among four very popular cross platform tools focusing on the availability of application programming interfaces, programming languages, supported mobile operating systems, licences, and integrated development environments. Similarly, Juntunen et al. [6] , and Karl Andersson and Dan Johansson [9] have discussed about the burgeoning interest on HTML5 and its potential to shape the future of mobile development.

The following provides a *literature survey* concerning cross-platform mobile development. The purpose of the survey is to gain insights about the trends and technologies used to devise cross-platform mobile solutions over the last few years.

4.1   Method

A *protocol*, inspired from Biolchini et al. [16], was developed according to which the systematic literature survey was conducted. We first identified the *keywords* to search for the relevant literature. Then after selecting the digital library sources, we used the keywords in *query string* and performed automated search in these libraries to retrieve the literature. Thereafter, only the papers meeting our *inclusion criteria* were filtered for the survey purpose. More description about the method and the steps involved have been presented hereunder.

Choice of Keywords.
The two keywords used to identify the literature were "*mobile*" and "*cross-platform*".

Sources Selection.
The literatures for survey were located from well renowned digital libraries IEEE Xplore Digital Library and ACM Digital Library.

Search Method.
Using the keywords in the query string, we searched for the papers in the digital archive.

Following query was performed in the ACM Digital Library:

```
(Title: "mobile", AND Title:"cross-platform")
Years 2000-2012
Found 15 within The ACM Guide to Computing Literature
```

Similarly, in IEEE Xplore Digital Library:

```
("Document Title":"mobile" AND "Document Title":"cross-platform")
Years 2004-2012
Found 14 within The IEEE Xplore Digital Library
```

The query retrieved 15 papers from ACM and 14 papers from IEEE. By reading titles, abstracts and keywords 10 papers were selected further from ACM while 13 papers made through in case of IEEE. Then on eliminating the duplicates from both, total of 19 papers remained. These 19 papers were read fully and thoroughly and in the end 17 papers were selected. The method has been depicted in the fig. 1. below.

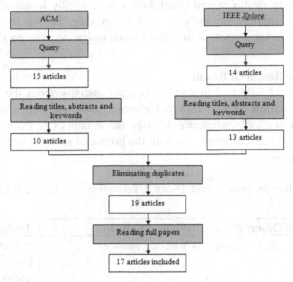

Fig. 1. Search method using the query string

Inclusion Criteria

Criterion 1. Firstly, the papers should be about the mobile devices meaning that the devices should be portable enough to carry around easily, such as mobile/smart phones, PDAs, etc. Albeit not always in the pocket like tablets but not laptops.

Criterion 2. Secondly, the papers should mainly focus on the cross-platform mobile development issues or mostly be relevant to it. Papers raising different kinds of issues related to mobile domain other than the cross-platform are not included.

Criterion 3. Thirdly, the papers should strictly adhere to the problem definition of this paper. It should be able to address the research question(s) raised above by providing the concrete recommendations and solutions. Papers that do not tackle the research questions, or at the best, propose insubstantial solutions with insufficient amount of information, papers providing vague analytic discussion rather than tangible solutions and recommendations are also excluded.

4.2   Outcomes

Timeline of the Literature.
After applying the inclusion criteria, we are left with 17 articles. From the papers selected we could clearly see the cross-platform development gathered momentum in recent years (2009 onwards). The survey does not have any articles through the years

2005 to 2008. During those years, the concept of cross-platform mobile applications or even mobile applications was not so much in prominence as it is now after the emergence of Apple's iOS and Google's Android. Also the hardware aspects of the mobile devices were fairly mediocre and smartphones were hardly heard of then. However there is couple of articles from the year 2004, where they also had the similar problem description albeit in the context of different mobile operating systems then.

Overview of the Research Results

Each article is read fully and thoroughly to gain insights about the problems it has raised and the solutions proposed to tackle those problems. Below we have summarized the results in a table where we describe the details of problem description that these work tried to address, combined with the proposed solution and the technologies used for this purpose.

Table 1. Survey Results Summarized. (Note: * Technologies that have been discontinued by their creators.)

No.	Problem Description	Proposed Solution	Technologies Used
1.	Portability and offline execution.	Web-based applications.	Yahoo! Mobile Widget*, Google Gears*, Google Gadget, Apache Shindig.
2.	Choice of platform and software for cross-platform development.	Cross-platform web-based development environment.	Cabana*(cross-platform web-based mobile development system), JavaScript.
3.	Cross-platform mobile development challenges.	A frame of component-based cross-platform mobile web application development. Application development divided into hierarchy of Application Layer, JS Engine Layer, Component Layer and OS Layer.	Not Applicable.
4.	Secured cross-platform access control for mobile web applications using privacy sensitive JavaScript APIs.	webinos platform, cross-device policy system for web applications on a wide range of web-enabled devices.	XACML (eXtensible Access Control Markup Language).
5.	Mobile phone game development tools for cross-platform development.	Swerve Studio, X-Forge.	C/C++, Java.
6.	Cross-platform mobile application with consistent user experience and real-time dynamic content delivery.	Cross-platform mobile development frameworks like Mobile Web Framework (MWF) released by UCLA and other device-agnostic approaches.	Web development technologies, native app wrappers.

Table 1. (*continued*)

7.	Device fragmentation and consistent user experience.	A hybrid (private/public) model cloud based enterprise mobile application.	Cloud based XML specification.
8.	Lack of standard for graphics on handheld devices.	GapiDraw platform.	C++.
9.	Non-uniform standards and the security of payment restricting the development of Mobile-Commerce.	Mobile payment standard CUPMobile of China UnionPay, customized mobile payment middleware CUPFace.	Web technologies like HTML, CSS and JavaScript for app development.
10.	Problems for existing mobile instant messaging systems regarding exchange of information across different platforms as they use the private protocols without the interconnective capability.	XMPP (the Extensible Messaging and Presence Protocol) achieves the unity of various IM protocols across different platforms.	Openfire server based on XMPP; XML, MySQL, Java ME.
11.	Unreliable network connection, applications adaptive to network conditions, consistent user experience across different platforms, battery life conservation.	A conceptual mobile application architecture which is adaptive, cross-platform, multi-network.	Not Applicable.
12.	Alternatives to Java ME when writing medical applications for mobile devices across multiple platforms.	An HTML-based medical information application.	HTML, PHP, MySQL.
13.	Adapting the user interface (UI) across to the actually mobile devices and mobile computing platforms.	A mobile cross-platform client-server architecture, where a set of 4 layers is defined that allows a plastic dynamic deployment of user interfaces.	PhoneGap; Web technologies like JavaScript, XML and HTML5, and native libraries.
14.	A mobile dictionary app of the English-Czech automatic control terms for the Department of Control Systems and Instrumentation with the view of the fragmented mobile landscape.	The use of HTML5 mobile frameworks.	HTML5, jQuery Mobile.

Table 1. (*continued*)

15.	Challenges in cross-platform development of mobile widget, and implementation of cross-platform API.	A conceptual MWPDL (Mobile Widget Portable Development Library) architecture is a replaceable approach to providing cross-platform support for development of mobile widgets.	Standard web technologies such as HTML, CSS, JavaScript and XML.
16.	Challenges for IoT based applications regarding orienting different intelligent terminals because of the heterogeneous platforms, which results in a problem of duplicated development.	OpenPlug Studio, based on Flex, is an appropriate cross-platform solution which realizes the conception of once development and multi-deployment.	Flex, ActionScript, MXML, CSS, C++.
17.	Requirements for the design and development of an end-user programming software system that supports the creation and deployment of cross-platform mobile mashups.	The *Mobile Mashup Editor* web application consisting of a client representing web browser based GUI editor and a server component representing storage. The *Mobile Mashup Viewer* based on the cross-platform mobile framework.	Web technologies; Google Web Toolkit, XML, Titanium Mobile Development Platform.

A more detailed outcome of the survey mentioning each paper is available here[1]. Wide range of problems regarding portability, native platform and device features, consistent user experience, choice of appropriate platform and software, approach of development, non-uniform standards, etc have been raised in the survey papers. Similarly different kinds of cross-platform tools, technologies and approaches have been used to provide solutions for these problems.

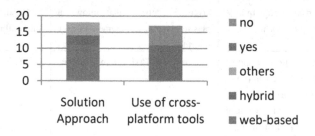

Fig. 3. Different solution approaches and use of cross-platform tools

---

[1] http://goo.gl/UbXZD

Based on the survey outcome, the fig. 3 shows web-based approaches have been the prevailing cross-platform solutions, however the hybrid approach is also catching up. The figure also shows some *others* solution approaches. This is because some papers from the early days differently tackled the issues and some are mainly concerned with the architecture and model for solutions. Also in the figure, we can see majority of solutions make use of some kind of cross-platform tools, SDKs, frameworks, libraries, platforms, etc.

5   Conclusions and Future Directions

Web-based solutions have been prevalent even from the initial days and have really gained momentum off late mainly due to recent developments and added functionalities of HTML5. Furthermore, as a result the web has moved towards becoming a software platform. The web-based mobile apps still have not achieved the level of performance the native code provides currently, but the recent developments are closing this gap on daily bases. Still the main criteria for choosing a web-based approach for development relies on the app objectives, and the business realities it tries to address [11].

Developments of hybrid approaches are now leveraging the web-based apps to a more serious alternative for native development. Access to native like features through a middleware such as PhoneGap or Titanium using the web based development tools brings a new dimension on the cross-platform development. The main advantages that this approach could bring are primarily on the unification of the development processes for different platform. All this combined with the saved developers' time thus directly affecting the overall cost for development. On the other hand, none of these middlewares provides a full and comprehensive support for all the mobile OS. Thus today there is a high probability that *"code once, deploy anywhere"* might most likely become *"code once, debug everywhere"*. Having the survey we conducted in mind, the main opportunities that cross-platform mobile development brings could be summarized as:

- Relying on web technologies for development purposes
- Shorter development time, thus reduced cost of development
- Suitable for rapid prototyping purposes and fast high fidelity mockups.
- Very good for educational purposes for mobile development.

Nevertheless despite the advantages that cross platform mobile development brings it can be said that they are still in the initial phases of development. The real opportunities for cross platform development will be more attractive once the HTML5 standard is fully developed and includes functions for access to native features of the phone. Though some features have already been well deployed, few are in experimental phases and limitedly implemented, and a lot more features still need to be incorporated. However works are underway to bring as many features as possible into the web. This trend of development is inline with the new development in the mobile OS market where we have Firefox OS and Ubuntu Phone that uses HTML5 as main development language. Thus in conclusion we can claim the opportunities of cross-platform development are closely embodied with the concepts of the web as a software platform.

References

1. International Data Corporation: Strong Demand for Smartphones and Heated Vendor Competition Characterize the Worldwide Mobile Phone Market at the End of 2012, IDC Says (January 2013), `https://www.idc.com/getdoc.jsp?containerId=prUS23916413&id=2126015`
2. Gartner, Inc.: Gartner Says Free Apps Will Account for Nearly 90 Percent of Total Mobile App Store Downloads in 2012, `http://www.gartner.com/newsroom/id/2153215` (September 2012), `http://www.gartner.com/DisplayDocument?ref=clientFriendlyUrl&id=2126015`
3. Cao, Y., et al.: Virtual Campfire - Cross-Platform Services for Mobile Social Software. In: Tenth International Conference on Mobile Data Management: Systems, Services and Middleware, MDM 2009, pp. 363–364 (2009)
4. Ramadath, S., Collins, M.: Mobile Application Development: Challenges and Best Practices (2012), `http://www.accenture.com/SiteCollectionDocuments/PDF/Accenture-Mobile-Application-Development-Challenges-Best-Practices.pdf`
5. Kaltofen, S.: Design and implementation of an end-user programming software system to create and deploy cross-platform mobile mashups. Student thesis. LinnaeusUniversity (2010), `http://urn.kb.se/resolve?urn=urn:nbn:se:lnu:diva-9300`
6. Juntunen, A., Jalonen, E., Luukkainen, S.: HTML 5 in Mobile Devices – Drivers and Restraints. In: 46th Hawaii International Conference on System Sciences, Hawaii, p. 1058 (2013)
7. Shah, N.: One Billion HTML5 Phones to be Sold Worldwide in 2013 (December 2011), `http://www.strategyanalytics.com/default.aspx?mod=pressreleaseviewer&a0=5145`
8. ABI Research.: 2.1 Billion HTML5 Browsers on Mobile Devices by 2016 says ABI Research (July 2011), `http://www.abiresearch.com/press/21-billion-html5-browsers-on-mobile-devices-by-201`
9. Andersson, K., Johansson, D.: Mobile e-Services Using HTML5. In: 2012 IEEE 37th Conference on Local Computer Networks Workshops (LCN Workshops), Clearwater, FL, pp. 815–816 (2012)
10. Korf, M., Oksman, E.: Native, HTML5, or Hybrid: Understanding Your Mobile Application Development Options, `http://wiki.developerforce.com/page/Native,_HTML5,_or_Hybrid:_Understanding_Your_Mobile_Application_Development_Options`
11. Charland, A., LeRoux, B.: Mobile Application Development: Web vs. Native. Queue - Data 9(4), 2–8 (2011)
12. Whinnery, K.: Appcelerator Development Center (May 2012), `http://developer.appcelerator.com/blog/2012/05/comparingtitanium-and-phonegap.html`
13. Seven, D.: What is a Hybrid Mobile App? (June 2012), `http://www.icenium.com/community/blog/icenium-teamblog/2012/06/14/what-is-a-hybrid-mobile-app-`

14. Heitkötter, H., Hanschke, S., Majchrzak, T.A.: Comparing cross-platform development approaches for mobile applications. In: 8th International Conference on Web Information Systems and Technologies, WEBIST 2012, pp. 299–311 (2012)
15. Palmieri, M., Singh, I., Cicchetti, A.: Comparison of Cross-Platform Mobile Development Tools. In: 2012 16th International Conference on Intelligence in Next Generation Networks, Berlin, pp. 179–186 (2012)
16. Biolchini, J., Mian, P.G., Natali, A.C.C., Travassos, G.H.: Systematic Review in Software Engineering. Technical Report RT-ES 679/05, Systems Engineering and Computer Science Department, COPPE/UFRJ, Rio de Janeiro (2005)

Virtual Studio Technology inside Music Production

George Tanev and Adrijan Božinovski

University American College Skopje, Computer Science Department, Skopje, Macedonia
{george.tanev,bozinovski}@uacs.edu.mk

Abstract. Music production in the 21st century is heavily based on the use of high performance computer-based systems and software applications that not only provide digital sound processing of the highest quality, but also offer pristine emulation of tons of simple and high-end hardware devices and instruments used in the music industry throughout history. One of the aims of this development process is to enable the integration of as much musical equipment as possible into one single device, i.e. to allow a massive number of tools and functionalities to be implemented inside simple interfaces designed for the popular computer platforms. As a major highlight and one of the leading achievements in this domain, this paper takes into account the eminent Virtual Studio Technology, by reviewing its development, widespread application, and creative potential in the music production industry as an integral part of the global music industry.

Keywords: music production, music software, digital signal processing, virtual instruments, VST plugins, Steinberg.

1   Introduction

About two decades ago, musicians would have laughed if they were told that classic synthesizers would be successfully re-created in software form and sold at a fraction of their original price. They would have been even less likely to believe that they might have half a dozen, or more of them, neatly integrated into their favorite MIDI (short for *Musical Instrument Digital Interface* [1]) & audio sequencer to be summoned at will. Yet VST instruments, a product of the renowned Virtual Studio Technology, do precisely this [2].

Ever since MIDI made its breakthrough in 1983, tracking and sequencing software was constantly utilized and developed [3]. Anyone making digital music prior to 1996 had been using their DAW (short for *Digital Audio Workstation* [4]) to control keyboards and samplers via MIDI and then routing all their external hardware through a traditional mixing desk. Steinberg, a musical software and hardware production company, had been around since 1984. With the release of Cubase 3.02 in 1996, this German-based establishment announced the Virtual Studio Technology (VST) interface specification which allowed a new breed of software developers to recreate all those bulky effects units as VST plugins [5].

V. Trajkovik and A. Mishev (eds.), *ICT Innovations 2013*,
Advances in Intelligent Systems and Computing 231,
DOI: 10.1007/978-3-319-01466-1_22, © Springer International Publishing Switzerland 2014

A few years later, in 1999, Steinberg updated their VST specification allowing VST plugins to receive MIDI data. This changed the game even further, as it was now possible to recreate keyboards, synths, and drum machines, too. This upgrade saw the birth of the Virtual Studio Technology Instrument, or VSTi for short.

With VST3, Steinberg released the next major revision of Steinberg's Virtual Studio Technology to the audio industry. VST3 marks an important milestone in audio technology with a completely rewritten code base, providing not only many new features, but also the most stable and reliable VST platform to date. Now VST plugins are able to process audio in 64-bit, accept audio inputs, allow multiple MIDI inputs and outputs, improve performance by applying processing to plugins only when audio signals are present on their respective inputs, and so on [6].

The latest upgrade to the VST interface specification – version 3.5 – was made in February, 2011. This update, among other things, included Note Expression where each individual note (event) in a polyphonic arrangement can contain extensive articulation information, which creates unparalleled flexibility and a much more natural feel of playing, say the people at Steinberg [7].

Fig. 1. Neon – The first VST instrument (bundled with Cubase VST 3.7) [8]

2   VST Plugins

The VST system was developed by Steinberg to enable a complete studio to be created in software. Even in its earliest incarnation, it allowed third-party developers to produce real-time effect modules that could "plug in" to the host application (initially Cubase). However, when Steinberg introduced the second version of the VST plugin standard, it also became possible to send MIDI data to and from such effects. This enabled developers to add more features, such as MIDI control of effect parameters and locking of effect settings to tempo. The inevitable result of this advance in the protocol was that this MIDI information was also used to run synth engines, rather than just effects processors. It is these synths, masquerading as effects plugins in order to fit directly into the sequencing environment, that are called VST instruments, or VSTi's.

Fig. 2 shows the user interface of Synthogy's Ivory II Grand Pianos, one of the most popular virtual grand piano collections. This VSTi relies on high levels of sampling and synthesis technology, incorporating Synthogy's exclusive, powerful 32-bit Sample Playback and DSP (short for *digital signal processing* [9]) engine, engineered specifically for recreating the acoustic piano sound. It includes a library of over 77 gigabytes of sampled world class grand piano instruments, with up to 18 discrete velocity layers with Sample Interpolation Technology for ultra-smooth velocity and note transitions. Ivory II employs complex modeling algorithms for half-pedaling, harmonic and sustain resonance, and comes with many customizable user controls and dozens of user-adjustable presets [10].

Fig. 2. Synthogy Ivory II VSTi – Sampled Grand Piano instrument collection [11]

A screenshot of Native Instrument's (NI) Guitar Rig 5 VST is shown in Fig. 3. The Guitar Rig series of plugins are NI's attempt at recreating high-end guitar amplifiers, cabinets, microphones, and effect-units, previously available in hardware form only. The software, tailored for guitar and, somewhat less so, bass players, uses amplifier modeling to allow real-time digital signal processing in both standalone and virtual studio environments.

2.1   Requirements

In order to make use of VST plugins it is essential to have a suitable VST-compatible host application. Just about every music production package available nowadays supports VST technology, and many come with their own built-in plugins and virtual instruments. In order to control things more easily, a good keyboard or controller is a must (see Fig. 4). A keyboard with weighted keys would give more control over expression when playing the virtual instruments, and having a controller with plenty of knobs and faders will allow more control over all plugin settings.

Relevant to mention here is that some VST plugins come with a standalone version and can be used independently from any host application. This limits their capabilities, however.

Fig. 3. NI Guitar Rig 5 VST – Guitar multi-effects processor [12]

An important thing to take into consideration is that VST plugins and virtual instruments are software, whereas their ancestors were all hardware. The computer and, more importantly, the sound card are now playing the major part of the hardware. And it basically boils down to common sense to make sure the user makes the most of their working setup. If a cheap computer with a built-in sound card is set to run too many VSTs, it will drag to a halt. A cheap sound card will result in a noticeable delay (latency) between playing a note on the keyboard and hearing the sound come out of the VSTi. For this reason, it is advisable to use a sound card with very low or *zero* latency, and to adjust the settings of the card and the host application for best performance.

In addition to the conventional PCI/PCI Express (referring to the *Peripheral Component Interconnect* local computer buses for attaching hardware devices) or built-in sound cards found in desktop and laptop computers (nowadays even tablets and smartphones), there are external audio interfaces being made and sold on the market. These can be hooked up to most any computing device via USB (the *Universal Serial Bus* connection standard) or FireWire (Apple's name for the *IEEE 1394* communications interface), and they offer all the same functionalities, each with its advantages and disadvantages [13].

Another important consideration to have is that in order to run VSTs the system has to have a fairly powerful CPU (*central processing unit*, i.e. processor). Since the sounds are created in real time using some of the processing power of the computer, each extra note or MIDI signal needs more calculations, and thus consumes additional CPU cycles. In order to be able to have more than just a few VST plugins running at once, it is also important to have sufficient RAM (*random-access memory*) – the more the better. From a hardware standpoint, the CPU and RAM pose the only practical limitations in regards to *which* (in terms of complexity) and *how many* VST plugins can run on a single machine at one time.

Fig. 4. M-Audio Axiom 25 MK2 USB MIDI Controller [14]

2.2   Platforms

Most musicians rely on Windows or Mac OS X as their main operating system. While Windows users tend to utilize applications that have native VST support, many Mac OS X users seem to choose some of Apple's applications, like Logic or Garage Band, which support Audio Units instead [15]. The Audio Unit (AU) may be thought of as Apple's architectural equivalent to Steinberg's VST. Because of the many similarities between these two formats, several commercial and free wrapping technologies are available (e.g. Symbiosis [16] and FXpansion VST-AU Adapter [17]). Other companies have developed their own virtual plugin formats, like DigiDesign and their RTAS (*Real-Time AudioSuite* [18]), for use with their proprietary software applications.

In order to provide cross-platform compatibility, plugin adapters have been developed for most operating systems and applications that lack native VST support. And to practically deal with this issue, most plugins nowadays are equipped with multi-format installation files.

2.3   A Programmer's Perspective

Steinberg have developed a VST Software Development Kit (SDK) – a set of C++ classes based around an underlying C API, which can be downloaded from their website [19]. There are several ports available, such as a Delphi version by Frederic Vanmol [20], a Java version from the jVSTwRapper project at Sourceforge [21], and two .NET versions – Noise [22] and VST.NET [23].

In addition, Steinberg have developed the VST GUI, which is another set of C++ classes, which can be used to build a graphical interface. There are classes for buttons, sliders, displays, etc. Nevertheless, these are low-level C++ classes and the look and feel still need to be created by the plugin manufacturer.

A large number of commercial and open-source VSTs are written using the Juce C++ framework [24] instead of direct calls to the VST SDK, because this allows multi-format (VST, AU, and RTAS) binaries to be built from a single codebase.

A notable language supporting VST is Faust [25], considering that it is especially made for making signal processing plugins, often producing code faster than hand-written C++.

2.4   In Practice

"*Quantum Leap* instruments have become an important part of my sonic template. I am really looking forward to using *EW/QL Pianos* and *Silk* in my future scores and I am using *STORMDRUM 2* right now on *Terminator 4*," said Danny Elfman about the EastWest Quantum Leap series of virtual instruments [26]. Elfman is an accomplished film score composer and music producer. Some of his notable works include *The Simpsons* Theme, his Grammy-winning score for *Batman*, the *Spider-Man 2* Main Title, and the scores for *Mission: Impossible* and *Alice in Wonderland* [27]. And he's not the only respectable or "credible" source of critique when it comes to the success of the penetrating Virtual Studio Technology. Many successful people in the music industry, regardless of whether they're just musicians, composers, producers, audio engineers, or some combination thereof, have reported their amazement and success at utilizing VST plugins and instruments in their work.

The VST technology has proven to be extremely practical and reliable, and its usage has become an inevitable routine in music production. Flagship musical instruments and effect processors have been successfully replicated, and a lot of "proprietary" VST plugins have been designed from scratch. The functionalities of massive hardware units traditionally used for *audio mastering* have been fastidiously recreated, often bundled into a single, multi-functional VST plugin (see Fig. 6). However, professional audio mastering engineers still tend to rely on both analogue, as well as digital audio equipment and outboard hardware gear, not just software [28].

3   On Hardware vs. Software Synthesizers

The near infinite possibilities of software would seem to render synths and workstation keyboards obsolete, yet hardware synths persist [29].

Hardware synthesizers, as the term implies, are tangible synthesizers that give users direct feedback between tactile input and audio output. There are many different brands, models, and styles of analog and digital hardware synths on the market, with some retaining or increasing monetary value long after their initial period, because of a particular type of sound delivered. Due to their tangible nature, they can be sold with relative ease. Of course, the more synthesizers one acquires, the more weight that needs to be carried from the studio to the gig and back [31].

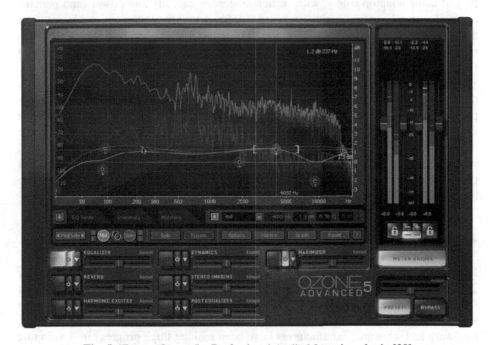

Fig. 5. iZotope Ozone 5 – Professional Audio Mastering plugin [30]

There is a school of thought that hardware synthesizers – especially analog ones – tend to sound "better" than their software counterparts. Here, consideration can be taken for the material resources and engineering used to go into constructing the "hardware" portion of the hardware synthesizer. The hardware synthesizer is usually built to do one or a few things, and to do that or those things well – including the faithful reproduction of modeled or sampled sound.

In part, due to component cost, storage is typically more limited for samples, program sets, and other data internal to the on-board programming and user-saved content. Thus, hardware synthesizers are limited in feature-set when compared to software synthesizers, but some can be expanded internally with user-upgradable modules, and most modern synthesizers and workstations include at least two traditional MIDI connectors so one can control another, or control a different piece of hardware using the MIDI protocol over said MIDI connector or USB or, if one goes far back, through a module that converts MIDI into CV (*control voltage*) signals that can be used by analog synthesizers such as an ARP 2600.

Software synthesizers and workstations (VST plugins and instruments being a large subset in this category) are programs or application suites that are installed on a personal computer, and sometimes, in this modern age, can even be loaded piecemeal from the so-called "cloud" (i.e. Internet). Just as with the hardware variety, software synths have their own sets of strengths and weaknesses.

The cost of the software can be very minimal, or can easily pile up as more workstations or synthesizers are acquired and installed. Next, there's the weight reduction: if using a laptop solely, with sufficient storage capacity and power (and perhaps a small control surface, mini-keyboard, audio interface and/or microphone), one might have a virtual studio weighing a total of ten to fifteen pounds in a backpack – as opposed to disconnecting a rack of modules weighing upwards of 75 pounds, with full-stage keyboards, stands, and so forth.

Considering the power and storage capacities of modern computer hardware these days, one can easily purchase a computer – desktop or laptop, even a tablet or smartphone – and install music production software on said computer system – and have such a multitude of options available at the outset that it may be hard to choose where to start in terms of what brand or "model" of software to choose, and, from there, where to start in terms of workflow within the software package itself. Also, there is the aspect that computers and/or the software running on them tend to "crash" more often than their hardware counterparts.

4   Conclusion

Music production technology has come a long way since its beginnings in the early 20th century. Thanks to the advancements in computer science and engineering, what took loads of time and expensive gear in the past can now be accomplished using a simple computer with a few peripherals attached to it. To simplify things further, hardware and software manufacturers have been making huge progress in developing new technologies and making them easily available to consumers.

The Virtual Studio Technology initiated by Steinberg has become a prominent field in audio software engineering, utilizing digital signal processing to simulate traditional recording studio hardware with software. There are probably tens of thousands of VST plugins around these days, varying both in capability and in character. Some are made by one person and given away for free, while others are developed by large companies and cost an arm and a leg. The benefits of using these plugins are of such extent that the word "plugin" doesn't do them any justice at all. "Real" musical instruments, racks of effect-processors, modules, etc. tend to take up a lot of physical space and require lots of effort for maintenance and relocation. The advantages of having these bulky and expensive devices at your disposal as software, obtained for a symbolic price, are obvious. The experience of using or playing these software devices or instruments might not always be the same or as good as working with the originals, due to the fact of ones being real and the others virtual. What VST developers have tried to do here is to recreate the sound of these originals as closely as possible, and in most cases than not, the results fulfill the expectations and the benefits surpass the costs. In addition, the digital nature of these plugins offers possibilities to users that are impossible to pull off with the originals.

All functions of a VST effect processor or instrument are directly controllable and automatable through a hardware controller. VST allows easy integration of external equipment, allowing tailor-made systems to be put together for specific purposes. And by being an open standard, the possibilities offered by VST have steadily been growing over the past decade. New virtual effect processors and virtual instruments are constantly being developed by Steinberg and dozens of other companies. Leading third-party VST instrument creators include renowned software companies such as Native Instruments, Arturia, and Spectrasonics, as well as known hardware manufacturers like Korg, Waldorf, and Novation. Companies like Waves, Sonnox, Antares, and TC Works have brought their virtual effect processors to the table.

The Virtual Studio Technology has totally revolutionized the world of music production. There are plugins and instruments that could never exist in the physical world but have come about thanks to the unique features and possibilities offered by this technology. The emerging creative possibilities have opened a new chapter in the history of making music, and with the ever-growing pool of developers and incremental technological advances, it is safe to presume that things can only get better.

References

1. Swift, A.: A brief Introduction to Midi. SURPRISE 97, Imperial College of Science Technology and Medicine (1997), http://www.doc.ic.ac.uk/~nd/surprise_97/journal/vol1/aps2/ (accessed February 18, 2013)
2. Walker, M.: Using VST Instruments. Sound On Sound Magazine (2000), http://www.soundonsound.com/sos/dec00/articles/vst.asp (accessed February 14, 2013)
3. Bozinovski, A.: Interaction with Music Information Systems using Modular Trackers. In: Chu, H.-W., Savoie, M., Sanchez, B. (eds.) Proceedings of the International Conference on Computing, Communications and Control Technologies, vol. 2, pp. 72–75. IIIS, Texas (2004)
4. Reyniers Audio. DAW Computer: description of Digital Audio Workstation. ReyniersAudio.com. (2012), http://daw-computers.com/ (accessed February 19, 2013)
5. Looperman. What is a VST plugin or VST instrument. Looperman (2012), http://www.looperman.com/blog/detail/55/what-is-a-vst-plugin-or-vst-instrument (accessed February 14, 2013)
6. Steinberg. VST3: New Standard for Virtual Studio Technology. Steinberg (2013), http://www.steinberg.net/en/company/technologies/vst3.html (accessed February 16, 2013)
7. Steinberg. 2011. VST 3.5 a milestone in VST development. Steinberg, http://imsproav.com/wp/2011/03/vst-3-5--a-milestone-in-vst-development/ (accessed February 14, 2013)
8. KVR Audio. Neon by Steinberg. KVR Audio Plugin Resources (2013), http://www.kvraudio.com/product/neon-by-steinberg (accessed February 14, 2013)
9. Smith, S.W.: The Scientist and Engineer's Guide to Digital Signal Processing. California Technical Publishing (1997), http://www.dspguide.com/pdfbook.htm (accessed March 17, 2013).

10. Synthogy. Ivory II Grand Pianos. Synthogy LLC (2010),
    `http://www.synthogy.com/products/ivorygrand.html` (accessed March 17, 2013)
11. PROCOM. Synthogy Ivory Italian Grand II. Procom Music (2013),
    `http://www.procom.no/softsynther/piano/synthogy-ivory-ii-italian-grand.html` (accessed February 15, 2013)
12. Hitsquad. Guitar Rig 5.1.0 for Windows XP. Hitsquad Pty Ltd. (2013),
    `http://www.hitsquad.com/smm/programs/GuitarRigWin/` (accessed March 17, 2013)
13. Sweetwater. FireWire, PCI or USB? Sweetwater Sound Inc. (2013),
    `http://www.sweetwater.com/feature/daw/firewire_pci_or_usb.php` (accessed February 19, 2013)
14. Djsuperstore. M-Audio Axiom 25 MK2 USB Midi Controller. djsuperstore (2013),
    `http://www.djsuperstore.co.uk/M-Audio-Axiom-25-MK2` (accessed February 14, 2013)
15. MHC. About VST Plugins. MHC (2010),
    `http://www.mhc.se/software/plugins/vst-plugins.php` (accessed February 15, 2013)
16. Google. symbiosis-au-vst – AU / VST Symbiosis. Google Code; Google Project Hosting (2011), `http://code.google.com/p/symbiosis-au-vst/` (accessed February 18, 2013)
17. FXpansion. FXpansion – VST-AU Adapter v2.0. FXpansion™ Audio UK Ltd. (2013),
    `http://www.fxpansion.com/index.php?page=5` (accessed April 15, 2013)
18. Avid. Introduction to Pro Tools Plug-ins. Avid Technology, Inc. (2013),
    `http://www.avid.com/US/resources/introduction-to-pro-tools-plug-ins` (accessed March 17, 2013)
19. Steinberg. 3rd Party Developer Area. Steinberg (2013), `http://www.steinberg.net/en/company/developer.html` (accessed February 18, 2013)
20. Vanmol, F.: VST. Axi world (2011), `http://www.axiworld.be/vst.html` (accessed February 18, 2013)
21. SourceForge. jVSTwRapper – Java-Based Audio Plug-Ins. SourceForge (2013),
    `http://jvstwrapper.sourceforge.net/` (accessed February 18, 2013)
22. Google. Noisevst – A .Net wrapper for Steinberg's VST plugin API. Google Code, Google Project Hosting (2007), `http://code.google.com/p/noisevst/` (accessed February 18, 2013)
23. Microsoft. VST.NET – Home. Microsoft CodePlex (2012),
    `http://vstnet.codeplex.com/` (accessed February 18, 2013)
24. Raw Material Software. C++ programming tools. Raw Material Software Ltd. (2010),
    `http://rawmaterialsoftware.com/juce.php` (accessed March 17, 2013)
25. Smith III, J.O.: Audio Signal Processing in Faust. Center for Computer Research in Music and Acoustics (CCRMA), Stanford University Department of Music (2013),
    `https://ccrma.stanford.edu/~jos/spf/aspf.pdf` (accessed March 18, 2013)
26. EastWest. Home – Sounds Online. Sounds Online A division of East West Communications, Inc..(2013), `http://www.soundsonline.com/` (accessed March 23, 2013)
27. Clemmensen, C.: Filmtracks: Danny Elfman. Filmtracks Publications (2013),
    `http://www.filmtracks.com/composers/elfman.shtml` (accessed March 23, 2013)

28. Gardner, B.: What is audio mastering | What is mastering. whatisaudiomastering.com. (2011), http://www.whatisaudiomastering.com/ (accessed March 23, 2013)
29. Snow, V.: Synths – Hardware vs. Software. Victor Snow > > Ramblings on Music (2013), http://victorsnow.com/?p=618 (accessed June 10, 2013)
30. iZotope. iZotope Ozone 5 and Ozone 5 Advanced | Complete Mastering System. iZotope, Inc. (2013), http://www.izotope.com/products/audio/ozone/features/otherHighlights.asp (accessed February 15, 2013)
31. Kupka, D.: On Hardware vs. Software Synthesizers / Workstations. Quora (2013), http://www.quora.com/Dan-Kupka/Posts/On-Hardware-vs-Software-Synthesizers-Workstations (accessed June 10, 2013)

End-User Software Reliability in the Case of the Inventory Tracking System in University of Prishtina

Isak Shabani[1], Betim Cico[2], Dhurate Hyseni[2], and Fatos Halilaj[1]

[1] University of Pristina, Faculty of Electrical and Computer Engineering, Pristina, Kosovo
{isak.shabani,fatos.halilaj}@uni-pr.edu
[2] South East European University, Faculty of Contemporary Sciences and Technologies,
Tetovo, Macedonia
{b.cico,dh11752}@seeu.edu.mk

Abstract. University of Pristina, a public University of Kosovo, has transferred all their assets management to an online Inventory Tracking System (namely e-Pasuria). We outline techniques to improve software reliability by providing mathematical analysis using probability, binominal distribution and for increase of performance we use data synchronization mechanisms to help system remain stable even when there are defects on the network. The Inventory Tracking System we discuss here is widely used by Kosovo Institutions and it is part of e-Government. Our focus will be isolated to a single institution, University of Pristina and its faculties. We provide a variety of results and samples showing the increase of reliability to the end users on the application of the provided techniques. Software reliability is hard to be calculated precisely because of its nature, but we have put up a lot of time and effort to evaluate this outcome. Reliability plays an important role in the usage of your application; it determines the length of application life.

Keywords: Software Reliability, Web Application, e-Pasuria.

1 Introduction

Often, when multi-user systems are developed to be used nation-wide by the government and citizens, reliability of the software should be taken under high considerations. Involvement of the end-users in this process is very valuable. The aim of the reliability is to make a system reliable and consistent with no-faults to the end-users. Thus, measurements and information of a reliable software application are very important. For every complex system that includes multi-users and data-manipulations, it will be really hard to achieve a consistent level of reliability. Detailed study and analysis and research on software application reliability is provided in this paper. This paper is structured in 6 Chapters: Introduction, Software Reliability Measurements, Practical Evaluation of Reliability Relating to e-Pasuria, Data Synchronization, Assets Management System and Conclusion.

V. Trajkovik and A. Mishev (eds.), *ICT Innovations 2013*,
Advances in Intelligent Systems and Computing 231,
DOI: 10.1007/978-3-319-01466-1_23, © Springer International Publishing Switzerland 2014

2   Software Reliability Measurements

Reliability $R(t)$ of a system at time t is the probability that the system operates without failure in the interval $[0,t]$, given that the system was performing correct at time 0 [7]. High reliability is required in situations when a system is expected to operate without interruptions.

Reliability is a function of time. The way in which time is specified varies considerably depending on the nature of the system under consideration. For example when a system is required to do some job in a short period of time, time is specified in units such as minutes/seconds/milliseconds.

To exactly define system reliability we should rely on the concepts related to software application. Based on these software applications needs we act on the same way, by operating in different point and in different time; this way the failure can be expressed only in terms of probability. Thus, fundamental definition for the software reliability depends on concepts of the probability theory. These concepts, provide basics of a software reliability, allow comparisons among systems and they provide fundamental logic for improvements of the fail rates, that will be reached during the application life cycle.

In general, a system may be required to perform different functions, each of which can have different reliability level.

In addition, in different time, software application can have different probability to perform required functions from the user under declared conditions.

Reliability represents probability of a success or possibility, that the software application has to perform it's functionality for the sake of the project under certain limits. More specifically, reliability is a probability that the product or part of it that operates in basis of predefined requirements for a defined period of time, under projects' conditions (i.e. number of transactions, bandwidth, etc.) works with no failure. In other words, reliability can be used as a measurement of a system's success for it to work as required. Reliability is a quality characteristic that customers demand from the producer of the product or better said a tool to evaluate safety of a software application.

Mathematically, reliability $R(K)$ is the probability that a system be successful in the time interval: [0-k]:

$$R(K) = 1 - P(K) \tag{1}$$

$$R(t) = \int_0^\infty \frac{1}{0} e^{-\frac{s}{0}} ds = e^{-\frac{t}{0}} \quad t \geq 0 \ . \tag{2}$$

The increase of R(k) decreases software application reliability (see Fig. 1).

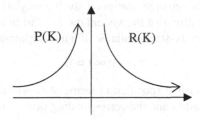

Fig. 1. Report between P(K) and R(K)

In order to have a complete reliability in a system, the prediction of mean time to repair (MTTR) is required for different conditions that can occur during system use. This is based from the system designers' experience in the past and experts' disposal for repair handlings.

Time system repair is composed of two specific intervals:

- Passive time for software application repair
- Active time for software application repair

Passive repair time is defined mainly from the time taken from the engineers that travel to the users' locations. In many cases, traveling time cost is exceeded by actual repair time cost. Active time for software application repair is related directly with system projection and is defined as follows:

Time between occurrences of a failure and user of the system is notified about it:

- Time required to explore the failure
- Time required to change the components
- Time required to verify that the problem has been solved and the system is fully functional.

Active time or software application repair can be correctly improved if the system is designed in such a way that errors can be identified easily and be corrected. The more complex the design of the system is made, the harder to find the errors and improve things in a system. Reliability is a measure that requires success for the system for a particular period of time and such as that failures are not allowed.

Availability $A(t)$ of a system at time t is the probability that the system will be functioning at the instant of time t.

$A(t)$ is also referred as point availability, or instantaneous availability. It is often necessary to determine the interval or mission available.

$$A(T) = \frac{1}{T} \int_0^T A(t)dt \ , \tag{3}$$

where $A(t)$ is the value of the point availability averaged over some interval of time T. This is the time required for the system to accomplish some task [7].

If a system cannot be repaired, the point availability $A(t)$ equals to the system's reliability, i.e. the probability that the system did not fail in between 0 and t. Thus, as T goes to infinity, the steady-state available of a non-repairable system goes to zero.

$$A(\infty) = 0 .$$

Steady state available is often specified in terms of downtime per year. Fig. 2 shows the values for the availability and the corresponding downtime.

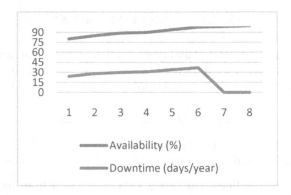

Fig. 2. Availability and corresponding downtime values per year

Availability is typically used as a measure for systems where short interruptions can be tolerated. Surveys have shown that nearly 60 percent of web users say that they expect a website to load in less than 3 seconds.

3   Practical Evaluation of Reliability Relating to e-Pasuria

Even though theoretical process of reliability evaluation doesn't seem to be very complex, safe development, complex, and exact for the reliability evaluation systems requires a lot of effort and research, because each product has its own properties, in concept, development and implementation. This for the fact that each product has its own properties, in concept, development and implementation. In the following we will examine the module of amortization in the system e-Pasuria, in details we will show the methodologies of reliability evaluation relating the application life-cycle. Mathematically, part of the failures are shown, all of these statistics will be presented in a tabular and graphical form.

Relating the Eq. (2) for the first three months (2190 days):

$$P(0 < T \le 2190) = \int_0^{2190} \frac{1}{2022} e^{-\frac{t}{2022}} dt \approx 0.334 .$$

Then from the Eq. (1):

$$D(T) = 1 - P(T) = 1 - 0.334 = 0.67 ,$$

where D(T) is the reliability of the software, whereas P(T) is time of failure for the software which is determined based on methodologies of evaluation of reliability relating the products' life-cycle (for the problem of amortization which is very sensitive). Main goal is the reliability of amortization of inventory accurately for each equipment. First case is presented in table 1 (see Fig. 3).

Table 1. First case with failure and reliability achievement against first version of the software

P(T)	D(T)
0.33	0.67
0.42	0.58
0.38	0.62

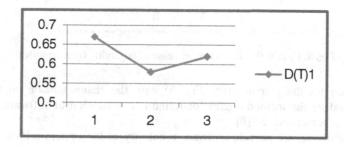

Fig. 3. Case with failure and reliability achievement against first version of the software application

After this, an upgrade (update) to the current version has been provided with amortization functionality provided allowing the user to automatically calculate the amount of amortization for the equipment based on categories provided in the system. Before the update, user could calculate the amount of amortization on monthly bases, and this was not accepted very well from officials/users because it did not fulfill the needs required by audits. Audit, required that the calculation of amortization be rectilinear, and should be provided in days, months, and years, which is now provided with the update. Initial results for the achieved reliability are too low.

Now we present the case when equipment amortization is well accepted by users and it is closely or better said very likely with rectilinear method. Now, we give statistics which are taken for the period of six months, after the release of the update to the software.

$$P(2190 < T \le 4380) = \int_{2190}^{4380} \frac{1}{2211} e^{-\frac{t}{2211}} dt \approx 0.10.$$

$$D(T) = 1 - 0.1 = 0.90.$$

Calculations of the reliability of the system after update are provided in Tab. 2 and Fig. 4.

Table 2. First case with failure and reliability achievement after the updates

P(T)	D(T)
0.10	0.90
0.04	0.96
0.01	0.99

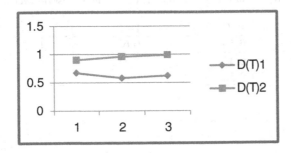

Fig. 4. Case with failures and increase of reliability by comparison

As is seen for the graphics (see Fig. 5) with the changes made on the update release, based on the methodologies for reliability evaluation of software life-cycle, reliability has been achieved [6].

In the following, we provide software reliability of the system e-Pasuria, relying the Eq. (4):

- Regular time for functioning is taken as 90% of full time
- Successful time is evaluated as an average of all values from different modules
- An average of 95% out of 100% resulted successful in e-Pasuria software:

$$R(90) = \sum_{90}^{100} (\frac{100}{90})(0.91)^{91}(0.05)^{91-5} \approx 0.877 .$$

It can be said that reliability of e-Pasuria software in general is approximately ~0.90, which indicates a high value of reliability.

4   Data Synchronization

Optimistic replication strategies are attractive in a growing range of settings where weak consistency guarantees can be accepted in return for higher availability and the ability to update data while disconnected [1]. These uncoordinated updates must later be synchronized (or reconciled) by automatically combining non conflict updates, while detecting and reporting conflict updates. The ability to support mobile and remote workers is becoming more and more important for organizations every day. It is critical that organizations ensure users have access to the same information they have when they are in the office. In most cases, these workers will have some sort of laptop, office desktop, Smartphone, or PDA. From these devices, users may be able to access their data directly through VPN connections, Web Servers, or some other connectivity method into the corporate networks. Synchronization gained great importance in modern applications and allows mobility in the context of information technology. Users are not limited to one computer any more, but can take their data with them on a laptop.

5   Assets Management System

5.1   The Concept of the System

e-Pasuria is an Asset Management System which is in used by all Ministries, Municipalities, Agencies in Kosovo. Through this application monitoring and controlling from the auditing agency is done for all integrated institutions in Kosovo. e-Pasuria provides the management of assets, their usage, expandable materials, equipment barcoding, amortization of equipment based on amortization percentage provided with categories, stock management, and online requests for officials when they require the usage of assets (see Fig. 5).

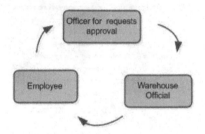

Fig. 5. Online request in e-Pasuria

5.2   Statistics on Data Transfers in e-Pasuria

Statistics with real data usage for the transactions made by users in e-Pasuria for each Faculty under University of Pristina are given as follows, Fig. 6:

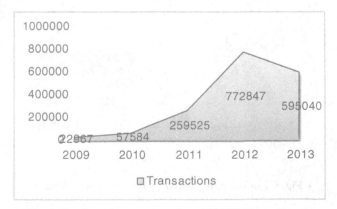

Fig. 6. Number of total transactions 2009-2013 in e-Pasuria

As can be seen from Fig. 8, number of transactions for the last four months (in 2013) is nearly equal to the number of total transactions from 2012. Number of transactions is greatly increased from 2009 – 2012. As number of transactions is growing, this is a good sign that user's reliability in the system, is also increased.

In the following (see Fig. 7), we provide number of transactions per period of 3 months for each Faculty in University of Pristina:

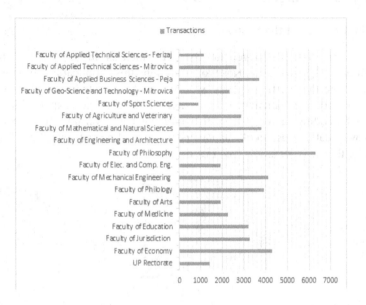

Fig. 7. User transactions for the last 3 months (2013) in e-Pasuria

As number of users is of a great value in the usage of the system, we provide statistics on number of users per year (see Fig. 8):

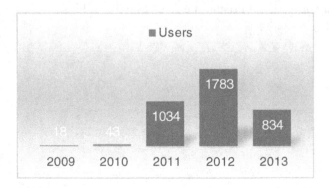

Fig. 8. Number of registered users per year in e-Pasuria

As can be seen on Fig. 8, number of new users registered in e-Pasuria has increased, the period for the year 2013 includes last for month as of today April 15th, 2013.

5.3   Technologies Used to Develop e-Pasuria

Application is built using Microsoft Technologies using Microsoft Visual Studio as a Development Environment, Microsoft ASP.NET, SQL Server 2008, ADO.NET, and Communication with other applications using Web Services and XML.

6   Conclusion

It is hard to calculate the exact reliability of software, but to come to those results we have included a bunch of factors involved with this process, such as availability of the software, number of transactions, users, downtime and we have shown mathematical methods on how to increase reliability on the software application.

Provided methods have increased the number of users in the e-Pasuria system, as a result of evaluation of reliability an incorporation of Data Synchronization algorithm into a system. The lack of regular power supply in Kosovo leads to many difficulties to the end user, making the software unreliable and hard to be used. As we have practically evaluated the reliability of e-Pasuria on the module of amortization this method could easily further be implemented in any module of any system.

Acknowledgments. We are grateful to all the people working at the University of Pristina that supported us on gathering of information and measurements on e-Pasuria software to make this research possible.

References

1. Shabani, I., Çiço, B., Dika, A.: Solving Problems in Software Applications through Data Synchronization in Case of Absence of the Network. IJCSI International Journal of Computer Science Issues 9, 1694–1814 (2012)
2. Mauher, M., Smokvina, V.: Municipal Asset and Property Management System for the Web Collaborative Environment. Multidimensional Management Consulting, Ltd. Zagreb Vinogradi 36 C, Zagreb, Croatia & City of RijekaKorzo 16, Rijeka, Croatia (2008), http://www.majorcities.eu/generaldocuments/pdf/ municipal_asset_and_property_management_system_for_the_web_ collaborative_environment.pdf
3. Çiço, B., Hajdini, A.,, Sadiku, S., Meha, S., Shabani, I.: The Increase of the Speed of Integration of Online Services for Citizens Through Standardization of Municipality Portals. In: 15th International Research/Expert Conference, Czech, TMT, pp. 501–504 (2011)
4. Kreutzkamp, J., Hagge, L., Deffur, E., Gellrich, A., Schulz, B.: Experience with an IT Asset Management System. DESY, Hamburg, Germany (2003), http://www.slac.stanford.edu/econf/C0303241/proc/ papers/TUDT001.PDF

5. Shabani, I., Çiço, B., Dika, A.: Web services oriented approach for data synchronization. In: 6th Annual South - East European Doctoral Student Conference Programme, DSC 2011, pp. 342–349 (2011)
6. Bailey, D., Frank-Schultz, E., Lindeque, P., Temple, J.L.: Three reliability engineering techniques and their application to evaluating the availability of IT systems: an introduction. IBM Syst. J. 47, 577–589 (2008)
7. Dubrova, E.: Fault Tolerant Design. An Introduction Department of Microelectronics and Information Technology Royal Institute of Technology. Stockholm, Sweden (2007)
8. e-Pasuria. Assets Management System, https://e-pasuria.rks-gov.net

Analyzing the Cryptographic Properties of the Sets of Parastrophic Quasigroups of Order 4

Vesna Dimitrova, Zlatka Trajcheska, and Marija Petkovska

Faculty of Computer Science and Engineering,
"Ss. Cyril and Methodius" University,
Skopje, Macedonia
vesna.dimitrova@finki.ukim.mk,
{trajcheska.zlatka,petkovska.marija}@students.finki.ukim.mk

Abstract. Quasigroups are algebraic structures that are suitable for cryptographic use and their cryptographic properties are intriguing. Looking into these properties we can classify the quasigroups based on different criteria and sort out the ones with best attributes for encryption and resistance to attacks. The Boolean representations of quasigroups allow us to find out more about their cryptographic properties. Some of them are already examined and determined. In this research we will use some previous conclusions in order to find out more about the cryptographic properties of the sets of parastrophic quasigroups .

Keywords: quasigroup, parastrophe, cryptography, algebraic immunity, nonlinearity, resiliency.

1 Introduction

The quasigroups are plain algebraic structures which are suitable for cryptographic and coding purposes due to their large, exponentially growing number, and also their specific properties. Formally, a groupoid $(Q, *)$, where $*$ is binary operation, is called a quasigroup if it satisfies:

$$(\forall\, a, b \in Q)(\exists!\, x, y \in Q)(x * a = b \wedge a * y = b) \tag{1}$$

In this paper we are interested only in the quasigroups of order 4. We use lexicographic ordering [2] of the quasigroups of order 4. This is done by representing the quasigroup as a string of length n^2 which is obtained by concatenation of the its rows. Then, this strings are ordered lexicographically and thus we acquire the ordering of quasigroups.

Each quasigroup belongs to a set of at most 6 quasigroups that are interdependent and connected through the specifics of their operations. These quasigroups are called parastrophes.

Another important issue that should be discussed before we pass to the main results of this paper is the Boolean representations of the quasigroups. Actually, we use the approach of representing the quasigroups as Boolean functions [2].

A Boolean function of n variables is defined as a function $f : \mathbb{F}_2^n \rightarrow \mathbb{F}_2$, where $\mathbb{F}_2 = \{0, 1\}$ is a two-element field [2][7]. Each Boolean function can be uniquely presented in its Algebraic Normal Form (ANF) [2][7], as a polynomial in n variables over the field $\mathbb{F}_2$ that has degree less or equal than 1 in each single variable, i.e.:

$$f(x_1, x_2, ..., x_n) = \sum_{I \subseteq \{1,...,n\}} a_I x^I \qquad (2)$$

where

$$x^I = \prod_{i \in I} x_i, \ x^{\emptyset} = 1 \ \text{and} \ a_I \in \{0, 1\}.$$

Every quasigroup $(Q, *)$ of order 2^n can be represented as a vector valued Boolean function $f : \{0, 1\}^{2n} \rightarrow \{0, 1\}^n$. The elements $x \in Q$ can be considered as binary vectors $x = (x_1, ..., x_n) \in \{0, 1\}^n$. Now, we represent the quasigroup in the following way: $\forall x, y \in Q$ we have that

$$x * y \equiv f(x_1, ..., x_{2n}) = (f_1(x_1, ..., x_{2n}), ..., f_n(x_1, ..., x_{2n}))$$

where

$$x = (x_1, x_2, ..., x_n), y = (x_{n+1}, x_{n+2}, ..., x_{2n})$$

and

$$f_i : \{0, 1\}^{2n} \rightarrow \{0, 1\}$$

are the Boolean function components of the previously defined f [2].

Because here we are considering only the quasigroups of order 4, it is clear that their Boolean representations will be composed of 2 Boolean functions each with 4 arguments. For each quasigroup we will analyze the cryptographic properties of both Boolean functions.

There is a classification of quasigroups by their Boolean representations that is already presented in previous research (see [2]) that we will use for comparison with the new results. In this classification the quasigroups are categorized as linear or non-linear by Boolean representations. A quasigroup is called linear by Boolean representation if all functions f_i for $i = 1, 2, ..., n$ are linear polynomials. On the other hand, a quasigroup is called non-linear by Boolean representation if there exist function f_i for some $i = 1, 2, ..., n$ which is not linear. There is also a specific subset of the non-linear quasigroups that are called pure non-linear quasigroups by Boolean representations. In these quasigroups all components are non-linear Boolean functions. This classification fully matches the one based on the algebraic immunity and nonlinearity of the Boolean representations of quasigroups of order 4 (see [1]).

2   Sets of Parastrophic Quasigroups

As mentioned before, each quasigroup belongs to a set of at most 6 quasigroups that are mutually dependent by their operations and are called parastrophes.

Namely, if Q is a quasigroup with operation $*$, than the other five parastrophic quasigroup operations are denoted by $/, \backslash, \cdot, //, \backslash\backslash$, respectively, and they are defined by the following rules [2] :

$$
\begin{aligned}
x/y = z &\Longleftrightarrow z*y = x \\
x\backslash y = z &\Longleftrightarrow x*z = y \\
x \cdot y = z &\Longleftrightarrow y*x = z \\
x//y = z &\Longleftrightarrow y/x = z \Longleftrightarrow z*x = y \\
x\backslash\backslash y = z &\Longleftrightarrow y\backslash x = z \Longleftrightarrow y*z = x
\end{aligned}
\tag{3}
$$

We used the lexicographic ordering of quasigroups of order 4 to represent the sets of parastrophic quasigroups in Table 1. We will assign a number to every set in order to facilitate the presentation of the results of our research.

Table 1. Cryptographic properties of the sets of parastrophic quasigroup

No.	Parastrophes	No.	Parastrophes
1	{1}	76	{428,548,571}
2	{2,3,25}	77	{32,42,122}
3	{5,18,121}	78	{36,79,114}
4	{7,9,49}	79	{39,44,76,89,104,106}
5	{11,51,57}	80	{48,128,140}
6	{12,13,55,65,75,105}	81	{88,120,143}
7	{15,19,52,59,81,99}	82	{94,95,112,118,136,141}
8	{28}	83	{154,162,217,265,434,457}
9	{35,53,58}	84	{156,221,437}
10	{47,84,102}	85	{159,165,219,267,436,459}
11	{61,68,131}	86	{168,270,463}
12	{63}	87	{202,225,300,363,441,483}
13	{96,117,134}	88	{205,227,307,409,449,529}
14	{144}	89	{208,231,312,415,455,535}
15	{145}	90	{210,273,298,361,465,482}
16	{148,170,171}	91	{214,277,304,367,471,488}
17	{155,195,291}	92	{216,279,310,412,473,533}
18	{167,244,317}	93	{346,369,489}
19	{172}	94	{350,356,372,421,498,541}
20	{174}	95	{352,358,375,418,501,538}
21	{176}	96	{360,423,545}
22	{178,218,433}	97	{4,26,27}
23	{183,187,198,245,294,316}	98	{8,10,31,41,73,98}
24	{185,242,314}	99	{14,77,113}
25	{189,246,318}	100	{21,83,100}
26	{190,248,320}	101	{22,23,90,103,123,130}
27	{191,268,460}	102	{24,126,139}
28	{201,299,341}	103	{30,33,50}

Table 1. (*continued*)

No.	Parastrophes	No.	Parastrophes
29	{215,255,309,329,386,387}	104	{37,43,54,60,82,101}
30	{232,456,559}	105	{64,87,119}
31	{233,250,322,362,443,481}	106	{69,86,107}
32	{236,278,446,472,509,512}	107	{70,71,93,110,132,133}
33	{257,327,344}	108	{72,135,142}
34	{259,331,388}	109	{146,147,169}
35	{260,332,390}	110	{150,151,173}
36	{261,283,333,410,475,530}	111	{153,161,193,241,290,313}
37	{263,335,392}	112	{157,163,196,243,293,315}
38	{286,478,558}	113	{179,197,292}
39	{345}	114	{180,200,295}
40	{376,502,564}	115	{181,220,435}
41	{379,394,493}	116	{182,223,438}
42	{382,422,496,518,524,542}	117	{186,266,458}
43	{403}	118	{192,272,464}
44	{405}	119	{203,305,385}
45	{406}	120	{207,311,391}
46	{407,429,549}	121	{209,251,297,323,337,340}
47	{432,552,574}	122	{212,252,302,324,342,343}
48	{514}	123	{226,442,505}
49	{516,522,565}	124	{229,451,553}
50	{519,525,562}	125	{234,253,325,365,444,484}
51	{520,526,566}	126	{235,275,445,467,506,507}
52	{528,568,570}	127	{237,254,326,368,447,487}
53	{572}	128	{240,280,454,474,554,555}
54	{575}	129	{262,284,334,414,476,534}
55	{576}	130	{264,287,336,416,479,536}
56	{359,399,401}	131	{282,470,508}
57	{6,17,29,34,74,97}	132	{285,477,556}
58	{16,20,78,115,125,137}	133	{348,354,395}
59	{38,45,56,66,124,129}	134	{351,357,396}
60	{62,67,85,91,108,109}	135	{370,490,513}
61	{149,152,175}	136	{374,500,563}
62	{158,164,199,247,296,319}	137	{377,397,491}
63	{177,194,289}	138	{380,398,494}
64	{184,188,224,271,440,462}	139	{381,420,495,517,523,540}
65	{204,306,389}	140	{384,424,504,546,567,569}
66	{211,249,301,321,338,339}	141	{404,427,547}
67	{230,452,557}	142	{408,430,550}
68	{238,276,448,468,510,511}	143	{426,527,544}
69	{239,256,330,413,453,532}	144	{431,551,573}
70	{258,281,328,366,469,486}	145	{40,46,80,116,127,138}

Table 1. (*continued*)

No.	Parastrophes	No.	Parastrophes
71	{288,480,560}	146	{92,111}
72	{347,353,393}	147	{160,166,222,269,439,461}
73	{373,499,561}	148	{206,228,308,411,450,531}
74	{378,419,492,515,521,539}	149	{213,274,303,364,466,485}
75	{383,400,402,425,503,543}	150	{349,355,371,417,497,537}

3   Cryptographic Properties of Boolean Functions

The Boolean functions have certain properties that are important of cryptological aspect. In a previous research [1] we have already explored these properties with the Boolean representations of the quasigroups of order 4. Now, we are more interested into examining these properties in the sets of parastrophic quasigroups and see if we can make a classification.

In this research we use the results from the previous research [1]. We are interested in several cryptographic properties of the quasigroups of order 4 and their parastrophes - algebraic immunity, nonlinearity, correlation immunity and resiliency.

We use the Boolean representations of quasigroups in Algebraic Normal Form. Besides that, we use the *boolfun* package [6], which is a convenient open source software that evaluates the cryptographic properties of Boolean functions. It is a package for the R language (a free software language and software environment for statistical computing and graphics).

3.1   Algebraic Immunity

In order to understand what is algebraic immunity we first need to define the term annihilator in B_n - the set of all Boolean functions that have n arguments. Namely, an annihilator of $f \in B_n$ is a function $g \in B_n$ such that $f(\overline{x}) \cdot g(\overline{x}) = 0$ for each $\overline{x} \in \mathbb{F}_2^n$. Now, the algebraic immunity of a Boolean function f, or $AI(f)$ is the smallest value of d such that $f(\overline{x})$ or $1 \oplus f(\overline{x})$ has a non-zero annihilator of degree d. [6][9].

The algebraic immunity is an indicator of the resistance to algebraic attacks for given Boolean function. Since each quasigroup of order 4 is represented as a pair of two Boolean functions (f_1, f_2), the algebraic immunity of each quasigroup is considered as a pair $(AI(f_1), AI(f_2))$ [1]. Quarter of the quasigroups of order 4 have the minimal value (1,1) as algebraic immunity, a quarter have the maximal (2,2) as algebraic immunity and one half have algebraic immunity (1,2) or (2,1).

Observing the algebraic immunity of the quasigroups in the sets of parastrophes we have identified:

- 37 sets that contain only quasigroups with algebraic immunity (1,1)
- 105 that contain some quasigroups with algebraic immunity (1,2) or (2,1) and some quasigroups with algebraic immunity (2,2)
- 8 sets that contain only quasigroups with algebraic immunity (2,2)

Figure 1 shows this distribution graphically.

An important observation is that the sets of parastrophes with bad cryptographic properties stand apart. Namely, these sets contain only quasigroups with bad cryptographic properties (that have algebraic immunity (1,1), nonlinearity (0,0) and that are linear by Boolean representation). In other words, if a quasigroup has bad cryptographic properties (algebraic immunity, nonlinearity and nonlinearity by Boolean representation), all of its parastrophes will have bad cryptographic properties, too.

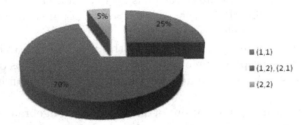

Fig. 1. Distribution of the algebraic immunity of the sets of parastrophic quasigroups

3.2   Nonlinearity

The term nonlinearity of a Boolean function in this subsection should be considered as follows [5].

Firstly, we need to define the term affine Boolean function. An affine Boolean function $h \in \mathcal{B}_n$ is a Boolean function in the form of:

$$h(x_1, x_2, ..., x_n) = \alpha_0 \oplus \alpha_1 x_1 \oplus \cdots \oplus \alpha_n x_n$$

where $\alpha_i \in \{0,1\}$ for $i = 0, 1, ..., n$. The set of affine Boolean functions with n arguments is denoted by $\mathcal{A}_n$.

Now, the distance between two Boolean functions $f, g \in \mathcal{B}_n$ is defined as $d(f, g) = |\{x \mid f(x) \neq g(x)\}|$.

The nonlinearity of f (denoted by $NL(f)$) is the minimal distance between f and any $h \in \mathcal{A}_n$ [5][8]. The nonlinearity of each quasigroup is considered as a pair $(NL(f_1), NL(f_2))$ [1]. The distribution of the sets of parastrophes in terms of nonlinearity is exactly the same as the distribution in terms of algebraic immunity, because the quasigroups that have algebraic immunity (1,1) also have nonlinearity(0,0), the ones that have algebraic immunity (1,2) or (2,1) have nonlinearity(0,4),(4,0) and the ones having algebraic immunity (2,2) have nonlinearity (4,4) [1].

3.3 Correlation Immunity and Resiliency

Correlation immunity is a property of Boolean functions that is an indicator of their resistance to correlation attacks. A Boolean function is correlation immune of order m, if the distribution of its truth table is unaltered while fixing any m inputs [4]. Resiliency is a property of Boolean functions that combines balance and correlation immunity. A function $f \in \mathcal{B}_n$ is m-resilient if f is balanced (has the same number of zeros and ones in the truth table) and its correlation immunity order is m [6]. It is related to the number of input bits that do not give statistical information about the output bit. Since all of the Boolean representations of the quasigroups of order 4 are balanced, this means that if the Boolean function has correlation immunity order m then it is m-resilient [1]. In other words, the distribution of quasigroups based on their resiliency matches the distribution of quasigroups based on their correlation immunity order.

In terms of the resiliency, we identified 5 classes of sets of parastrophic quasigroups:

- 56 sets that contain only quasigroups with resiliency (1,1)
- 20 that contain some quasigroups with resiliency (1,1) and some quasigroups with resiliency (1,2)
- 20 sets that contain some quasigroups with resiliency (1,1) and some quasigroups with resiliency (2,2)
- 48 sets that contain some quasigroups with resiliency (1,2) and some quasigroups with resiliency (1,3)
- 6 sets that contain only quasigroups with resiliency (2,2)

Figure 2 shows this distribution graphically.

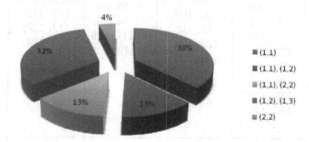

Fig. 2. Distribution of the resiliency of the sets of parastrophic quasigroups

3.4 Algebraic Immunity, Nonlinearity and Resiliency of the Sets of Parastrophic Quasigroups

The goal of this research is to classify the sets of parastrophic quasigroups based on their cryptographic properties. We have already discussed about the algebraic immunity, nonlinearity and resiliency individually, but it is also important to make a parallel between the results, as we know that algebraic immunity and nonlinearity are opposed to the correlation immunity and that we have to make a compromise between these properties. Using the numbers that denote each set of parastrophic quasigroups from Table 1, in Table 2 we represent the intersection between the mentioned properties.

Table 2. Parastrophic quasigroup and their cryptographic properties

AI/NL \ Res.	(1,1)	(1,1),(1,2)	(1,1),(2,2)	(1,2),(1,3)	(2,2)
(1,1)	1, 3, 14, 17, 26, 34, 37, 38			2, 4, 5, 6, 7, 9, 11, 12, 15, 16, 18, 19, 21, 23, 24, 25, 27, 28, 29, 31, 32, 33, 35, 36	8, 10, 13, 20, 22, 30
(1,2) and (2,2)	39, 40, 43, 44, 47, 50, 54, 56, 63, 64, 65, 66, 71, 75, 77, 79, 82, 85, 86, 87, 97, 101, 102, 103, 108, 110, 111, 114, 116, 122, 126, 129, 130, 134, 135, 138, 139, 140, 141, 142	41, 45, 52, 57, 67, 73, 78, 83, 89, 95, 100, 105, 106, 109, 115, 118, 125, 128, 131, 137	49, 51, 53, 55, 61, 62, 70, 72, 74, 76, 88, 90, 92, 94, 96, 98, 117, 119, 121	42, 46, 48, 58, 59, 60, 68, 69, 80, 81, 84, 91, 93, 99, 104, 107, 112, 113, 120, 123, 124, 127, 132, 133, 136	
(2,2)	143, 144, 145, 146, 147, 148, 149, 150				

4 Conclusion

To conclude, we have made a new classification of the parastrophic quasigroups of order 4, using their Boolean representations and considering several cryptographic properties. These classification can assist in understanding the connection between the parastrophic quasigroups, since we can see that the sets of parastrophes hold similar properties, at least for the algebraic immunity, nonlinearity and resiliency.

These results are especially important for the parastrophic quasigroup string transformation [3], which uses these sets of parastrophic quasigroups. Therefore, the presented results can be used for selecting the sets that have good cryptographic properties and use them for this transformation. As further work, we can examine the cryptographic properties of the classes of isomorphic quasigroups and see if there is a match with the results of this research. Also, we can apply these results for some further cryptographic primitives.

References

1. Dimitrova, V., Trajcheska, Z., Petkovska, M.: Analyzing the Cryptographic Properties of the Boolean Representation of Quasigroups of Order 4. In: 10th Conference for Informatics and Information Technology, CIIT 2013, Bitola (in print, 2013)
2. Dimitrova, V.: Quasigroup Processed Strings, their Boolean Representations and Application in Cryptography and Coding Theory. PhD Thesis, Ss. Cyril and Methodius University, Skopje, Macedonia (2010)
3. Dimitrova, V., Bakeva, V., Popovska-Mitrovikj, A., Krapež, A.: Cryptographic Properties of Parastrophic Quasigroup Transformation. In: Markovski, S., Gushev, M. (eds.) ICT Innovations 2012. AISC, vol. 207, pp. 235–243. Springer, Heidelberg (2013)
4. Dündar, B.G.: Cryptographic Properties of some Highly Nonlinear Balanced Boolean Functions. Master's Degree Thesis, The Graduate School of Applied Mathematics, The Middle East Technical University, Ankara, Turkey (2006)
5. Hirose, S., Ikeda, K.: Nonlinearity Criteria of Boolean Functions. KUIS Technical Report, KUIS-94-0002 (1994)
6. Lafitte, F.: The boolfun Package: Cryptographic Properties of Boolean Functions (2012)
7. Pommerening, K.: Fourier Analysis of Boolean Maps. A Tutorial, Mainz (2005)
8. Braeken, A.: Cryptographic Properties of Boolean Functions and S-Boxes. PhD Thesis, Katholieke Universiteit Leuven, Leuven, Belgium (2006)
9. Zhang, X.-M., Pieprzyk, J., Zheng, Y.: On Algebraic Immunity and Annihilators. In: Rhee, M.S., Lee, B. (eds.) ICISC 2006. LNCS, vol. 4296, pp. 65–80. Springer, Heidelberg (2006)

Optimizing Durkins Propagation Model Based on TIN Terrain Structures

Leonid Djinevski[1], Sonja Stojanova[2], and Dimitar Trajanov[2]

[1] FON University,
Av. Vojvodina, 1000 Skopje, Macedonia
leonid.djinevski@fon.edu.mk
[2] Ss. Cyril and Methodious University,
Rugjer Boshkovikj 16, 1000 Skopje, Macedonia
stojsonja@gmail.com, dimitar.trajanov@finki.ukim.mk

Abstract. In order to bring wireless ad hoc networks simulation a step closer to real-life scenarios, 3D terrains have to be taken into account. Since 3D terrain involves larger amounts of data, network simulations with heavy traffic load, requires compute intensive calculations. In this paper we evaluate the usage of efficient point location for network simulation in 3D terrains, in order to increase the performance of the overall simulation. Our experimental results show a reasonable speedup using the jump-and-walk point location algorithm when computing the propagation between two wireless nodes, as well as for the overall performance increase for a complete simulation scenario.

Keywords: Network Simulators, TIN, Point Location, Durkins Radio Propagation, Parallel.

1 Introduction

When researching new scenarios and protocols in a controlled and reproducible environment, network simulators are the main tools that allow the user to represent various topologies, simulate network traffic using different protocols, visualize the network and measure the network performances.

Many network simulators have been developed as open-source tools by the academia and proprietary tools by the industry. NS-2 network simulator stands out as the most popular and most widely adopted, thus it is considered as de facto standard regarding network simulation. Simulation for 802.11 ad hoc networks with mobile nodes, special routing protocol and fully specified network traffic is supported by the NS-2 network simulator.

Most of the commonly used propagation models when assessing performance of ad hoc networks, take into account the following mechanisms of reflection, diffraction, scattering, penetration, absorption, guided wave and atmospheric effects [10]. However, the available models in NS-2, do not take the terrain profiles in consideration, thus the obtained results for the received signal strength from a transmitter are not close to real life scenarios.

V. Trajkovik and A. Mishev (eds.), *ICT Innovations 2013*,
Advances in Intelligent Systems and Computing 231,
DOI: 10.1007/978-3-319-01466-1_25, © Springer International Publishing Switzerland 2014

Since the communication between nodes in wireless ad hoc networks is usually carried out in irregular terrains, the terrain profile should be taken into account. Authors in [13] layout such implementation called the Durkins propagation model, which is an extension of the NS-2 simulator [1], where the terrain profile is represented by Digital Elevation Model (DEM) [21] data structure.

An improvement of the extension of the Durkins propagation model is presented in [27], where the implementation uses Triangular Irregular Networks (TIN) [24] based terrains. This tool allows the simulation modeler to conduct more realistic simulation scenarios, and analyze the way the terrain profile affects the ad hoc network performances. Although the implementation in [27] is very useful, it does not scale well, thus starting a simulation for medium to large networks, results in few hours up to days, even weeks of execution time which is unsuitable for investigating protocols. Additional optimization was proposed by authors in [12] where they have speed up the point location of Durkins propagation model, by parallel execution using GPU devices.

In this paper, we present an analysis of three main efficient access algorithms implemented for point location in the Durkins propagation model based on TIN terrain profile. Additionally we investigate how the efficient point location method affects the overall execution time of a given simulation, for different detail levels of the terrain.

The obtained results help the simulation modeler to choose the appropriate point location algorithm for a terrain profile with a given detail level, in order to achieve faster simulations.

The remainder of this paper is organized as follows: In Section 2 we present a small introduction to mobile ad-hoc network (MANETs) and the Durkins terrain aware radio propagation model for the NS-2 simulator. We present the DEM and TIN data structure that describe a given terrain profile in Section 3. Theoretical analysis of point location methods are presented in Section 4, followed by a short presentation of the applied testing methodology in Section 5. The obtained results are analyzed in Section 6. We conclude in Section 7.

2   MANETs and the Durkin's Propagation Model

Communication among people on the move has evolved remarkably during the last decade. One of the possibilities for such mobility is the mobile radio communications industry growth by orders of magnitude, thus making portable radio equipment smaller, cheaper and more reliable [17]. The large scale deployment of affordable, easy-to-use radio communication networks has created a trend of a demand for even greater freedom in the way people establish and use the wireless communication networks [9].

One of the consequences to this ever present demand is the rising popularity of the ad hoc networks. A MANET does not rely on any infrastructure, thus, it can be established anywhere on the fly [26]. Wireless mobile nodes communicate directly with the absence of base station or an access point. Therefore, nodes establish the network environment by self organization in a highly decentralized manner. In order to achieve this goal every node has to support the so-called multihop paths. The multihop path concept is introduced to allow two distant nodes to communicate by the means of the intermediate nodes to graciously forward the packets to the next node that is closer to the destination. This is controlled by a special ad hoc routing protocol [8] that is concerned with discovery, maintenance and proper use of the multihop paths.

The independence of existing infrastructure, as well as the ability to be created instantly, that is, on demand, has made the ad hoc networks a very convenient and irreplaceable tool for many on-the-go situations like: rescue teams on crash sites, vehicle to vehicle networks, lumber activities, portable headquarters, late notice business meetings, military missions, and so on. Of course, every one of these applications demands a certain quality of service from the ad hoc network and usually the most relevant issue are the network performances in terms of end-to-end throughput. However, the tradeoff of having no infrastructure and no centralized manner of functioning has influenced the ad hoc networks performances greatly on many aspects.

3   Terrain Data Structures

Many applications utilize Geographical Information Systems (GIS) technology for which geographical referenced information of sea bottom or physical land surface is required [28][18][6][25][7]. This technology provides representation of terrain elevation data using two broad methods: raster images or vectors. The raster data format is consisted of rows and columns of cell, wherein each cell a numerical value describes an attribute of the surface. Therefore each cell is a single pixel, and the grid of all the cells with its width and length represent the resolution of a terrain. The most widely spread raster data format is described by the Digital Elevation Model (DEM) standard [23][5][3][4]. On the other hand, the vector data format uses geometrical units like points, lines and polygons to represent objects. Hence, compared to the raster data format, much less data is needed to represent the same object, resulting in smaller files for storing the data.

The most popular vector data format for representing elevation is the Triangular Irregular Networks (TIN) [2][29][20]. The basic building block of the TIN data structure is a record that is consisted of points that associate 3D coordinates, which are interconnected by lines, distributed in such way to form triangles, where no triangles are overlapping each other. Both of the data structures (DEM and TIN) for the same terrain are visualized on Figure 1.

The raster data format is more suitable for modeling surfaces, although the vector data format can produce much more precise representation of a given surface with the same amount of data. Beside the accuracy of the surface, the raster data format regarding hill peaks and ridges is reliable as much as the network resolution allows it, while for the TIN data format hill peaks and ridges are very precise.

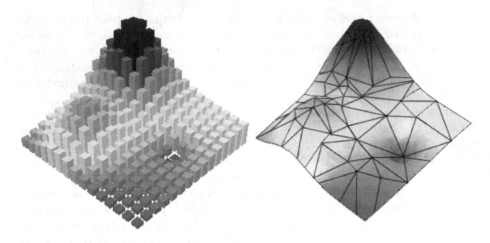

Fig. 1. Terrain representation using DEM data (left), and TIN data (right)

3.1   Optimal Access to the Network

Since the TIN networks are vector structures, there is no direct way to obtain an elevation of a point on the terrain. The most simple method for point location is to traverse a TIN network by visiting all triangles only once [15][14][16]. However, this method is ineffective when dealing with large-scale TIN networks. There are several point location strategies for the TIN model [19]. Some of these strategies include: hierarchical data structures, spatial indexing structures, quad trees, etc. Although, these solutions are very effective, the implementation effort for the data structures they use, can be nontrivial, so in practice more simple methods like walking algorithms are used. These algorithms solve the point location problem by traversing the triangles of the TIN, starting from a random triangle, until the triangle containing the query point is reached.

4   Theoretical Analysis

In this section we analyse few of the most popular and widely used algorithms for point location in TIN terrains: quad trees, jump-and-walk, and windowing.

4.1   Quad Trees

As the name of the algorithm implies, quad trees data structure is used, which is a regular search tree of degree 4 as presented in Fig. 2. The root square is decomposed in 4 subsquares recursevly in the same fashion as the quad tree expands. The decomposition continues until the subsquare data is simple enought. The algorithm is consisted of two steps:

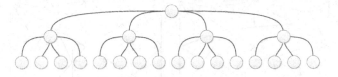

Fig. 2. Quad-tree algorithm

- Traverse down the tree unti the data structure terminates with a triangle.,
- Traverse the subsquare structure in order to find the triangle that contains the point that is queried.

4.2   The Jump-and-Walk Algorithm

The jump-and-walk [11] is one of the most competitive walking algorithms in which the traversal from the starting to the final triangle is determined from the line segment (p,q) where p is the starting and q is the query point (Fig. 3 on the left). The jump-and-walk algorithm consists of three main stages:

- Obtaining m random, but uniformly distributed starting points in the XY-plane, where m is around $n^{1/3}$,
- Determine the index i , for $i = 1..m$, of the starting point such that the Euclidian distance between p_i and q is minimal,
- Traverse the triangles that intersect the line segment (p_i, q) until the triangle that contains the query point is located,

The additional data structure required is for storing the index of the neighboring triangles for a given triangle in the TIN terrain.

4.3   The Windowing Algorithm

Windowing as an approach for the point location where the events take place in given squarearea, for example silnal propagation between two (Transiver-Receiver) nodes. The windowing algorithm is as folllows:

- in this step the position of the window needs to be located,
- and all triangles that intersect with the window need to be located.

5   Testing Methodology

All of our tests were performed on the same hardware infrastructure: (Intel i7 920 CPU at 2.67GHz, with 12GB RAM at 1333MHz, GPU NVIDIA Tesla C2070), with Linux operating system Ubuntu 10.10. The implementations were compiled with the NVIDIA compiler nvcc from the CUDA 4.2 toolkit.

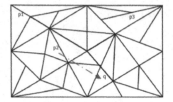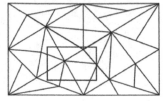

Fig. 3. Jump-and-walk algorithm (left), and windowing algorithm (right)

The testing methodology is based on 2 experiments which show inverse execution time dependence of problem size (number of triangles in the TIN terrain) for the jump-and-walk approach and overall performance increase in a simulation scenario.

Experiment 1 determines the inverse execution time compared to different terrain requirements. We increase the terrain size and measure the speed for the executions of jump-and-walk approach, standard sequential approach of traversing all triangles and the parallel implementation of the standard approach. Our hypothesis to be confirmed experimentally is to achieve the highest speed for the jump-and-walk approach.

In Experiment 2 we implement the jump-and-walk approach in given NS-2 simulations, by varying different node mobility (nodes moving with speed of 1, 2 and 5 m/s) and different traffic load (from 0.1 to 7 Mbps). We use a TIN terrain with dimensions of 1 million square meters, and highest relative point of 200m. We have 100 nodes that are uniformly dispersed in the simulation area. The node transmission range is set to the standard 250m given by the use of the IEEE 802.11b standard wireless equipment. The antenna height is set to 1.5m and it has no relative offset against the wireless node. For route discovery and path set up we adopt the AODV protocol [22]. The simulation time is set to 1.5 hours as to the average battery life of a notebook. During the mobile simulations, the nodes are moving according to the random direction model in the terrain boundaries. The average node speed is varied from 0 (static nodes) to 1, 2 or 5 m/s with standard deviation of 0.1 m/s. We measure the average execution time for all network traffic scenarios by varying the network load from 0.1 to 7 Mbps using UDP data packets with 1 KB size. Our hypothesis to be confirmed experimentally is to achieve higher speedup for the average execution time using the jump-and-walk approach, compared to the average execution time without the jump-and-walk approach.

6   Results of the Experiments

This section presents the results that show performance increase in wireless simulation using the jump-and-walk, quad trees, and windowing for point location algorithms.

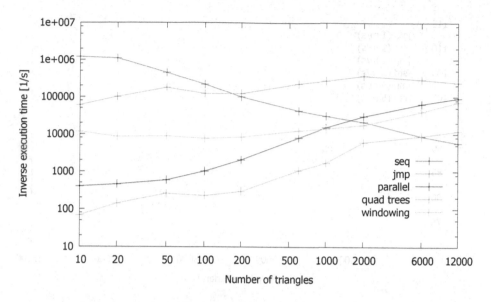

Fig. 4. Inverse execution time of different point location algorithms

6.1 Experiment 1

Figure 4 depicts the inverse execution time for different terrain requirements, where seq stands for the sequential implementation, parallel stands for the parallel implementation, jmp stands for jump-and-walk implementation, quad trees stands for the quad trees algorithm, and windowing for the windowing algorithm. Thus, it is easy to notice that the jump-and-walk implementation is the fastest for larger number of triangles. For smaller number of triangles the parallel implementation does not utilize the GPU resources, therefore there is slow down and the speed is significantly lower than the sequential implementation. Only for terrains with more than 1000 triangles, the parallel implementation expresses higher speed than the sequential implementation. The speed of the sequential implementation is the best for small number of triangles. However, by increasing the number of triangles, the speed of the sequential implementation is getting worse.

6.2 Experiment 2

In our second experiment we are evaluating the speedup of jump-and-walk implementation and the standard parallel implementation for compared to given terrain requirements, for average node velocity of 1, 2 or 5 m/s. The results are depicted in Figure 5 where the dotted lines represent the standard parallel implementation, and the solid lines represent the jump-and-walk implementation. For jump-and-walk implementation the results confirm our expectation to achieve higher performance.

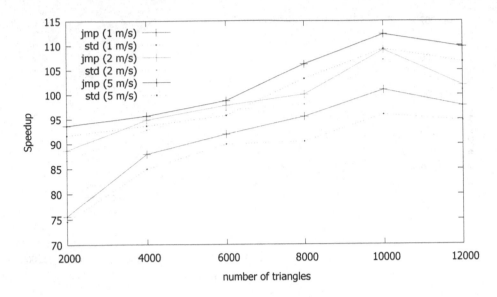

Fig. 5. Overall NS-2 network simulation speedup using different point location algorithms

7   Conclusion and Future Work

Implementations of three efficient access algorithms are presented in TIN terrain representation, as an alternative to DEM terrain representation discussed in previous work [13].

There are many proprietary and open-source network simulators that are available. However, NS-2 Simulator is the most widely used and adopted by the research community as de facto standard regarding network simulation. For NS-2 Simulator we developed our efficient access implementations.

We performed two experiment that evaluate the performance of the optimization using the jump-and-walk approach and overall performance increase in a NS-2 simulation.

The experiment 1 confirms that the jump-and-walk algorithm achieves best speed for terrains with more than 200 triangles. Since terrains with less than 1000 triangles do not describe the terrain well, the results point out that jump-and-walk algorithm is the fastest. Additionally, the experiment 2 confirmed that the effective point location implementation for the jump-and-walk algorithm achieves performance increase for overall network simulation scenarios of terrain aware radio propagation model.

Although there are theoretically more efficient algorithms for point location, we have chosen the jump-and-walk algorithm. In order the jump-and-walk algorithm to be effective, the assumption that the starting points are uniformly distributed across the terrain needs to be fulfilled, which is the case with the nodes in our simulation scenario.

Additionally, the jump-and-walk algorithm requires a very simple data structure, therefore there are very few requirements needed to modify the standard implementation of the terrain aware extension for the Durkins propagation model in NS-2 Simulation.

Acknowledgement. We gratefully acknowledge the Nvidia Corporation for support in graphics hardware as part of the CUDA Teaching Center program.

References

1. ns-2, network simulator (1989)
2. Unit 39-the tin model (2011),
   http://www.geog.ubc.ca/courses/klink/gis.notes/ncgia/u39.html
3. U.s. geological survey national mapping division: Part 1 general, standards for digital elevation models (2011),
   http://nationalmap.gov/standards/pdf/1DEM0897.PDF
4. U.s. geological survey national mapping division: Part 2 general, standards for digital elevation models (2011),
   http://nationalmap.gov/standards/pdf/2DEM0198.PDF
5. Usgs dem (2011), http://edc.usgs.gov/guides/usgs_dem_supplement.html
6. What is gis? (June 2011), http://www.gis.com/content/what-gis/
7. Abdul-Rahman, A., Pilouk, M., et al.: Spatial data modelling for 3d gis (2008)
8. Reddy, B.A.K., Hiremath, L.C., Performance, P.S.: comparison of wireless mobile ad-hoc network routing protocols. IJCSNS 8(6), 337 (2008)
9. Brown, J.S., Duguid, P.: Social life of information. Harvard Business Press (2002)
10. Cichon, D.J., Kurner, T.: Propagation prediction models. COST 231 Final Rep. (1995)
11. Devroye, L., Mücke, E.P., Zhu, B.: A note on point location in delaunay triangulations of random points. Algorithmica 22(4), 477–482 (1998)
12. Djinevski, L., Filiposka, S., Trajanov, D., Mishkovski, I.: Accelerating wireless network simulation in 3d terrain using gpus. Tech. Rep. SoCD:16-11, University Ss Cyril and Methodius, Skopje, Macedonia, Faculty of Information Sciences and Computer Engineering (June 2012)
13. Filiposka, S., Trajanov, D.: Terrain-aware three-dimensional radio-propagation model extension for ns-2. Simulation 87(1-2), 7–23 (2011)
14. Gold, C.M., Charters, T., Ramsden, J.: Automated contour mapping using triangular element data structures and an interpolant over each irregular triangular domain. ACM SIGGRAPH Computer Graphics 11(2), 170–175 (1977)
15. Gold, C., Cormack, S.: Spatially ordered networks and topographic reconstructions. International Journal of Geographical Information System 1(2), 137–148 (1987)
16. Gold, C., Maydell, U.: Triangulation and spatial ordering in computer cartography. In: Proc. Canad. Cartographic Association Annual Meeting, pp. 69–81 (1978)
17. Hekmat, R.: Ad-hoc networks: fundamental properties and network topologies. Springer (2006)
18. Kennedy, M.: Introducing Geographic Information Systems with ArcGIS: A Workbook Approach to Learning GIS. Wiley (2009)

19. van Kreveld, M.: Digital elevation models and tin algorithms. In: Algorithmic Foundations of Geographic Information Systems, pp. 37–78. Springer (1997)
20. Li, Y., Yang, L.: Based on delaunay triangulation dem of terrain model. Computer and Information Science 2(2), 137 (2009)
21. Maune, D.F.: Digital elevation model technologies and applications: The DEM users manual. Asprs Pubns (2007)
22. Meltzer, R., Zeng, C.: Micro-benchmarking the c2070 (January 2013),
    http://people.seas.harvard.edu/~zeng/microbenchmarking/
23. Peckham, R.J., Jordan, G.: Digital terrain modelling: development and applications in a policy support environment, vol. 10. Springer (2007)
24. Rognant, L., Goze, S., Planes, J., Chassery, J.: Triangulated digital elevation model: definition of a new representation. International Archives of Photogrammetry and Remote Sensing 32, 494–500 (1998)
25. Shi, W.: Principles of modeling uncertainties in spatial data and spatial analyses. CRC Press (2010)
26. Tonguz, O.K., Ferrari, G.: Ad Hoc Wireless Networks: A Communication-Theoretic Perspective. John Wiley & Sons, Ltd. (2006)
27. Vuckovik, M., Trajanov, D., Filiposka, S.: Durkin's propagation model based on triangular irregular network terrain. In: Gusev, M., Mitrevski, P. (eds.) ICT Innovations 2010. CCIS, vol. 83, pp. 333–341. Springer, Heidelberg (2011)
28. Wise, S.: GIS basics. CRC Press (2002)
29. Zeiler, M.: Modeling our world, environmental systems research institute Inc., Redlands (1999)

Hierarchy and Vulnerability of Complex Networks

Igor Mishkovski

Faculty of Computer Science and Engineering, Ss. Cyril and Methodius University – Skopje
igor.mishkovski@finki.ukim.mk

Abstract. In this paper we suggest a method for studying complex networks vulnerability. This method takes into account the network topology, the node dynamics and the potential node interactions. It is based on the PageRank and VulnerabilityRank algorithms. We identify the problem with these algorithms, i.e. they tend towards zero for very large networks. Thus, we propose another method to evaluate the amount of hierarchy in a given complex network, by calculating the relative variance of the system vulnerability. This measure can be used to express how much one network is being hierarchical, thus revealing its vulnerability. We use the proposed method to discover the vulnerability and hierarchical properties of four characteristic types of complex networks: random, geometric random, scale-free and small-world. As expected, the results show that networks which display scale-free properties are the most hierarchical from the analyzed network types. Additionally, we investigate the hierarchy and vulnerability of three real-data networks: the US power grid, the human brain and the Erdös collaboration network. Our method points out the Erdös collaboration network as the most vulnerable one.

Keywords: complex networks, hierarchy, vulnerability.

1   Introduction

The topological hierarchy of a complex network gives a fundamental characteristic of many complex systems. Complex networks serve as a backbone of the complex system and its dynamics, and often they have different topological organization. Many of the today's present systems, like the World Wide Web, the Internet, the semantic web, the actor network and the scientific collaboration network have strong hierarchy in their structure [1]. The hierarchy of these and other complex systems is closely related with their vulnerability because the parts of the system which are in the highest hierarchical levels of the system will have the highest impact on the performances of the system when being attacked. Thus, by finding how much one network (system) is hierarchical one can classify its vulnerability, and vice versa.

The hierarchy of the complex networks is often measured by the degree of the node, and this type of hierarchy is called explicit hierarchy [2]. In many systems the nodes with higher degrees are often more important and by that higher in hierarchy than other nodes in the network. For example, in the protein interaction network, proteins that have more bindings (physical reactions) are higher in the hierarchy. In

V. Trajkovik and A. Mishev (eds.), *ICT Innovations 2013*,
Advances in Intelligent Systems and Computing 231,
DOI: 10.1007/978-3-319-01466-1_26, © Springer International Publishing Switzerland 2014

the power grid network, the generators and substations linked with more high-voltage transmission lines would be placed higher in the hierarchy, meaning that they are the most vulnerable nodes in the power grid network. However, the explicit hierarchy measure using only node degree can sometimes be misleading placing unimportant nodes high in the hierarchy, while not taking into account the hierarchy level of its neighbors. In order to obtain realistic hierarchical representation of the network, the level of hierarchy of the neighbors and a given node must be in a similar range.

Thus, another approach, given in [2], is to quantify the implicit hierarchy of the network based on the system vulnerability measure introduced by Latora and Marchiori [3]. Latora and Marchiori calculate the point-wise vulnerability of the network by:

$$V(i) = \frac{E - E(i)}{E} \tag{1}$$

where:

$$E = \frac{1}{N(N-1)} \sum_{i \neq j} \frac{1}{d_{ij}} \tag{2}$$

represents the global efficiency of the network [4], N is the total number of nodes in the network, d_{ij} is the minimal distance between the i-th and j-th node, and $E(i)$ is the network efficiency after removal of the i-th node and all its edges. The maximal value V of the vector $V(i)$ corresponds to the vulnerability of the network as a whole [3]. The assumption that the authors used in [2] is that the most vulnerable nodes, the nodes with maximum point-wise vulnerability, are placed on the highest position in the network hierarchy. Also, when removal of certain nodes causes more serious disturbance of the network efficiency compared to the other nodes, the method assumes that the network has a strong hierarchical properties.

In [4] authors quantify the hierarchical topology of a network using the concept of hierarchical path (see also [5]). The hierarchy is measured as the fraction of shortest paths that are also hierarchical. They found that the hierarchy of random scale-free topologies smoothly declines with the power law coefficient γ.

In this paper we propose an algorithm for measuring the vulnerability of a complex network (system) based on the PageRank algorithm [6]. This paper is extension of the work presented in [7] where the authors introduce a new approach for finding the most vulnerable nodes in the network, called VulnerabilityRank. Based on the above algorithm we introduce a new approach for measuring the hierarchy of a complex network. The proposed algorithm takes into account not only the network topology, but also the node dynamics and potential node interactions. The dynamics of each node is implemented using the node-degree binary influence model.

Another contribution of this paper is that all of these algorithms are tested on four different network types: random (Erdös-Renyi) network, geometric random network, scale-free network and small world networks as well as different real data networks: human brain network [8], US power grid [9] and Erdös collaboration network [9].

The rest of the paper is organized as follows. In Section 2 we explain the algorithm for measuring the vulnerability of the network. Section 3 explains the newly proposed algorithm for measuring the hierarchical properties of a network. Section 4 describes the generic network types and real data networks which are used in the simulations and Section 5 presents the results for the vulnerability and the hierarchy of the observed networks. Section 6 concludes the paper.

2 Algorithm for Measuring Vulnerability of Complex Network

In [7] the authors propose a measure for the vulnerability of a node using an influence matrix D which corresponds to an arbitrary graph G. If G is a finite simple undirected connected graph with N nodes and $A=(a_{ij})$ is its adjacency matrix then the node-degree influence matrix $D=(d_{ij})$ can be defined to be binary, such that $d_{ij} = a_{ij} / \sum_i a_{ij}$.

Then the nodes' vulnerability, according to [8], is equal to the stationery distribution π of the network influence matrix D, which is the normalized left eigenvector of the influence matrix associated with the eigenvalue 1. We identify the network vulnerability with the maximum value of the vector π. This measure can serve for comparing the vulnerability of different types of networks. As toy example, we can consider three networks with different simple topologies, such as: the star-like network, ring-like network and tree-like network, see Fig.1.

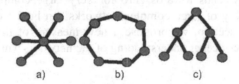

a)   b)   c)

Fig. 1. Example Networks: a) star-like, b) ring-like, c) tree-like

The adjacency matrix of the star-like network (Fig 1.a) is:

$$A = \begin{pmatrix} 0 & 0 & 0 & 0 & 0 & 0 & 1 \\ 0 & 0 & 0 & 0 & 0 & 0 & 1 \\ 0 & 0 & 0 & 0 & 0 & 0 & 1 \\ 0 & 0 & 0 & 0 & 0 & 0 & 1 \\ 0 & 0 & 0 & 0 & 0 & 0 & 1 \\ 0 & 0 & 0 & 0 & 0 & 0 & 1 \\ 1 & 1 & 1 & 1 & 1 & 1 & 0 \end{pmatrix}$$

And its raw matrix D is:

$$D = \begin{pmatrix} 0 & 0 & 0 & 0 & 0 & 0 & 1 \\ 0 & 0 & 0 & 0 & 0 & 0 & 1 \\ 0 & 0 & 0 & 0 & 0 & 0 & 1 \\ 0 & 0 & 0 & 0 & 0 & 0 & 1 \\ 0 & 0 & 0 & 0 & 0 & 0 & 1 \\ 0 & 0 & 0 & 0 & 0 & 0 & 1 \\ 1/6 & 1/6 & 1/6 & 1/6 & 1/6 & 1/6 & 0 \end{pmatrix}$$

This matrix D is now stochastic, but it's reducible, so it cannot have a unique positive stationery distribution. To force irreducibility, the following transformation is used:

$$\bar{D} = \alpha D + (1-\alpha)ee^T / n \tag{3}$$

where n is the order of A and α is the dampening factor. Now the matrix $\bar{D}$ is both stochastic and irreducible, and its stationery vector (the PageRank and the VulnerabilityRank vector) for $\alpha = 0.85$ is:

$$\pi^T = (0.0882, 0.0882, 0.0882, 0.0882, 0.0882, 0.0882, 0.4710)$$

The maximum value of the stationary vector, 0.4710, can now be used to describe the vulnerability of the whole network. As expected, this network is the most vulnerable from the toy example networks given in Fig. 1. Using the algorithm on the rest of the networks one can calculate that the tree-like network has 0.2413 vulnerability index, while the most robust is the ring-like network with vulnerability of 0.1429.

This method can take into account not only the network topology but also the node dynamics and potential node interactions. The dynamics of each node can be additionally implemented by an arbitrary (finite) Markov chain. In this case the method can be extended if the node-degree binary influence model is replaced with a heterogeneous influence model.

The only problem that this method might encounter is that the measure for the network vulnerability tends towards zero for very large complex networks. Thus, measuring vulnerability of large complex networks can be a tedious problem. As a remedy, in the next section, we propose a new measure of the complex networks hierarchy that provides deeper understanding of the networks' vulnerability.

3   Algorithm for Measuring the Hierarchy of Complex Networks

The vulnerability of the network can be used to quantify the hierarchical properties of a complex network.

First the point-wise vulnerability $V(i)$ is calculated using the VulnerabilityRank algorithm [7] instead of using the global efficiency of the network, given in (1) and (2). After that, the parameter of the fluctuation level (i.e. the hierarchical properties of the network) can be calculated using:

$$h = \frac{<\Delta V^2>}{<V>^2}, \tag{4}$$

where V stands for the network vulnerability,

$$<\Delta V^2> = \frac{1}{N}\sum_{i=1}^{N}(V(i) - <V>)^2,$$

and

$$< V >= \frac{1}{N} \sum_{i=1}^{N} V(i)$$

is the mean vulnerability. Because of the underlying VulnerabilityRank algorithm the mean vulnerability can be calculated as $<V>=1/N$.

The parameter h gives the hierarchical properties of the observed networks. For our toy examples from the previous section, the hierarchy of the star-like network is 0.8824, there is no hierarchy at all in the ring-like network and the tree-like network shows hierarchy of 0.2157. The important property of this hierarchical parameter is that it does not tend to be zero when $N \rightarrow \infty$. Because this method is solely based on the VulnerabilityRank it does take into account the network topology, node dynamics and potential node interactions (for the proof please see [7]). Again, the dynamics of each node can be additionally implemented by an arbitrary (finite) Markov chain. Thus, by extending the VulnerabilityRank we can also extend the proposed method for finding hierarchy in networks.

4   Complex Networks

Complex network is a complex graph-based structure made of nodes (which can be individuals, computers, web pages, power grid plants, organizations, cities, proteins in the human body, etc.) that are connected by one or multiple types of interdependence (i.e. friendship, network links, power transport network, trade, roads, chemical reactions, etc.) These graphs or networks have certain properties which limit or enhance the ability to do things with them [10]. For example, small changes in the topology, shutting down only small number of nodes, may lead to serious damage to the network capabilities.

4.1   Generic Network Types

As a reference point, we study four generic network types. Number of the nodes in all generic networks is 500 and the average node degree is around 6. All the networks were connected (i.e. there are no islands) and the connectivity is checked by investigating the value of the second eigenvalue of the Laplacian matrix [11].

The first network type, ER network [12] is the classic unweighted Erdös-Renyi random network. The nodes are randomly connected with a probability of a link between nodes i and j of 0.012.

The second network type is the geometric random network (GR) [13]. The nodes in this network are randomly scattered along a 1 m² square terrain and their connectivity radius was calculated as:

$$r = \sqrt{\frac{< k >}{N \pi}} \qquad (5)$$

where $<k>$ is the average node degree.

The scale-free network (SF) was based on the Barabasi-Albert (BA) power-law random graph [14]. At the beginning the network was consisted of 4 entirely connected nodes and the new nodes connected to the existing ones using linear preferential attachment.

The last generic network type is the small world network (SW). The network was generated using the Watts-Strogatz model [15] with a probability of reconnection of 0.1.

Table 1 analyzes the topological and connectivity dependent properties of the observed generic networks types. It examines the average node degree (AND), the average clustering coefficient (CC) and the average normalized node betweenness (AnNB) of the networks.

Table 1. Topology dependent properties of the observed generic complex networks types

Measure/Network	ER	GER	SW	SF
AND	6.27	6.26	6.00	5.98
CC	0.014	0.627	0.447	0.055
AnNB	0.521	3.352	0.894	0.445

4.2   Real Data Networks

The proposed vulnerability and hierarchy methods were used on several real data networks: the brain, the US power grid and the Erdös collaboration network.

The first network represents the structural connectivity of the entire human brain. The used data was originally obtained by a diffusion magnetic resonance imaging scan with the approach described in [16]. The network has two layers: physical and logical. The logical layer consists of connections in the gray matter in the brain, while the physical layer reflects the axonal wiring used to establish the logical connections.

The second real data network represents the power grid in the US. The network that we used in our simulation is provided by [17]. This network has 4941 nodes and 13188 edges.

The third real data network represents a network whose edges are the collaboration between Paul Erdös and other mathematicians. The Erdös network [17] has 472 nodes and 2,628 edges (collaborations). Additionally, in table 2 the topological and connectivity dependent properties are shown for the real data networks: the Erdös collaboration (EC), logical network of the brain (LB), physical network of the brain (PB) and the US power grid network (USPG).

Table 2. Topology dependent properties of the observed real data complex networks

Measure/Network	EC	LB	PB	USPG
Number of nodes	472	1013	4445	4941
Number of links	2628	30738	41943	13188
AND	5.568	30.343	9.436	2.669
CC	0.347	0.456	0.373	0.107
AnNB	0.531	0.182	0.186	0.364

5   Vulnerability and Hierarchical Analysis

In our simulations we measure the vulnerability and the hierarchical properties of the above mentioned networks. Please keep in mind that the network vulnerability investigated in this paper is related to node vulnerability as opposed to link vulnerability which is not considered.

The first analysis was related to the vulnerability of the generic network types. As expected, the scale free topology is the most vulnerable network (index: 0.0190). The most robust network is the small world network (index: 0.0028). The random network has vulnerability index of 0.0044 and the geometric random 0.0033.

As a next step we investigate the vulnerability of the US power grid, the logical and physical human brain network. The results shows that the nature's complex network, the human brain (index: 0.0005), is more robust than the man-made network, the US power grid (index: 0.0012). Also one can see that the logical functioning of the brain is the most vulnerable to node failure (index: 0.0042).

In the next part of this section we analyze the hierarchy of the observed networks. From the generic network types, the most hierarchical network is the scale free network which has hierarchical value of 0.8454 and the least hierarchical is the small world network with value of 0.0117. The random network has hierarchical value of 0.1133 and geometric random 0.0620. For the real data networks, as expected, the most hierarchical network of the observed real data networks is the Erdös collaboration network with hierarchy of 0.7462, while the US power grid (h=0.2786) and the logical brain network (h=0.2685) show similar hierarchical properties. The least hierarchical network is the physical brain network with hierarchy of 0.1378.

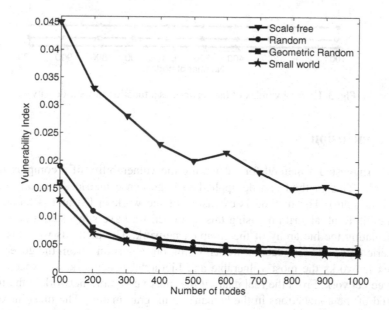

Fig. 2. Vulnerability index of the network as a function of their density

Next we analyze how the vulnerability index depends on the density of the network. We used the generic networks types and change the number of nodes in the network from 100 to 1000. In addition the average degree of the nodes remained unchanged for all networks. As expected the vulnerability index is decreasing as the number of the nodes in the network increases, as shown in Fig. 2.

Another analysis (see Fig. 3) shows how the calculated hierarchy depends on the network density for the generic networks. Fig. 3 shows that it remains almost constant for all type of generic networks, except for the scale-free topology. In the case of scale-free topology the hierarchy increases as the number of the nodes in the network grows.

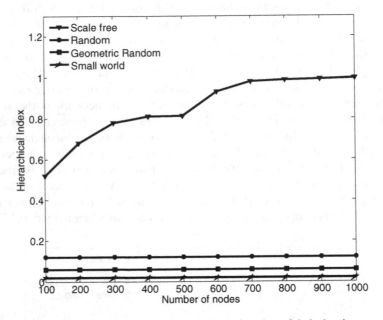

Fig. 3. Hierarchy index of the network as a function of their density

6   Conclusion

We have suggested a method for calculating the vulnerability of a complex network as a whole. This method can be applied to huge dense matrices and any complex network (system). The method is extension of the work in [7] and is based on the VulnerabilityRank algorithm. Using this approach we have suggested a new method for calculating the hierarchy of the complex network. The methods were tested both on generic types and real data networks. Using our algorithm, from the generic set of complex networks the most vulnerable and hierarchical network, as expected is the scale free network. From the real data set the most vulnerable network is the network consisted of the connections in the human's brain gray matter. The most hierarchical network is the Erdös collaboration network.

Additionally, in this paper, we have analyzed how the network density affects the network vulnerability and hierarchy. The network vulnerability tends to zero as the density of the network increases. The speed of the vulnerability decrease is different for different network types. The hierarchy of the network remains constant for all types of generic networks except for the scale-free model where this measure increases as the number of the nodes in the network grows.

References

1. Ravasz, E.: Evolution, hierarchy and modular organization in complex networks.Ph. D. thesis, University of Notre Dame (2004)
2. Gol'dshtein, V., Koganov, G.A., Surdutovich, G.I.: Vulnerability and hierarchy of complex networks. cond-mat/0409298 (2004)
3. Latora, V., Marchiori, M.: Economic Small-World Behavior in Weighted Networks. The European Physics Journal B-Condensed Matter and Complex Systems 32(2), 249–263 (2002)
4. Trusina, A., Maslov, S., Minnhagen, P., Sneppen, K.: Hierarchy Measures in Complex Networks. Physical Review Letters 92(17), 178702 (2008)
5. Tangmunarunkit, H., Govindan, R., Jamin, S., Shenker, S., Willinger, W.: Tech. Rep 01-746. Computer Science Department, University of Southern California (2001)
6. Brin, S., Page, L.: The anatomy of a large-scale hypertextual Web search engine. In: Proceedings of the 7th International Conference on World Wide Web, vol. 7, pp. 107–120 (1998)
7. Kocarev, L., Zlatanov, N., Trajanov, D.: Vulnerability of networks of interacting Markov chains. Phylosophical Transactions of the Royal Society A: Mathematical, Physical and Engineering Sciences 368(2010), 2205–2219 (1918)
8. Kurant, M., Thiran, P., Hagmann, P.: Error and Attack Tolerance of Layered Complex Networks. Physical Review E76, 5 (2007)
9. US power grid, http://www.cs.helsinki.fi/u/tsaparas/MACN2006/data-code.html
10. Filiposka, S., Trajanov, D., Grnarov, A.: Survey of Social Networking and Applications in Ad Hoc Networks. In: ETAI VIII National Conference 2007 (2007)
11. Briggs, K.: Graph eigenvalues and connectivity, http://keithbriggs.info/documents/graph-eigenvalues.pdf
12. Karonski, M., Rucinski, A.: The Origins of the Theory of Random Graphs. In: The Mathematics of Paul Erdös I, pp. 311–336. Springer (1997)
13. Penrose, M.: Random Geometric Graphs. Oxford University Press (2004)
14. Barabasi, A., Albert, R.: Emergence of scaling in random networks. Science 286, 509–512 (1999)
15. Watts, D.J.: Small Worlds: The Dynamics of Networks between Order and Randomness. Princeton University Press (2003)
16. Hangmann, P., Kurant, M., Gigadent, X., Thiran, P., Weeden, V.J., Meuli, R., Thiran, J.P.: Mapping Human Whole-Brain Structural Networks with Diffusion MRI. PLoS One 2, e597 (2007)
17. UF Sparse Matrix Collection, http://www.cise.ufl.edu/research/sparse/matrices/Pajek/

Asymptotic Performance of Bidirectional Dual-Hop Amplify-and-Forward Systems

Katina Kralevska[1], Zoran Hadzi-Velkov[2], and Harald Øverby[1]

[1] Department of Telematics, Faculty of Information Technology,
Mathematics and Electrical Engineering, Norwegian University of Science and
Technology, Trondheim, Norway
{katinak,haraldov}@item.ntnu.no
[2] Faculty of Electrical Engineering and Information Technologies,
Ss. Cyril and Methodius University, Skopje, Macedonia
zoranhv@ukim.edu.mk

Abstract. This paper considers a bidirectional amplify-and-forward (AF) relaying system where two nodes communicate through an intermediate node due to the lack of a direct path. The communication is realized over Rayleigh fading channels in two phases: multiple access and broadcast phase. We derive closed form and asymptotic expressions for the outage probability of two-way wireless relaying system. The validity of our performance analysis for moderate and high signal-to-noise ratio (SNR) is verified with Monte Carlo simulations.

Keywords: Two-way relaying, amplify-and-forward, outage probability, Rayleigh fading.

1 Introduction

Cooperative relaying over wireless fading channels has been shown to be a practical technique to enhance the capacity and the coverage by allowing user cooperation [8]. Two-hop channels where a relay assists in the communication between a source and a destination have attracted a lot of interest. Traditional relay systems operate in a half-duplex mode, hence the transmission of one information block from the source to the destination occupies two channel uses. This leads to a loss of spectral efficiency due to the pre-log factor $\frac{1}{2}$ [4]. In order to overcome this drawback a two-way relaying is proposed [7]. Due to the lack of a direct path, a bidirectional connection between a source and a destination is established by using a half-duplex relay that mitigates the loss in the spectral efficiency.

There are two main methods for relaying, amplify-and-forward (AF) and decode-and-forward. The simplest relaying technique is based on the AF method where the relay amplifies and forwards the signal to the receiving nodes. In the recent years, many variants of AF relaying systems have been studied, depending on the number of relays [9] and the fading channels. The authors in [6] investigate the performance of practical physical-layer network coding schemes for two-way

V. Trajkovik and A. Mishev (eds.), *ICT Innovations 2013,*
Advances in Intelligent Systems and Computing 231,
DOI: 10.1007/978-3-319-01466-1_27, © Springer International Publishing Switzerland 2014

AF relaying systems. In [5] the performance of cooperative protocols for the two-way relaying assited by a single half-duplex relay is studied. Later, Duong et al. [2] derived the exact closed-form expressions for the main performance parameters of bidirectional AF systems in independent but not necessarily identically distributed (i.n.i.d.) Rayleigh fading channels. Yang et al. [11] present a performance analysis of two-way AF relaying systems over i.n.i.d. Nakagami-m channels. Part of these works present a relatively complex analysis for the outage probability. Therefore, studying the systems for high signal-to-noise ratios (SNR) is especially important since it results in relatively simple and accurate expressions. These high SNR approximations explain how the system behaviour depends from some system parameters [10].

In this paper, we focus on a two-way relaying AF system where in the first phase the nodes send information to the relay and in the second phase the relay broadcasts the signal to the nodes. We do not neglect the conditions in the wireless environment, therefore we consider a half-duplex AF relaying system that operates over Rayleigh fading channels. An accurate and novel expression for the outage probability for high SNRs is derived.

The rest of the paper is organized as follows: Section II presents the system model. Section III presents an exact and asymptotic analysis on the outage probability for the considered system. The exactness of the derived asymptotic expression is verified with Monte Carlo simulations in Section IV. Section V concludes the paper.

2   System Model

We consider a two-way wireless relaying system where the two nodes, T_1 and T_2, communicate with each other through the intermediate node R. The node R is a half-duplex relay which utilizes amplify-and-forward strategy. There is no direct path between the two receiving nodes T_1 and T_2; therefore, the communication is with the help of the intermediate node. As it is presented in Fig. 1, the communication between T_1 and T_2 is realized in two phases: multiple access phase (MA) and broadcast phase (BC). In the first phase, the nodes T_1 and T_2 transmit their information blocks towards the relay R. In the second phase, the relay R amplifies the received signal according to its available average transmit power and sends it to T_1 and T_2.

During the first phase, we assume that node T_1 transmits the information block $x_1[k]$ with an average power P_1 and node T_2 transmits the information block $x_2[k]$ with an average power P_2. The fading is assumed to be slowly varying, so the random fading amplitudes remain static for the duration of the information block. The relay R receives the following information in time slot k

$$y_R[k] = h_1[k]x_1[k] + h_2[k]x_2[k] + n_R[k] \tag{1}$$

where h_1 and h_2 are the Rayleigh fading amplitude of the T_1–R and T_2–R links, respectively, with mean squared values $E[h_1^2] = \Omega_1$ and $E[h_2^2] = \Omega_2$. $n_R \sim \mathcal{CN}(0, \sigma_3{}^2)$ is the additive white Gaussian noise at the relay. In order to

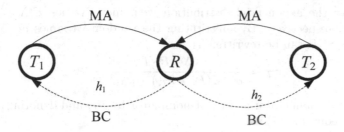

Fig. 1. Wireless communication system where the two receiving nodes T_1 and T_2 communicate through the full-duplex relay R

meet the average transmit power constraint, the relay scales the signal presented with equation (1) by

$$g[k] = \sqrt{\frac{P_3}{P_1|h_1[k]|^2 + P_2|h_2[k]|^2 + \sigma_3{}^2}}. \tag{2}$$

In the next time slot, the relay R transmits the amplified signal to the nodes T_1 and T_2. The communication in the direction T_1–R–T_2 is given by

$$y_2[k+1] = h_2[k+1]g[k]h_1[k]x_1[k] + h_2[k+1]g[k]h_2[k]x_2[k] +$$
$$+ h_2[k+1]g[k]n_R[k] + n_2[k+1] \tag{3}$$

where $n_2 \sim \mathcal{CN}(0, \sigma_2{}^2)$ is the AWGN at T_2. Whereas, the communication in the direction T_2–R–T_1 is given by

$$y_1[k+1] = h_1[k+1]g[k]h_2[k]x_2[k] + h_1[k+1]g[k]h_1[k]x_1[k] +$$
$$+ h_1[k+1]g[k]n_R[k] + n_1[k+1] \tag{4}$$

where $n_1 \sim \mathcal{CN}(0, \sigma_1{}^2)$ is the AWGN at T_1.

We assume that the nodes T_1 and T_2 can subtract the back-propagating self-interference in (3) and (4), since they know their own transmitted information blocks and they have the perfect knowledge of the corresponding channel coefficients. By substituting (2) in (3) and (4), we can calculate the instantaneous SNR at the nodes T_i, $i = 1, 2$, $i \neq j$ as

$$\gamma_{T_i} = \frac{P_{T_j}|h_{T_i}|^2|h_{T_j}|^2 P_3}{P_3|h_{T_i}|^2\sigma_3{}^2 + P_{T_i}|h_{T_i}|^2\sigma_{T_i}{}^2 + P_{T_j}|h_{T_j}|^2\sigma_{T_i}{}^2}. \tag{5}$$

Without loss in generality, we assume that $\sigma_{T_i}{}^2 = \sigma_{T_j}{}^2 = \sigma_3{}^2$ and in the following analysis we use the notation $\sigma_0{}^2$. Let introduce the random variables

$$X = |h_{T_i}|^2$$
$$Y = |h_{T_j}|^2 \tag{6}$$

which follow the exponential distribution with mean values $E[X] = \Omega_i$ and $E[Y] = \Omega_j$, respectively. By substituting the random variables in (5), the instantaneous SNR can be rewritten as

$$\gamma_{T_i} = \frac{XY P_{T_j} P_3}{\sigma_0^2 X (P_3 + P_{T_i}) + \sigma_0^2 Y P_{T_j}}. \tag{7}$$

By dividing the numerator and the denominator by P_{T_j} and denoting $\frac{P_3}{\sigma_0^2}$ as γ_0, we simplify equation (7) as

$$\gamma_{T_i} = \gamma_0 \frac{XY}{kX + Y} \tag{8}$$

where k is a ratio from the transmit powers of the nodes in the system, i.e., $k = \frac{P_3 + P_{T_i}}{P_{T_j}}$.

3   Analysis of the Outage Probability

3.1   Exact Outage Probability

The outage probability is defined as the probability that the instantaneous SNR at the receiving nodes is below a predefined threshold γ_{th}. In terms of the instantaneous SNR γ_{T_i}, the outage probability is actually the cumulative distribution function (CDF) of γ_{T_i} evaluated at γ_{th},

$$P_{out} = F_{\gamma_{T_i}}(\gamma_{th}) = Pr\{\gamma \le \gamma_{th}\} = Pr\{\gamma_0 \frac{XY}{kX + Y} \le \gamma_{th}\} \tag{9}$$

which is actually a sum of the two probabilities γ_{T_i}, the outage probability is actually the cumulative distribution function (CDF) of γ_{T_i} evaluated at γ_{th},

$$P_{out} = Pr\{y < \frac{kxC}{x - C}|x < C\} + Pr\{y < \frac{kxC}{x - C}|x \ge C\} \tag{10}$$

where $C = \frac{\gamma_{th}}{\gamma_0}$, γ_0 is the transmitted SNR. The result for the first part of (9) is $1 - \exp\left(-\frac{C}{\Omega_i}\right)$ and the second part is calculated as

$$Pr\{y < \frac{kxC}{x - C}|x \ge C\} = \int_C^\infty F_Y(\frac{kxC}{x - C}) f_X(x) dx. \tag{11}$$

The closed form expression for the outage probability of the two-way relaying system is given by

$$P_{out} = 1 - 2\sqrt{\frac{k\gamma_{th}^2}{\gamma_0^2 \Omega_i \Omega_j}} \exp\left(-\frac{\gamma_{th}}{\gamma_0}(\frac{k}{\Omega_j} + \frac{1}{\Omega_i})\right) K_1(2\sqrt{\frac{k\gamma_{th}^2}{\gamma_0^2 \Omega_i \Omega_j}}) \tag{12}$$

where $K_1(\cdot)$ is the first-order modified Bessel function of the second kind [3, Eq. 8.432]. In this section we derived the closed form expression for a general case related to the transmit powers of the nodes in the system. In [2], the closed form of CDF is derived when the nodes have equal transmit power. The authors use the Jacobian transformation and the Laplace transform to derive an expression for a specific case. Additionally, the authors in [12] present an expression for the outage probability for bidirectional AF relaying for certain power allocation.

3.2 Asymptotic Outage Probability

In this section, we present an asymptotic expression for the outage probability of our proposed two-way amplify-and-forward relay network. The asymptotic analysis typically results in relatively simple and accurate approximate expressions for moderate and high SNRs from which the system behavior is easier to deduce. To derive the high outage probability approximation, we use the power series expansion of a Bessel function $K_n(\cdot)$ [1, Eq. 9.6.11]

$$K_n(z) = \frac{1}{2}(\frac{1}{2}z)^{-n} \sum_{k=0}^{n-1} \frac{(n-k-1)!}{k!}(-\frac{1}{4}z^2)^k + (-)^{n+1}\ln(\frac{1}{2}z)I_n(z) +$$

$$+ (-)^n \frac{1}{2}(\frac{1}{2}z)^n \sum_{k=0}^{\infty}\{\psi(k+1) + \psi(n+k+1)\}\frac{(\frac{1}{4}z^2)^k}{k!(n+k)!}. \tag{13}$$

In our case $n = 1$.

With the help of [1, Eq. 9.6.7] for $\nu = 1$ and $z \to 0$ due to $\gamma_0 \to \infty$, and [1, Eq. 6.3.2] the function $K_1(\cdot)$ is rewritten. Note that in high SNR regime

$$\lim_{\gamma_0 \to \infty} \exp(-\frac{\gamma_{th}}{\gamma_0 \Omega_i}) = 1 \tag{14}$$

and

$$\exp(-\frac{\gamma_{th}}{\gamma_0 \Omega_i}) \approx 1 - \frac{\gamma_{th}}{\gamma_0 \Omega_i}. \tag{15}$$

Applying these approximations, the expression for the outage probability for high SNR regime is

$$P_{out} \approx \frac{\gamma_{th}}{\gamma_0}(\frac{1}{\Omega_i} + \frac{k}{\Omega_j}) - \frac{k\gamma_{th}^2}{\gamma_0^2\Omega_i\Omega_j}(\ln(\frac{k\gamma_{th}^2}{\gamma_0^2\Omega_i\Omega_j}) - 1 + 2\mu) \tag{16}$$

where μ is the Euler's constant.

To the best of authorsknowledge, this approximation is novel. These results are validated in the next section by Monte Carlo simulations.

4 Numerical Results

In this section we verify the tightness of our approximation through Monte Carlo simulations. Fig. 2 and Fig. 3 present the OP vs. transmit SNR of the considered two-way relaying system under Rayleigh fading for different transmit powers of the nodes. The curves are obtained from our asymptotic expression and via simulation for two thresholds γ_{th} (0 dB and 5 dB). We can note that if the average SNR increases, the outage probability decreases. Additionally, the outage probability increases proportionally with γ_{th}. Moreover, they show an excellent match for moderate and high SNRs. This proves the validity of our derived expressions in the previous Section.

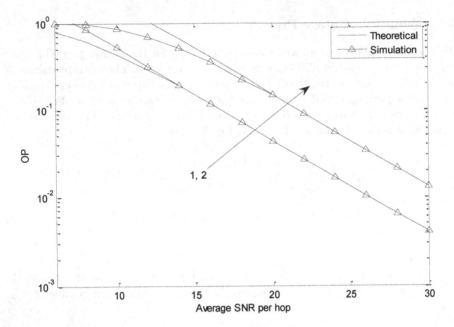

Fig. 2. OP for two-way relaying system for γ_{th}=0dB (1) and 5dB (2) and $P_3 = P_{T_i} = 0.5P_{T_j}$

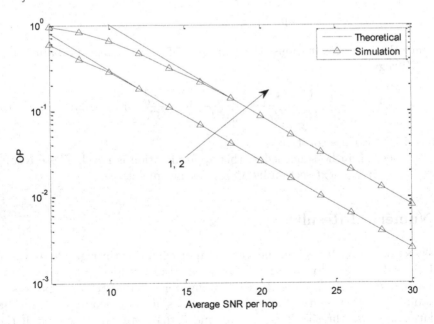

Fig. 3. OP for two-way relaying system for γ_{th}=0dB (1) and 5dB (2) and $P_3 = P_{T_i} = 2P_{T_j}$

5 Conclusion

In this paper, we study the exact and the asymptotic behaviour of the outage probability for a two-way AF relaying half-duplex system in the presence of Rayleigh fading. We derive expressions for this performance parameter. The validity of the derived expression for moderate and high average SNRs is verified by Monte Carlo simulations.

References

1. Abramowitz, M., Stegun, I.A.: Handbook of mathematical functions with formulas, graphs, and mathematical tables. Dover publications (1972)
2. Duong, T.Q., Hoang, L.N., Bao, V.N.Q.: On the performance of two-way amplify-and-forward relay networks. IEICE Transactions 92, 3957–3959 (2009)
3. Gradshteyn, I.S., Ryzhik, I.M.: Table of Integrals, Series, and Products. Elsevier Science (2000)
4. Laneman, J.N., Tse, D.N.C., Wornell, G.W.: Cooperative diversity in wireless networks: Efficient protocols and outage behavior. IEEE Transactions on Information Theory 50, 3062–3080 (2004)
5. Li, Q., Ting, S.H., Pandharipande, A., Han, Y.: Adaptive two-way relaying and outage analysis. IEEE Transactions on Wireless Communications 8, 3288–3299 (2009)
6. Louie, R.H.Y., Li, Y., Vucetic, B.: Practical physical layer network coding for two-way relay channels: performance analysis and comparison. IEEE Transactions on Wireless Communications 9, 764–777 (2010)
7. Rankov, B., Wittneben, A.: Spectral efficient protocols for half-duplex fading relay channels. IEEE Journal on Selected Areas in Communications 35, 379–389 (2007)
8. Sendonaris, A., Erkip, E., Aazhang, B.: User cooperation diversity. part I. system description. IEEE Transactions on Communications 51, 1927–1938 (2003)
9. Song, L.: Relay selection for two-way relaying with amplify-and-forward protocols. IEEE Transactions on Vehicular Technology 60, 1954–1959 (2011)
10. Wang, Z., Giannakis, G.B.: A simple and general parameterization quantifying performance in fading channels. IEEE Transactions on Communications 51, 1389–1398 (2003)
11. Yang, J., Fan, P., Duong, T.Q., Lei, X.: Exact performance of two-way AF relaying in nakagami-m fading environment. IEEE Transactions on Wireless Communications 10, 980–987 (2011)
12. Zhang, Y., Ma, Y., Tafazolli, R.: Power allocation for bidirec- tional AF relaying over rayleigh fading channels. IEEE Communications Letters 14, 145–147 (2010)

Smartphone Traffic Review

Boban Mitevski and Sonja Filiposka

Faculty of Computer Science and Engineering, Ss. Cyril and Methodius University - Skopje
bobi_mit@yahoo.com, sonja.filiposka@finki.ukim.mk

Abstract. Because of rapidly growing subscriber populations, advances in cellular communication technology, increasingly capable user terminals, and the expanding range of mobile applications, cellular networks have experienced a significant increase in data traffic. Smartphone traffic contributes a considerable amount to Internet traffic. The increasing popularity of smartphones in recent reports suggests that smartphone traffic has been growing 10 times faster than traffic generated from fixed networks. Understanding the characteristics of this traffic is important for network design, traffic modeling, resource planning and network control. This paper presents an overview of the methodology for collecting data using smartphone logging and discusses traffic characteristics from some recent researches. The main goal is to explore some important issues related to smartphone traffic in order to prepare for further, deeper analysis.

Keywords: mobile, traffic, measurement, comparison.

1   Introduction

Because of the emergence of user-friendly smartphones and the advances in cellular data network technologies, the volume of data traffic carried by cellular networks has been experiencing a phenomenal rise. Smartphone sales already surpass desktop PCs [1], and this year is expected for the tablet sales to also surpass PCs [2].

Due to the fact that smartphones and tablets are already becoming more prevalent as members of the wireless infrastructure and ad hoc networks, which are trying to meet the needs of applications and users, there is a need to define a model that will reflect the traffic they generated in the network. This model will allow for realistic consideration of the behavior of the network built by this kind of devices, by creating simulation scenarios that will reflect realistic use of the network.

To make the modeling of traffic and movement in smart phones, the first step is to identify the characteristics of the traffic generated by smartphones. When considering 3G terminals, all traffic is based on packages. By studying this traffic it may be possible to identify models. Herein, models denote "regular, observable sequences of events that are repeated over time." Event in this context is sending or receiving a packet. If it is possible to identify traffic patterns and find correlations between events, newly developed algorithms and generators for network simulation scenarios can be made such that take advantage of the knowledge of these models and the correlations of network and user side.

V. Trajkovik and A. Mishev (eds.), *ICT Innovations 2013*,
Advances in Intelligent Systems and Computing 231,
DOI: 10.1007/978-3-319-01466-1_28, © Springer International Publishing Switzerland 2014

Technology can easily be used for logging events on these devices with access to real traffic data in real environments. However, when using this technique it must be noted that logging can affect the behavior of users. Similar to traditional methodologies, reactivity (behavioral modification because of measurement) can happen if real steps for planning of selection, causes, saving and coding behaviors and preferences in a way that preserves natural behaviors are not taken into account. This can seriously affect the validity of the data that is obtained from these devices. By carefully designing, logging in smart devices can be used for advanced research to establish the empirical models, the development of theories and test some hypotheses.

One of the main observations of smartphone logging data is to understand the impact of location on the use of smartphones [9]. It is important if participants change their usage patterns when they are in different locations. Another point of view is concerned with the impact of time periods of the day of smartphone usage [8]. Understanding dependencies of certain access times of the day can be useful for determining the social context of the user on the network. It will provide an opportunity for development of intelligent techniques for reducing the response time for individual users.

One of the additional observations is to make a comparison of traffic generated via Wi-Fi and cellular networks [6]. Such comparisons are useful for predicting usage patterns in mixed Wi-Fi/mobile service contexts (4G). Many cellular providers plan to develop such mixed networks to address the capacity constraints of mobile networks.

The overview in this paper also explores the possible relationships between the proximity of users and patterns of smartphone usage. Such analysis is useful for improving the basic knowledge for use in different circumstances. It is important whether smart phones are primarily used when the users are alone or they are more often used in companionship of more people. Such basic understanding can help in the understanding of the location dependencies in usage (for example, whether participants searched more in locations where they are likely to be alone) and eventually understand the types of applications that participants prefer in different scenarios [3].

Relationship between movements of users and the use of smart phones are also important in order to understand usage patterns when they are moving or stationary [9].

2   Methodologies for Data Collection

There are several methodologies for collecting smartphone data, such as traditional methodologies that require user input, which reduces accuracy. They can interrupt users or bore them with constant requests for report, which they enter according to their memory of the events.

Login methodologies have resolved most of these problems by adding surveillance technology. These methodologies provide access to data that can be gathered without supervising being present or reporting needs from users. Therefore, data collected from loggers are typically considered more objective, accurate, and realistic.

There are several important factors for implementing naturalistic approach to logging smartphone usage [4]:

— Variables - Variables of interest, have a major impact on the nature of the methodology. For example, communications through smart devices contain text

messages, which are considered more private than e-mail. By collecting these data, users can change their normal communication behaviors. However, researchers can collect the number of words, or part of the content, for example used emoticons.
— Privacy - Privacy must be considered at several levels in this methodology. Researchers can collect data from communication with people who are not involved in the research, but it can affect the behavior of users, avoiding too personal communications because their privacy is not guaranteed [7]. Several limitations of privacy should be implemented in the research, in order to adapt to the methodology. Thus, users should be aware of how their data will be used. Principles of research and the anonymization process should be explained in detail before the start of the study. Furthermore, participants should be assigned numbers because of the anonymity and connecting the data with names should be avoided. Data should be encrypted and should not contain content of communications or contact information. If such data is needed, then before they are taken, they should be granted unique alphanumeric codes.
— Participants - Several researches have received data from users who were not aware that their behavior is being monitored. For implementing a naturalistic methodology, participants should be fully informed of the data collected from their phones. Also, a careful selection of participants is needed.
— Duration of the research - Longer surveys influence the effect of monitoring to fade. Also, some events by their nature are rare. Small frame times may miss important events. Longer collecting of data has the potential to collect richer information about the cycles and trends that may not be so obvious in the shorter studies. Typically for such researches, longer is considered better.
— Obtrusiveness - Measurement obtrusiveness increases participant reactivity. There are several ways in which researchers or logging technology can impose and remind participants that they are being followed or hinder the normal behavior. For example, the requirement of participants to respond to messages or perform a data upload procedure to collect data can increase their reactivity. These activities can provide important information such as the current context where users use their device, but they come at the cost of interruption of normal activities. Also, the constant need for participants to perform any actions related to research are unnatural actions and may lead to additional activities that do not occur normally. Naturalistic logging methodology should not require user actions to record the data. A minimum number of meetings must be scheduled with participants to collect data. The optimal implementation of the research methodology is related meetings to be scheduled before and after the survey.
— Interface - Another factor that is important to preserve realistic and generalized behaviors are the types of interfaces that are implemented in the technology used in the research. Using new interfaces (eg, custom search engine) or change technologies during the research (eg, change device) can affect the validity of the data by giving false installments behaviors, increased variability and lead to many other problems. Participants are getting used to monitoring over time through a stable interface.
— Tasks - Naturalistic approach allows participants to carry out those tasks they normally perform on their devices. To implement this methodology, researchers should avoid influencing what users do with their smartphone.

— Technology - One challenge in smartphone logging is the design of constraints to encourage participants to use the instrumented technology as if it was their own. This can be difficult because smartphones are typically not used in isolation from other technologies. Many actions that can be performed on smartphones, can be made on other technologies such as laptop or another phone. Therefore, researchers should provide incentives to participants to encourage the use of these devices. Such incentives are promoting the latest offers and unlimited packages.

3   Overview of Recent Results

In this section, we give an overview of the recent studies and obtained results concerning smartphone traffic that can be used as a baseline for future research and pattern behavior discovery.

Falaki et al. [5] collected traces from 255 smartphone users of two different smartphone platforms, with 7-28 weeks of data per user and performed a detailed traffic analysis. Two sets of data were obtained, the first from 33 Android users, and the other from 222 Windows Mobile users. The authors concentrated on the application usage, session characteristics and energy consumption.

Shahriar Kaisar [3] carried out a research on 39 students from one University for a period of 5 weeks. Data was collected through a custom logger installed on their Android devices that recorded sent and received traffic.

Falaki et al. [6] collected traffic data from 43 users who used two platforms. Data was collected through loggers on devices that recorded sent and received traffic in windows of two minutes. The first data set was from 8 users on Windows Mobile and 2 Android users, and the second from 33 users of Android smartphones. Data was collected from 1 to 5 months for each user.

Diversity among users comes from the fact that users use their smart phones for different purposes and with different frequencies. For example, users who use games and maps often, usually have longer interactions. The results show that browsing contributes over half of the traffic, while email, media, and maps each contribute roughly 10% [3].

In addition to quantitative differences, there are qualitative similarities between users, which facilitates the task of learning user behavior. For several key aspects of the use of smartphones, the same model can describe all users, but with different parameters for each. For example, the time between user interactions can be presented with a Weibull distribution (Figure 1) as defined with (1). For each user, the parameter of this model is less than 1, indicating that the longer the elapsed time since the last user interaction, the less likely is to start the next interaction. It has also been obtained that the relative popularity of applications declined quickly for all users.

$$f(x; \lambda, \kappa) = \begin{cases} \frac{\kappa}{\lambda}\left(\frac{x}{\lambda}\right)^{\kappa-1} e^{-(x/\lambda)^k} & x \geq 0 \\ 0 & x < 0 \end{cases} \tag{1}$$

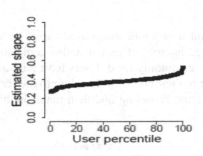
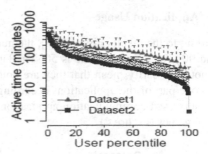

Fig. 1. Weibull distribution [3]   **Fig. 2.** Daily interaction [5]

3.1   Interaction Time

First data set users [5] had longer interaction times because they had longer interaction sessions with the same number of sessions. However, there were major differences between the users, so in the first data set, the smallest value was 30 minutes per day, and the largest 500 minutes per day (one third of a day). The other users were not centered around these two extremes, but are distributed between them (Figure 2).

3.2   Interactive Sessions

Interactive sessions provide a detailed look at how the user uses the phone. Their characteristics are also important because the use of energy depends not only on how long your phone is used, but also by the distribution of use. Very short interactions consume more energy compared to some long actions, because of the need to wake up the phone and radio.

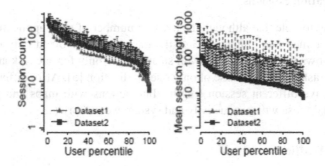

Fig. 3. Session analyzes (a)Number of sessions (b)Session length

It has been obtained that users had interacted with their smart phones from 10 to 200 times a day [5]. The duration of a session varied from 10 to 250 seconds (Figure 3). It has also been obtained that for every user most sessions are short, but some were very long.

3.3   Application Usage

According to the results given in [5], the number of applications used varied between 10 and 90. The average value is 50. The large number of user installed applications, does not necessarily mean that they are both commonly used. Users have paid attention to one part of the applications according to their choice. According to the results, most users used communication applications and browsing and then maps, multimedia, and games (Figure 4).

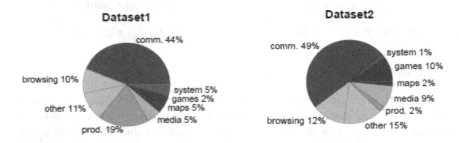

Fig. 4. Application popularity [5]

According to Kaisar [3], most of the logged traffic was a result from application use, and then from images, text and video. Viewed by usage, the most popular applications were browsing, maps and social networking. However, most of the traffic was from the applications for news and the Android Market.

On the other hand, from the results given in [6], browsing and email were dominating in use, while multimedia and maps were major contributors.

3.4   Application Sessions

It is interesting to note that although the average number of used applications was 50, it has been obtained that almost 90% of the interactions involved only one application, which shows that users interact with smartphones only for one task at a time, and most of these tasks require the use of only one application [5]. Also, different types of applications have different session lengths. Interactions with maps and games were the longest, and those with productivity and system were shortest.

3.5   Daily Traffic

Unlike interactive events and use of applications, data traffic is not intentional user action but a consequence of these actions. The analyses given in [5] include information only from the first set of data, with logging data traffic (sent and received) of cellular networks and 802.11 devices.

Traffic received varied from 1 to 1000 MB, and transmitted from 0.3 to 100 MB [5]. Averages were 30 MB sent and 5 MB traffic received. These results show that the traffic generated by smart phones is comparable to the traffic generated by computers several years ago. This high level of traffic has major implications for the provisioning of wireless carrier networks as smartphone adoption increases.

To examine which types of applications are favored more from customers that generate more traffic, users were divided into two equal classes according to their cumulative daily traffic. As expected, applications for communication were more popular among the users that generate more data traffic.

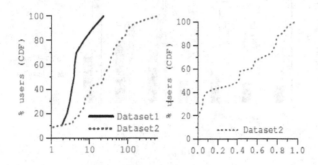

Fig. 5. Traffic [6]

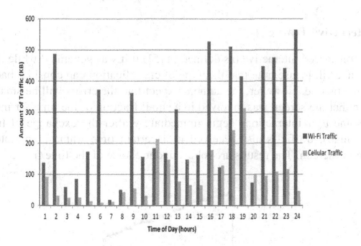

Fig. 6. Daily cyclic patterns [3]

The daily traffic observed in [6] moved from 2 to 20 MB in the first data set, and from 1 to 500 MB in the second set (Figure 5a). Received traffic was 10 times greater than the transmitted traffic. Compared with residential broadband traffic, this was one size smaller. Two factors could explain the differences in the two sets of data. The first factor was that the second set of data was dominated by Android users, who have had much more interaction from Windows Mobile users. The second factor, which was associated with the first, was that many users in the second set of data have often used WiFi. Although 20% of users do not use WiFi at all, in 20% of users, 80% of their traffic was via WiFi (Figure 5b).

Kaisar [3] examined the daily cyclic patterns of use, and obtained that the bulk of the traffic was between 4 pm and midnight (Figure 6). Traffic was also examined in different days of the week (Figure 7).

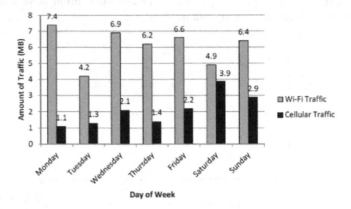

Fig. 7. Weekly cyclic patterns [3]

3.6   Interactive Traffic

Traffic is considered interactive (as defined in [5]) if it was generated while the screen is turned on. Still, please note that this type of classification can consider background traffic as interactive. However, the authors expect that the errors will be small since it was shown that the screen was involved in a small fraction of the time for most users. Because some user interactions begin immediately after the exchange of traffic (eg receiving email), traffic is also received in a certain time frame (1 minute) before turning on the screen. The results are robust under choice of the time frame.

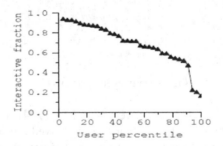

Fig. 8. Interactive Traffic [5]

Results obtained in [5] show that for 90% of users, more than 50% of the traffic was interactive, but for the others almost all of the traffic was not interactive (Figure 8). It can be concluded that for different users, almost all to almost none of the traffic was generated by applications in the background. It was also obtained that 80% of customers generate twice the traffic than the average during their main peak hour of usage.

3.7   Transfers

According to Kaisar [3], when using WiFi, the average size of the transfers was 5,31 KB, and the average access time was 86,54s, while for mobile networks the average size of transfers was 2.36 KB, and the average access time was 23,97 s.

In [6], the authors have gone one step further where individual transfers were identified using TCP flows. TCP flows have been identified by the means of IP addresses and ports. Although the average size of the transfer was 273 KB sent and 57 KB received, most transfers have been proven to be extremely small (Figure 9). 30% of transfers contained less than 1 KB and 10 packets.

Small sizes of the transfers have multiple implications. Due to the high amount of energy that is consumed by the 3G radio, it is possible to have much higher energy consumption when going from a state of sleep in a ready state and from inactive to a state of sleep. Another implication is that the large excess bytes from lower protocols can dominate, and the longer time required for "handling".

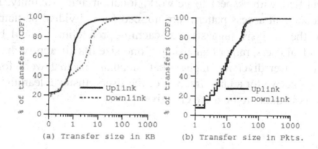

Fig. 9. Transfer sizes [6]

Retransmissions are identified using sequence numbers and provide a good estimate of path loss rate. Across all transfers, the retransmission rate was 3.7% for sent data, and 3.3% for received data (Figure 10). These rates were much higher than those in wired paths. It was concluded that the loss of TCP packets was the main bottleneck for transmission capacity.

3.8   Power Consumption

Energy consumption depends on two factors: 1) user interactions and applications and 2) hardware and software. If the second factor dominates, the energy consumption of users with identical phones would be similar. Otherwise, the power consumption would be varied according to user behavior. According to the study, heavy users have only 250 mAh and lightest 10 mAh. If the battery capacity is 1200 mAh, this leads to a time of use from 4 to 120 h.

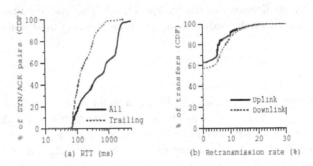

Fig. 10. Retransmission rate [6]

4   Conclusion

This paper gives an first order overview of the smartphone traffic characteristics and analysys of the relationships between smartphone usage and user context. Heterogeneity is found to be the main theme of the obtained results. Strong heterogeneity has been observed among usage patterns of different participants. Heterogeneity has also been identified with respect to network, location and proximity. Location and time-of-day dependent access patterns are noticed for individual participants. Results obtained from the analysis suggests that caching/prefetching should be tailored to habits of individual users, rather than using a `one size fits all' approach.

This extent of user diversity implies that mechanisms that work for the average case may be ineffective for a large fraction of the users. Instead, learning and adapting to user behaviors is likely to be more effective.

References

1. Smartphone Sales Overtake PCs for the First Time [STUDY],
   http://mashable.com/2012/02/03/smartphone-sales-overtake-pcs/
2. When Will Tablet Shipments Overtake PCs? http://www.tech-thoughts.net/
   2013/02/when-will-tablet-shipments-overtake-pcs.html
3. Kaisar, S.: Smartphone traffic characteristics and context dependencies, PhD Thesis, University of Saskatchewan, Canada (2012)
4. Tossell, C.C., Kortum, P., Shepard, C.W., Rahmati, A., Zhong, L.: Getting Real: A Naturalistic Methodology for Using Smartphones to Collect Mediated Communications. Advances in Human-Computer Interaction (2012)
5. Falaki, H., Mahajan, R., Kandula, S., Lymberopoulos, D., Govindan, R., Estrin, D.: Diversity in smartphone usage. In: Proceedings ofthe 8th International Conference on Mobile Systems, Applications and Services, pp. 179–194. ACM (2010)
6. Falaki, H., Lymberopoulos, D., Mahajan, R., Kandula, S., Estrin, D.: A First Look at Traffic on Smartphones. In: Proceedings of the 10th ACM SIGCOMM Conference on Internet Measurement, pp. 281–287. ACM (2010)

7. Bal, G.: Revealing Privacy-Impacting Behavior Patterns of Smartphone Applications, Mobile Security Technologies (2012)
8. Xu, Q., Erman, J., Gerber, A., Mao, Z., Pang, J., Venkataraman, S.: Identifying diverse usage behaviors of smartphone apps. In: Proceedings of 2011 ACM SIGCOMM Conference on Internet Measurement, pp. 329–344. ACM (2011)
9. Becker, R., Caceres, R., Hanson, K., Isaacman, S., Loh, J.M., Martonosi, M., Rowland, J., Urbanek, S., Varshavsky, A., Volinsky, C.: Human Mobility Characterization from Cellular Network Data. Communications of the ACM 56(1), 74–82 (2013)

Author Index